The Complete Home Video Book

Dr. Peter Utz has produced and directed more than 500 instructional-TV productions for the City University of New York. He has also published *Video User's Handbook* and more than 35 articles in media and television journals, and presently supervises the media department at the County College of Morris in Randolph, New Jersey. He also teaches professional and home video courses.

Peter Utz

Volume I
THE COMPLETE HOME VIDEO BOOK

A Source Book of Information Essential to the Video Enthusiast

A SPECTRUM BOOK

PRENTICE-HALL, INC., Englewood Cliffs, New Jersey 07632

Library of Congress Cataloging in Publication Data

Utz, Peter.
　　The complete home video book.

　　"A Spectrum Book."
　　Bibliography: p.
　　Includes index.
　　1. Home video systems.　　I. Title.
TK9960.U88　　1983　　　778.59　　82-10129
ISBN 0-13-161349-9 (pbk. : v.1)

© 1983 by Prentice-Hall, Inc., Englewood Cliffs, New Jersey 07632.
All rights reserved. No part of this book may be reproduced in any form
or by any means without permission in writing from the publisher.
A Spectrum Book. Printed in the United States of America.

10　9　8　7　6　5　4　3　2　1

ISBN 0-13-161349-9 {PBK. VOL. I}

Editorial/production supervision: Marlys Lehmann
Page layout: Marie Alexander
Cover design © 1983 by Jeannette Jacobs
Manufacturing buyer: Cathie Lenard

This book is available at a special discount when ordered in
bulk quantities. Contact Prentice-Hall, Inc., General
Publishing Division, Special Sales, Englewood Cliffs, N.J. 07632.

Prentice-Hall International, Inc., *London*
Prentice-Hall of Australia Pty. Limited, *Sydney*
Prentice-Hall Canada Inc., *Toronto*
Prentice-Hall of India Private Limited, *New Delhi*
Prentice-Hall of Japan, Inc., *Tokyo*
Prentice-Hall of Southeast Asia Pte. Ltd., *Singapore*
Whitehall Books Limited, *Wellington, New Zealand*
Editora Prentice-Hall do Brasil Ltda., *Rio de Janeiro*

To Barbara, whose tlyping is GETTIng bettre.

Contents

Preface *xiii*

Chapter 1
ALL ABOUT TV SETS *1*
   Operating a TV Receiver *1*
   Setting Up
   a TV Receiver for Use *3*
   How TVs Work *4*
   Adjusting the Controls
   on a TV Set *7*
   TV Monitors
   and Monitor/Receivers *11*
   Common TV Ailments
   and Cures *19*
   New Developments *21*
   TVs in a Nutshell *23*

Chapter 2
MORE ABOUT TV ANTENNAS *25*
   Monopole Antenna *25*
   Dipole or
   Rabbit-Ear Antenna *25*
   Loop and
   Bow Tie Antennas *26*
   Portable TV Placement *27*
   Portable TV Antennas *27*
   Improvised TV Antennas *28*
   Bigger Antennas *29*
   How TV Antennas Work *29*
   Kinds of TV Antennas *30*
   Installing a TV Antenna *32*
   Aiming the Antenna *32*
   Antenna Wire *34*
   Matching Transformers *36*
   Antenna Joiners *37*
   Signal Splitters *38*
   TV Couplers *40*
   Antenna Preamplifiers *41*
   Line Equalizers *41*
   Ghost Eliminators *41*
   Attenuaters *41*

Common TV-Antenna
Ailments and Cures   42

Antennas in a Nutshell   47

Chapter 3
CABLE TV   49

How Cable TV Works   49

Decoding Cable's
Signal for Your TV   52

Two-Way Cable   65

Future Developments   67

Cable TV in a Nutshell   68

Chapter 4
VIDEOCASSETTE RECORDERS   70

Basic Operation of
a Videocassette Recorder   71

Connecting Up Your VCR   79

Other Controls and
Features Found on VCRs   85

Player/Recorder
Compatibility   90

Getting More
Out of Your VCR   93

Portable VCRs   104

Ways to Wire
Your Video System   113

All About Tape   115

Common VCR
Ailments and Cures   127

Future Developments   134

VCRs in a Nutshell   135

Chapter 5
TV CAMERAS AND LENSES   138

How TV Cameras Work   138

Operating a TV Camera—
the Basics   141

Features and Controls
on the TV Camera   143

More About
TV Camera Lenses   157

Tripods and Dollies   168

Camera Cable
Compatibility   170

Common TV Camera
Ailments and Cures   171

TV Cameras in a Nutshell   175

Chapter 6
**ADVANCED CAMERA
TECHNIQUES AND EFFECTS**   178

Handling the Camera   178

Camera Angles
and Picture Composition   189

Creative Camera Angles,
Techniques, and Tricks   200

Surreptitious Recording   207

Shooting Sports Events   208

Camera Techniques
in a Nutshell   208

Chapter 7
RECORDING AUDIO   210

The Basic Basics   210

The Microphone   214

Choosing and Using
the Proper Microphone
for a Recording   217

Automatic Volumes Controls   223

Recording
Stereo Simulcasts   224

Mixers   225

Sound Mixing Techniques   231

Special Audio Devices   235

Common Audio
Ailments and Cures   237

Audio in a Nutshell   241

Chapter 8
LIGHTING 244
- The Basic Basics 245
- The Kind of Light a Camera Needs 247
- Basic Lighting Techniques 249
- Creative Lighting Strategies 256
- Lighting for Color 258
- Care of Lighting Instruments 260
- Lighting in a Nutshell 261

Chapter 9
GRAPHICS 263
- The Basic Basics 263
- Fitting Your Graphic to the TV Screen 264
- Boldness and Simplicity 267
- Gray Scale 267
- Typography 269
- Displaying Graphics 274
- Backgrounds for Titles 277
- Making Graphics Come Alive 278
- Film-to-Tape Transfer 278
- Graphics in a Nutshell 283

Chapter 10
COPYING A VIDEO TAPE 285
- The Basic Basics 285
- Making the Best Possible Signal for Copying 286
- Connecting Up Your Equipment 287
- Making the Copy 288
- Copying Commmercially Prerecorded Programs 290
- Scan Conversion 291
- Copying Other Video Signals 293
- Photographics Copying 293
- Copying in a Nutshell 297

Chapter 11
EDITING A VIDEO TAPE 298
- Assemble Editing 300
- Recording Something Over 301
- Insert Editing 302
- Using Scan Conversion for Glitch-Free Edits 305
- Audio Dubbing 306
- Editing from Another Video Tape 308
- Don'ts and Do's of Editing 313
- Transitions 318
- Editing Strategies 325
- Editing Techniques 326
- Electronic Editors 327
- Editing in a Nutshell 329

Index 331

Preface

EXCITING THINGS TO COME

So you want to buy a videocassette recorder, and maybe a camera, but you don't know what kind to buy or if it'll be too difficult to operate. Or maybe you're thinking of putting up a big dish in your back yard to pull in a zillion TV stations off the satellites. Or maybe you just want to connect several TV sets to the same antenna or perhaps to "The Cable" and don't know how.

Perhaps you tried reading some of those "home video" magazines to learn more about this new technology, only to discover that they seem to be written for people already familiar with video, not for beginners. Or maybe you're looking for the answer to a specific question and you don't feel like digging through forty video magazines for the information.

Or maybe you have a videocassette recorder and would like to clean the heads and do other minor maintenance tasks yourself. Perhaps you also have a TV camera and would like some tips on composing shots for professional-looking TV shows of your own.

If any of the above is true, then you're clutching the right book. Not only will *The Complete Home Video Book* cover these and many other topics, it will do so simply and painlessly (except for paper cuts).

This will be a gentle first step into the world of video. You'll need no prior electronic or photographic knowledge to read this book. Each topic is covered, starting with the basics and working gradually up from there. You'll find no electronics mumbo jumbo that only technicians care about. Nor will you find a lot of video jargon that is of interest only to academicians.

Instead, what you will find is a lot of up-to-date information, presented simply and sometimes humorously. You'll also discover that this information, written by an educator, will be easy to remember and apply.

For instance, as you page through this book you may notice that certain terms are printed in SMALL CAPITALS. As you deal in video, you'll find everyone using these words again and again. The special capitals will help you learn and remember these terms.

Acknowledgments

Thanks first go to Florence Hahn, who provided me with the most essential tools of the author trade: pencils.

Thanks also are due my models, Theresa Ciccone, Donna Insinga, and Dorothy Van Duyne.

The illustrations are by Olive Volsky and many of the photographs by Chuck Brehm. The teeth marks in the corners of some of the pictures are by my cat, Tinkle. Thanks, Tink.

Cover Notes for Volume I

Unless you're holding two books in your hand, you're holding only half a book. The other half is Volume II, still sitting on the shelf. Why did we divide this book into two parts? So the author could tell his friends he'd written *two* books? Nope. To make the books handier for propping up three-legged sofas of various heights? Nope again. To trick you into buying Volume I cheap knowing you'd read it and get hopelessly involved, and then sock you with an inflated price when you went to buy Volume II to see how the story came out? Wrong again.

We're figuring that while most of you will buy both volumes to have a *complete* reference book on home video, some of you may be interested only in TVs, videocassette recorders, and cameras. Why pay for all those extra pages on satellite receivers, TV projectors, and video maintenance and accessories if you don't give a flying quack about those things? Thus the honey-hearted publishers went to extra trouble and made it possible for you to buy half a book and save a little wampum. Thoughtful, huh?

Believe that and you'll buy anything. Might as well take Volume II.

Chapter 1
All About TV Sets

HIS MASTER'S FACE

In the Preface, I promised to start with the easy stuff first, but then, who ever reads a preface?

According to a recent study, the average three-year-old spends four hours a day watching TV. We all learned how to operate TVs long ago and need no further instruction, right? Then why is it that half the TV sets you see are displaying green or magenta faces? And why do so few folks watch UHF TV stations?

Considering that billions of dollars go into producing and broadcasting crisp pictures to our homes, and considering that most of this book will involve how to produce or record crisp pictures of our own, it would be a shame to have all that work turn to mush in a horribly adjusted TV set. So here come the basics (skim through them if you wish) followed by some lesser-known secrets of The Tube.

Operating a TV Receiver

Since there are many kinds of TV receivers, the descriptions in this chapter may not fit your particular set exactly. Usually the difference will be in the labeling and/or location of the knobs and switches. When in doubt, read your TV's instruction manual (which I'll bet you can't find).

Okay, let's turn the set on, assuming you can find the knob. The knob you are looking for may be called: POWER, ON, ON/OFF, OFF, VOLUME, or it may have no markings at all. If you can't find the switch easily, try turning, pulling, or pushing a few. There are no external knobs on a TV receiver that, when maladjusted, can permanently harm the set, so go ahead and explore.

Now that you've switched the set on, you'd like to know if the set is really working. You can wait a minute for the TV to warm up and come on, or you can immediately tell that the set is on by (1) seeing the light illuminating the channel numbers come on (most receivers except the small ones have channel selectors that light up), or (2) hearing a faint, high-pitched squeal that tells you the TV circuits are working. If your set has an instant-on feature, you will also get a picture immediately, but sets without this capacity require a warm-up period before the picture appears.

1

If you have waited about 30 seconds and nothing happens, turn to Common TV Ailments and Cures on page 19 at the end of this chapter and start troubleshooting to find out why the set is not getting power.

Common controls on a TV set

ON/OFF, VOLUME
Gives power to the set and controls loudness of the sound.

CONTRAST
Makes the whites whiter and the blacks blacker.

BRIGHTNESS
Makes blacks and whites both whiter. In brightly lit rooms, the brightness should be turned up to make the picture more visible; in dimly lit rooms it should be turned down, thus affording greater picture detail.

VERTICAL HOLD (or just VERTICAL)
Stabilizes the picture and keeps it from "flipping" or "rolling."

HORIZONTAL HOLD (or just HORIZONTAL)
Centers the picture and keeps it from twisting into a bunch of diagonal lines. Sometimes this control is in the back of the set; sometimes it is even hidden behind a hole in the cabinet and requires adjustment with a screwdriver. (However, a Vodka Collins will do.)

CHANNEL SELECTOR
Picks the channel you want to watch.

FINE TUNING
Tunes in a channel exactly, to improve picture and sound.

AUTOMATIC FINE TUNING or AFT or AFC
When pressed, this control automatically FINE TUNES your channel and sometimes activates other automated circuits.

COLOR INTENSITY (or just COLOR)
Adjusts how saturated or pastel the colors become.

HUE or TINT
Adjusts the greens and reds to balance, not overpower, each other.

Some sets also have the following:

SHARPNESS or DETAIL
Makes the picture sharper; it is usually left in the sharpest position.

> TONE
> Makes the sound sharper or more bassy.
> DISTANT/LOCAL
> A switch, usually near the antenna, which adjusts the antenna sensitivity. Unless you're very near a TV station, leave the switch on DISTANT for best results.

Setting Up a TV Receiver for Use

A TV set falls off the back of a truck and into your arms. You tried your best to stop the driver, but to no avail. You might as well take it home and try it out.

After plugging it in and turning it on, your next endeavor is to get the TV signal into the set so that you have a clear picture and good sound. For this you need an antenna, unless you are receiving your TV signal via The Cable, in which case you connect the cable to the back of your set in the place your antenna would go. If you are in a neighborhood where TV signals are strong (such as 20 miles from the TV station with no mountains or tall buildings between you and the TV transmitter), you can usually get a satisfactory TV signal from the set's own built-in antenna if it has one.

Because a TV set will sometimes use a separate, external antenna and sometimes use its own antenna, there is a place behind the set where the antennas are connected and disconnected. To use the set's own antenna, you must check to see that it is connected. This is sometimes achieved by flipping an ANTENNA switch to the INT (or INTERNAL) position, which means that the built-in antenna is automatically connected to the TV's circuits when the switch is in that position. Most times, however, you connect the set's antenna to its circuits by finding the appropriate wires (see Figure 1-1), slipping the metal ends of the wires under the two screws, and tightening the screws with a screwdriver. It doesn't matter which wire goes under which screw, so long as the wires from the rabbit ears (the VHF antenna, which is the one that telescopes) go to the antenna connection marked VHF. The little bow-tie or loop-shaped antenna connection is the UHF antenna and connects to the screws marked UHF. With this done, you're ready to watch TV.

If, however, you are distant from a TV transmitter, the rabbit ears will give you a grainy or snowy picture. To improve the picture, you need to disconnect the rabbit ears and connect the wires from a larger, more sensitive antenna in place of the rabbit ears. It doesn't matter which wire goes

Figure 1-1
Connecting the internal or "rabbit ear" antenna

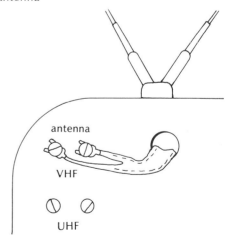

under which screw as long as the two bare wire ends are not touching each other or any other metal. Once the antenna is connected (as in Figure 1-2), you're ready to use the set. For receivers with the INT/EXT switch, be sure the switch is set to EXT. (There is more about antennas in Chapter 2.)

Here are some handy tips for setting up a TV receiver for a group of people to watch:

1. Avoid having bright lights on in the room, or tilt the set slightly forward to reduce reflections and glare on the screen's surface.

2. If the room is cursed with an unshaded window (or other bright light source), place the TV in front of the window. That way the screen will be shaded from the light and easier to see.

3. If the building is made of brick and steel and the TV has its own built-in antenna, place the TV near a window. TV signals travel poorly through brick and steel.

4. A 9-inch screen (TV screens are measured diagonally) can be comfortably seen by five or fewer viewers. A 21-inch screen is good for up to 20 viewers. For more than 30 viewers, definitely use more than one receiver.

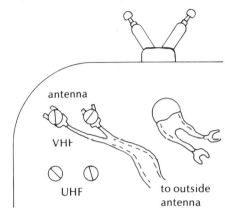

Figure 1–2
Disconnecting the internal antenna and connecting an outside antenna wire

5. If echoes are a problem, angle the set out from a corner of the room. Perhaps turning the TONE control to "sharp" will help, especially with the sounds of speech.

6. Television shows are most effective and are most easily remembered when preceded and followed by some discussion of the subject shown.

How TVs Work

This is just a quickie science lesson so you'll understand what's happening later when things get more complicated.

Words to know

AUDIO
The sound portion of a TV presentation.

VIDEO
The picture portion of a TV presentation. This electrical signal usually comes directly from a TV camera or video tape machine.

SYNC
Short for synchronizing. Another electronic signal that holds the picture steady on your TV screen.

> **RF**
> Radio Frequency. The kind of signal that goes through your TV-antenna wire.
>
> **VHF**
> Very High Frequency. Channels 2 through 13 are called the VHF channels.
>
> **UHF**
> Ultra High Frequency. Channels 14 through 83 are called the UHF channels and are selected on a separate tuning knob from the VHF channels. The UHF channel selector is usually activated when the VHF channel selector is turned to the position marked "UHF" or "U." UHF and VHF channels each have their own fine tuning knobs.
>
> **AFT**
> Automatic Fine Tuning. Pushing this button helps tune the set to the station and sometimes turns on other automatic circuits in the receiver.

VIDEO, AUDIO, SYNC, AND RF

When a television program is produced, whether "live" or by video tape recording, a camera takes the picture, changing it into an electrical signal called VIDEO. A microphone takes the sound and makes another electrical signal called AUDIO. And a special device called a SYNC GENERATOR creates a third electrical signal called SYNC that keeps the picture stable. When SYNC and VIDEO are electronically combined into a single electrical pulse, the signal is called *composite video*, but most TV people refer to them both simply as "VIDEO."

The TV broadcaster combines the AUDIO and VIDEO and SYNC using a device called a MODULATOR, which codes the three into another signal called RF. The RF is transmitted, travels through the air, gets picked up by your antenna, and goes into your TV receiver. By tuning your TV receiver to a particular channel (the same one that was broadcasted), a circuit in the TV set decodes the RF signal and breaks it up into VIDEO, AUDIO, and SYNC, as shown in Figure 1-3. The VIDEO goes to the TV screen, the AUDIO goes to the speaker, and the SYNC goes to special circuits which hold the picture steady. Incidentally, by adjusting BRIGHTNESS, CONTRAST, HUE, and COLOR INTENSITY, you adjust the TV's VIDEO circuits. By manipulating the VOLUME control, you adjust the AUDIO circuits. By moving the VERTICAL or HORIZONTAL controls, you adjust the SYNC circuits in your TV.

HOW THE PICTURE IS MADE

Inside your TV is an electronic gun, something like a machine gun for electrons. It's hidden beneath that bump in the middle of the back of every TV, and it's part of the TV's picture tube. This gun shoots a beam of electrons at the inside face of the TV screen, which is covered with phosphor dust. When the electrons hit the phosphor it glows. To make a picture, this gun sweeps across the screen from side to side, much as your eyes sweep across each line on this page and eventually cover the entire page. The SYNC circuits tell the gun how fast to

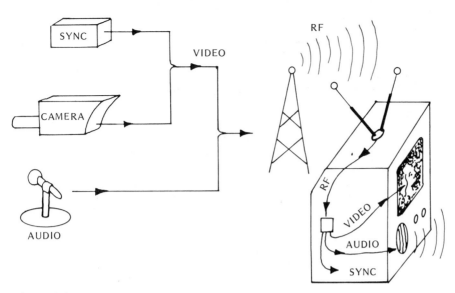

Figure 1–3
AUDIO, VIDEO, SYNC and RF

sweep and when to sweep, as well as when to stop shooting at the end of a line (just as you stop reading at the end of a line and zip your eyes back to the beginning of the next line). SYNC also tells the gun to stop shooting when it reaches the bottom of the TV screen, then reaims it at the top and starts it shooting again. As the beam zigzags across the screen the VIDEO signal tells the gun to shoot harder or weaker, depending on whether it is tracing a lighter or darker part of the picture. By turning up your BRIGHTNESS control, you tell the gun to shoot harder at the screen, thus lighting up the phosphor brighter.

This gun zips its beam across the screen 15,750 times each second, and it starts at the top 60 times each second. The phosphors keep glowing until the next time the beam comes around and zaps them again. Thus the screen appears smooth and flicker-free. European TVs retrace themselves only 50 times per second rather than 60, and as a result European TV flickers more noticeably than American TV. If you look very closely at your TV screen, especially at a light part of the picture, you will be able to see all those tiny horizontal sweep lines that make up the picture.

COLOR TV PICTURES

In a darkened room, aim a red flashlight at a white wall and you'll see red. Turn it off and you'll see black. Shine a blue light on the wall and you'll see blue. Shine green and you'll see green. Shine red and blue together and the colors will mix to create a new color, magenta. Shine red and green and you'll see yellow. Shine all three and you'll get white. All the other colors can be made from shining various proportions of these three primary colors.

A monochrome TV set makes its picture by electronically projecting black (which is really just the absence of white) and white onto a screen. A color TV makes its color picture by creating three pictures on its screen: one red-and-black, another green-and-black, and the third blue-and-black.

Where only red is created, you see only red. Where the red and green pictures overlap on the screen, you see yellow. Where all three pictures are black, you see black. Where all three colors converge with equal strength, you see white.

In a color picture tube there are three electron machine guns, one for each color. The face of the color TV screen is made of dots or bars of phosphor which shine red, blue, or green when hit by electrons. Look really closely at your TV screen while it's displaying a white picture and you'll be able to see the discrete colored dots or stripes which make up the whole colored picture. The blue gun is arranged so that it can only hit the blue phosphors, the red can only hit red, and the green only green. The three guns independently scan a picture onto the screen, and the three pictures overlap to create dazzling views of space shuttles and clogged sink drains alike.

In the back of some TV sets are adjustments for these three electron guns. They are used by TV repairmen to weaken or strengthen a certain color, balancing all three so that one doesn't overpower the others and tint all your pictures.

Adjusting the Controls on a TV Set

CHANNEL SELECTOR

Television and radio signals travel through the air as invisible electromagnetic waves. Some waves vibrate very fast and are said to have a high FREQUENCY, while others oscillate more slowly and have a lower FREQUENCY. As you tune your radio up from the lower numbers to the higher numbers on the dial, you are tuning in stations with higher FREQUENCIES.

Unlike radio dials, which show you the actual FREQUENCY you're tuned to, television dials just give you channel numbers. Channels numbered 2 through 13 are called the VHF, or Very High Frequency channels. Channels 14 through 83 are called the UHF, or Ultra High Frequency channels. UHF channels are usually selected on a separate tuning knob from the VHF channels, and the UHF channel selector is usually activated when the VHF channel selector is turned to "U" or "UHF," a spot on the dial between channels 13 and 2. The UHF and VHF channels have their own separate FINE TUNING knobs.

To tune in a station properly, assuming your antenna is correctly aimed (more on that later), you adjust FINE TUNING as follows:

1. First switch off any automatic controls, if your set has them, as they will confuse the issue. They may be labeled AUTO, AUTO COLOR, or AUTOMATIC FINE TUNING (or AFT), or AFC.

2. Next adjust the FINE TUNING control, which is probably a knob concentric with the channel selector (as in Figure 1-4). On some sets you have to push it in and turn it to make the dial work. When adjusted in one direction, the picture will look soft and a little fuzzy. When adjusted in the other direction the sound may become raspy and cause wavy lines in the picture while the picture becomes rough-edged and grainy (as in Figure 1-5). If you adjust to this second position and then back up just a little until the picture is sharp and the sound is good, you will have a well-tuned signal.

3. Now turn on the AFT, or AUTO control (or whatever), and this will reactivate automatic circuits which will further tune the set to the station. Your picture should now look the best it can be.

Later, in the chapter on cable TV, we'll examine some TV channels that aren't listed on your dial and can only be received through a converter box.

Figure 1–4
Channel selector and concentric FINE TUNING dial

Figure 1–5
FINE TUNING at one extreme, showing a grainy picture with slight wavy lines in it

BRIGHTNESS AND CONTRAST

Black-and-white TVs don't make black-and-white pictures. They just make pictures of nothing-and-white. The *nothing* part is the same darkness you find on the screen when the set is turned off. A TV that's turned off and sitting in your windowless King Tut's basement will have a darker screen than the same TV if sitting on your sun-drenched patio. Thus, when the set's turned on, its basement picture will have rich, black blacks while the patio picture is washed out and gray. Since you can't make the blacks on the screen blacker (without throwing a heavy blanket over you and your patio TV) you make the whites whiter by turning up the BRIGHTNESS and CONTRAST. This makes the blacks *look* blacker by comparison. But when the environment is dark again, it pays to turn these controls back down. Not only is the excess contrast unnecessary, but it robs sharpness and subtlety from your picture.

All of the above is true for color TVs too; their pictures are made of nothing-and-color and their blacks are only as black as the face of the picture tube when the set's off. If you examine a number of color sets in a TV showroom you may notice that some have light-faced picture tubes and others dark-faced ones. The dark-faced ones will yield blacker blacks than will their counterparts in a brightly lit room, so they are more desirable TVs to have in this respect. Some color sets have a PICTURE control which adjusts the CONTRAST and BRIGHTNESS together to simplify this day-night accommodation. Others have an electric eye sensor which measures the brightness in the room and automatically adjusts the picture accordingly.

COLOR AND HUE

To adjust the colors on your TV correctly, follow these steps: First make sure your TV set has "warmed up" for 2 to 3 minutes before you start adjusting, unless it is an "instant-on" type. "Cold" TVs have a tendency to change colors while warming up. Next tune in a station as best you can. Turn off the automatic controls and turn down COLOR all the way and you will display a black-and-white picture on your color set.

8 All About TV Sets

Next adjust your BRIGHTNESS and CONTRAST so that you have black blacks, white whites, and smooth grays in between. Now that you have a good black-and-white picture we're ready to put the color back.

Incidentally, if a station is weak, or you're playing a defective tape through the TV, or your antenna signal is fouled up, your TV's color picture may flash color on and off or do other distracting things. By turning the COLOR all the way down as described above, you will get rid of the distracting color mess and may enjoy the show more, even if it can only be seen in black-and-white.

Turning up the COLOR control increases the SATURATION or vividness of the color. Turning the HUE (or TINT) control changes all the colors from a greenish cast to a reddish cast. To have a good color picture you want *enough* color and the *right* color. Let's work on the *right* color first—HUE.

Dresses, sunsets, and pizzas can be any color. You can't use them as a guide to proper color adjustment because you never know if you've made them the color they are really supposed to be. Somehow, though, you can always look at a Caucasian face and tell if the color is right. Therefore, we use "flesh tones" to guide our HUE adjustments. To do this, first tune to a station with lots of flesh. Daytime soap operas always have a lot of facial close-ups, so stay home from work one day to adjust the color on your TV using the soaps. Your boss will understand. Next turn COLOR *way up*, too far. This is so you can see the colors better. Then adjust HUE until the face colors look right. And last, turn COLOR down until the overall picture looks good. You're done. The theory here is that if the faces look good, everything else will look good.

Newer TV's have a special circuit which makes this adjustment for you. This feature may be called various names (like Colortrack), but what it does is lock the circuits onto a signal broadcast by most stations as part of their TV picture. This signal, called Vertical Interval Reference Signal (VIRS), makes your TV circuits strive to reproduce color and contrast exactly the way they left the studio.

VERTICAL HOLD

This knob adjusts the SYNC circuits in your TV so that your picture doesn't roll or flop over, or jitter. Normally, the electron gun stops shooting at the bottom of your picture and starts again at the top. Misadjust SYNC and the gun will errantly think it's at the end of its picture while it's right in the middle. That black bar you see rolling across your screen (like in Figure 1-6) is the gun ending one picture, turning off, and starting another, but in the wrong place.

Figure 1–6
Vertical roll—adjust VERTICAL control

VERTICAL HOLD usually needs adjustment when the TV is getting a garbled signal. Video tape players, weak TV stations, or other interference sometimes mess up the SYNC signal, requiring you to adjust VERTICAL.

HORIZONTAL HOLD

This adjustment, hidden on many sets, moves the picture left and right, centering it on your screen, and keeps it from tearing into diagonal lines (like in Figure 1-7). It adjusts the part of the SYNC circuit that tells the electron gun that it's at the end of a line, should turn off, zip back, and start shooting a new line across the screen.

What throws the HORIZONTAL HOLD off is the same thing that messes up VERTICAL HOLD: a bad signal from somewhere (unless the TV set is defective).

Figure 1–7
Diagonal lines—adjust HORIZONTAL HOLD

Words to know

TV RECEIVER
A television set like the one you have at home. It connects to an antenna and shows channels 2–13 and channels 14–83.

TV MONITOR
A television set without a speaker, so it makes no sound. Instead of picking up channels off an antenna, it gets its picture from a direct connection to a TV camera or a video tape machine.

TV MONITOR/RECEIVER
A television set that does both jobs listed above. It can either get its signal off an antenna or from a direct connection to a TV camera or video tape machine.

MONITOR
The verb "monitor" by itself means "to check quality." You can monitor someone's heartbeat with a stethoscope. An AUDIO monitor could be a loudspeaker to let you know a microphone is working. Thus a TV MONITOR displays the kind of VIDEO signal you are receiving. Often people who deal with TV use the word MONITOR to mean TV MONITOR.

TV Monitors and Monitor/Receivers

THE TELEVISION MONITOR

In contrast to a TV RECEIVER, a television MONITOR does not play AUDIO and does not change channels. All it does is display a picture that it receives directly from a TV camera or recorder via a wire or cable. Like TV RECEIVERS, MONITORS have ON/OFF switches and controls for CONTRAST, BRIGHTNESS, VERTICAL, and HORIZONTAL HOLD, and perhaps also for picture height and width, but they have no AUDIO controls.

Who would ever use such a thing as a TV without sound and without tuner? Professional TV people, security companies, people who *only* are interested in a picture from a camera, etc., would. Why should they pay to have a speaker and tuner take up space in the box when they have no intentions of using them? Also, TV MONITORS, being specialized equipment, generally have better circuits and give sharper pictures than regular TV sets. They also cost more.

But there's another reason why professionals prefer MONITORS to regular TV receivers. To get VIDEO into a home TV RECEIVER, you have to convert it to a channel number using a MODULATOR (something like a radio transmitter) so you can feed it into the antenna of the TV. Once this RF signal goes in the TV's antenna, the TV reconverts it back to VIDEO (review Figure 1-3). All this converting and reconverting makes the picture fuzzier. It would be nice to skip all this RF foolishness and send the VIDEO from the videocassette machine or camera *straight* into the TV MONITOR. Thus no picture degradation.

To feed a signal to the MONITOR, simply plug the VIDEO cable into the socket in the back of the set marked VIDEO IN, and a picture should appear on the TV screen (if there's a VIDEO signal in the cable).

CONNECTING THE TV MONITOR

How do you know when a cable has VIDEO in it? If you can't see where the wire is coming from, you can make a good guess as to what it carries just by its looks. If it is ¼" in diameter, round, and looks like a stiff rubber hose, it probably contains AUDIO, VIDEO, or RF. If the plug on the end is an AUDIO plug (you'll see how to recognize these later), the wire most likely contains AUDIO. If the cable has an F CONNECTOR (see Figure 1-8), it probably carries RF from an antenna. If the plug is the BNC type or the PL259 type (see Figure 1-8 again), it probably carries VIDEO. It generally will not harm anything to plug the plug in anyway (if it fits) and see if you get a picture. Incidentally, it wouldn't hurt to learn the names of these connectors and how to recognize them.

Most home videocassette recorder manufacturers have really complicated the issue by choosing to use a PHONO (sometimes called RCA) plug and socket (see Figure 1-9) to carry their VIDEO signals. Up to recently, RCA plugs were only used for AUDIO. When you see such a plug, you'll have to trace it back to its source to find out whether it's carrying AUDIO or VIDEO.

USING SEVERAL TV MONITORS TO SHOW THE SAME PICTURE

You may notice a lot of other sockets and switches on the back of the MONITOR. Some of these are for connecting several MONITORS together. Before discussing how this is done, let's consider the gas company. It runs a big gas pipe to your house, by your house, and to the next house. You use a tiny amount of gas and the remainder goes to your neighbor. He uses a bit and the

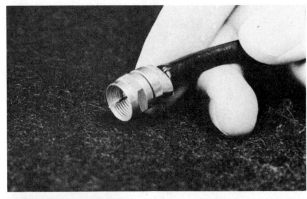
F connector

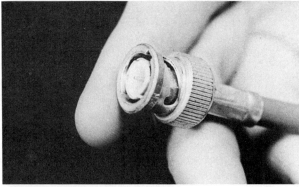
BNC plug

PL 259 connector
(the PL stands for plug)
Also called UHF connector

Figure 1–8
Common TV connectors

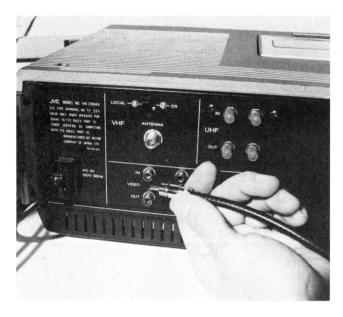

Figure 1–9
PHONO (or RCA) plug and socket. Normally used for AUDIO, most home videocassette recorders use these connectors for VIDEO.

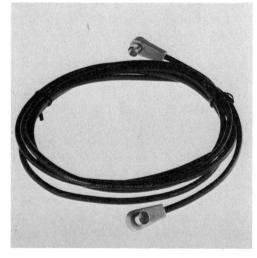

TV cables. Left, VIDEO cable with RCA plugs on ends for use with most home videocassette recorders. Special COAXIAL wire is used for carrying VIDEO signals efficiently. Common PATCH CORDS used for your home stereo equipment also may have RCA plugs and can work in a pinch, but degrade the VIDEO signal. Right, RF cable with push-on F plugs on ends for use with cable TV connections, TV ANTENNA INPUTS, videocassette recorder ANTENNA INPUTS, anywhere RF signals need to go. (Photos courtesy of Comprehensive Video Supply Corp.)

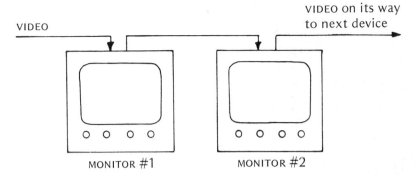

Figure 1-10
VIDEO LOOPED through two MONITORS

remainder goes to his neighbor. At the end of the line, the gas company puts a cap on the end of the pipe to plug it up. They don't want any unused gas leaking out.

Similarly, VIDEO is run to your MONITOR by plugging a VIDEO cable into the VIDEO input of the MONITOR. The MONITOR samples a tiny bit of the signal to make the picture. If the signal is to go to another MONITOR after that, it is LOOPED or BRIDGED, which means that another cable is connected to the back of the first MONITOR and runs to the VIDEO input of the second MONITOR, as shown in Figure 1-10. Most of the VIDEO signal enters the first MONITOR and then exits through these wires in the "looping through" process. It is possible to show the same picture on five or so MONITORS this way, with each taking a bit of the signal and passing the rest on.

Similar to the gas company which had to plug the gas main after the last house, something has to be done to the last MONITOR on the line. If it is not looping through to somewhere, the last MONITOR must be TERMINATED. This is a sort of electronic "plugging up."

There are two ways to terminate this line at the end. The first way is to plug a 75-OHM TERMINATOR into the back of the MONITOR in the place where the cable looping to the next set would have gone, as shown in Figure 1-11. This 75-OHM TERMINATOR looks like a PL259 connector without a cord. TV studios keep them around because every time they use a piece of VIDEO equipment with a VIDEO input that can be BRIDGED or LOOPED, they have to TERMINATE the device (unless they are LOOPING elsewhere).

The second way to terminate is to use the switch provided for this purpose (if the equipment has one). It will probably be near the socket and be marked 75Ω (Ω is the symbol for ohm) or 75-OHM TERMINATOR. The switch positions will be marked 75Ω and HI Z. Use the HI Z position when the signal is to LOOP THROUGH to another piece of equipment. In the 75Ω position, the switch TERMINATES the signal. Don't use both the switch and the terminator plug—use one or the other. Using both is called "double terminating" and causes a picture problem.

How do you tell when a MONITOR or another piece of equipment has the capability of being LOOPED or BRIDGED? Usually the sockets come in pairs set close together and have the words LOOP or BRIDGE printed near them.

In short, the law reads like this: *If the MONITOR (or other device) is made so that it can LOOP THROUGH, it must be either LOOPED to somewhere else or TERMINATED.*

So if, for instance, you are running VIDEO to only one MONITOR, and if it has a terminating switch, the switch must be in the 75Ω position as in Figure 1-12. If, conversely, you LOOP a MONITOR to something, the MONITOR's switch should then go to the HI Z position.

VIDEO from sources

75ω TERMINATOR

Figure 1–11
MONITOR properly terminated

Figure 1–12
MONITOR properly terminated

VIDEO ADAPTERS

What if the cable has a PL259 plug but the socket in the back of the MONITOR is a BNC type? Such incompatibility occurs all the time because different manufacturers use different kinds of sockets. The PL259 and BNC are used by TV professionals and serious amateurs. More common for home use is the PHONO plug. If you plan to do much serious TV work, buy some adapters, like those shown in Figure 1-13, that will permit you to convert from one kind of connector to the other. Also, learn the names of these connectors and how to recognize them—it will pay off.

VIDEO connectors are either male or female. Examine a few and you will readily deduce the origin of the sexual connotation. Almost always, the plugs will be male and the sockets in the TV devices will be female. The PL259 (male plug) goes into a SO259 (female socket). The BNC male plug goes into a BNC female socket. The "barrel" connector allows two males to connect and could be considered the homosexual of VIDEO connectors; the T CONNECTOR appears to be suited for a "ménage à trois."

Figure 1-13
VIDEO adapters

TV Monitors and Monitor/Receivers 15

More VIDEO adapters, from left to right; Female F to male mini, female F to male F push-on, female F to right angle male F push-on, female BNC to right angle male BNC (Courtesy of Comprehensive Video Supply Corp.)

MONITOR/RECEIVER

A TV MONITOR/RECEIVER (or receiver/monitor) does the job of either a TV MONITOR or a TV RECEIVER. It either accepts RF from which it derives picture and sound, or it accepts VIDEO and AUDIO separately and displays them.

The most common MONITOR/RECEIVERS may look exactly like home TV sets except for one switch and one socket on the side or back. The switch changes the TV from a MONITOR to a RECEIVER.

The simplest MONITOR/RECEIVERS have eight-pin sockets on them rather than the traditional connectors. The eight-pin plug that goes into this is called, strangely enough, an 8 PIN (Figure 1-14), and usually a video tape machine is at the other end of the cable. Instead of connecting a VIDEO cable from the tape machine and then connecting a separate AUDIO cable (to get the picture and sound from the tape player to come out of the TV set) between the two, this 8-PIN connecter is used. It has both VIDEO and AUDIO wires in it. One convenient cable does the whole job. Incidentally, Sony calls this cable a "monitor connecting cable." Panasonic calls this a "VTR-monitor connection cable" or "TV control cable." Informally, some call it a "Jones Cable."

Figure 1–14
8-PIN plug

To attach the cable, line up the pins on the plug with the holes on the corresponding socket on the MONITOR/RECEIVER; then push the plug in until it clicks. To unplug it, squeeze together the two buttons on either side of the plug to unlatch the connection, and then withdraw the plug as shown in Figure 1-15.

If someone has been using the MONITOR/RECEIVER as a regular TV receiver, and you now wish to use it to play a video tape, all you have to do is:

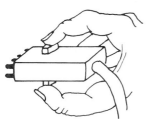

Figure 1-15
Grasping the 8-PIN plug

1. Turn the MONITOR/RECEIVER power on. Doing this assures that the device will be "warmed up" and ready to go by the time you finish the other steps.

2. Plug in the 8-PIN plug and be sure the other end of the cable is plugged into the video tape player as shown in Figure 1-16.

3. Find the switch on the MONITOR/RECEIVER that says AIR/LINE or TV/VTR and switch it to the LINE or VTR position. This switch converts the TV RECEIVER into a TV MONITOR with sound.

shown in Figure 1-17. Just run the AUDIO cable from the video tape player to the socket that says LINE IN or AUX (for auxiliary) IN, or EXT (for external) AUDIO IN and plug it in. Then run the VIDEO cable to the socket that says VIDEO IN or EXT IN. If you are using the popular 9-inch Sony CVM 920U MONITOR/RECEIVER (shown in Figure 1-18), the label reads simply EXT IN and Sony leaves it to your good sense to tell which socket is audio and which is video from the socket shape. Professional MONITOR/RECEIVERS, however, are well labeled, like the one shown in Figure 1-19, so you can't go wrong deciding where to plug things in.

Once your machines are connected properly, set the terminating switch to 75Ω unless you are LOOPING THROUGH to another MONITOR or MONITOR/RECEIVER.

Have you noticed that it seems strange to be taking a signal out from a place called EXT IN? The IN seems to imply that signals go in, not out. Rest assured that your suspicions are still pretty much correct. All of the

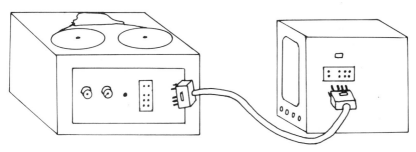

Figure 1–16
Connecting a TV MONITOR/RECEIVER to a video tape machine via an 8-PIN cable

This is all there is to it.

If you wish to use the MONITOR/RECEIVER as a straight RECEIVER again, simply flip the switch back to TV or AIR. Done. You don't even have to disconnected the 8-PIN plug.

What if your boss borrowed your 8-PIN cable to tie down the trunk lid of his car when he took off for that two-week vacation in Quebec? If your MONITOR/RECEIVER is blessed with separate VIDEO and AUDIO sockets in the back, you can use them as

sockets labled EXT IN are actually inputs. Any of them will accept a signal into the TV set. The reason you get a signal coming out of the adjacent IN socket is that the sockets are BRIDGED, or electrically connected together. As described earlier, almost everything that goes in one of those sockets is free to come out the adjacent socket to be used elsewhere. The TV set isn't *creating* any of the signal that is LOOPING to the next set; it is only sampling a little bit and allowing the

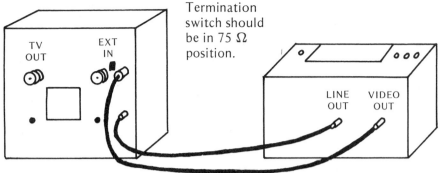

Figure 1–17
Wiring separate AUDIO and VIDEO to your MONITOR/RECEIVER

Figure 1–18
Rear view of a Sony CVM 920U TV MONITOR/RECEIVER

Figure 1–19
Rear view of a professional TV MONITOR/RECEIVER

remainder to pass on. In fact, you can turn off, unplug, or throw darts at the first set and it will still pass the signal on to the next set if the two are properly connected.

Some MONITOR/RECEIVERS have additional AUDIO and VIDEO sockets labeled TV OUT and AUDIO OUT (see Figures 1-18 and 1-19 again). Here's what they do. RF (remember RF?) comes down the antenna wire or distribution cable if you have "Cable TV." Depending on where the channel selector is set, the TV decodes the RF signal into VIDEO, SYNC, and AUDIO. The combined VIDEO and SYNC come out the VIDEO socket labeled TV OUT, while the AUDIO comes out the adjacent AUDIO socket. These two signals could be sent to a video tape recorder or to another MONITOR, allowing you to watch a TV program on two sets at the same time. So to convert a broadcast TV program into VIDEO and AUDIO for recording it or redisplaying it, just connect to the TV OUT sockets.

Now back to the 8-PIN connector for a moment. Remember how the 8-PIN plug carried AUDIO and VIDEO into the set, only without the mess of connecting separate wires to the VIDEO IN and AUDIO IN? The 8 PIN does more than that. It not only sends signals *to* the MONITOR/RECEIVER, but it will receive AUDIO and VIDEO signals *from* the MONITOR/RECEIVER. Use the 8 PIN and you don't have to bother with the TV OUT sockets. You'll see the convenience of this later when we discuss recording TV programs off the air.

In conclusion, a TV RECEIVER uses its antenna to pick up an RF signal and convert it to picture and sound. The process is simple, but in the conversion about 10% of the picture quality is lost. A TV MONITOR can display a picture from a VIDEO signal. It's not much use unless you have the VIDEO signal to start with, but if you *do* have such a signal, it will display a cleaner picture than a TV that has to mess around with RF. A TV MONITOR/RECEIVER, though about 150 percent as expensive as a TV RECEIVER, is the best of both worlds. It will display an RF signal as well as a straight VIDEO signal, and it will convert RF to VIDEO-and-AUDIO signals for you.

Common TV Ailments and Cures

Nothing happens when you turn the set on

1. Make sure it is plugged in.

2. Or, unplug it and plug it in somewhere else.

3. Or, make sure the wall plugs aren't controlled by a wall switch that's been turned off.

4. Check the CIRCUIT BREAKER (an electronic, resettable fuse), usually indicated by a red button on the back of the set. If the set comes on shortly after you push the button, fine. If it blacks out again, however, something is wrong. Don't use the set.

5. If the set is a portable type, designed to run on batteries, look for a power selector switch labeled BATTERY/AC or CHARGE and be sure it is set to the AC position, or *away* from CHARGE. Incidentally, any power switch that says DC means Direct Current. That is the kind of power you get from batteries. If the switch says AC, that means Alternating Current, which is the kind of power we in the U.S. get from a wall socket. If you get no power in the DC mode, your batteries may be dead. If you get no power in the AC mode, your wall outlet may be at fault.

6. If none of this works, read the instruction booklet.

Now that your set is on, you still may not get good picture and sound. Here's what can be wrong and what to do.

Good picture, no sound

1. Check the volume. Is it turned up enough?

2. Make sure there are no headphones plugged into the set (they may automatically cut off the sound).

3. Flip channels. If none of the channels has sound, the problem is in the receiver. If only the one channel has no sound, the problem is in the transmission from the station, or you're watching a silent opus with Theda Bara and Rudolph Valentino. (Some television programs have no sound.)

Sound but no picture

1. If the picture is black, turn up the BRIGHTNESS control.

2. If the picture is gray and washed out, turn up the CONTRAST.

3. Again, flip channels. Something may have temporarily stopped the picture transmission from the station.

Both picture and sound, but the picture . . .

1. Is not centered on the screen (as in Figure 1-20). Adjust the HORIZONTAL control. If you can't find it on the front of the set, look on the side, in back, or behind a trap door below the screen. Sometimes it's hidden behind a hole in the cabinet and must be adjusted with a long screwdriver.

2. Has diagonal black lines through it (as in Figure 1-7). Again, adjust the HORIZONTAL control.

3. Flips or rolls (as in Figure 1-6). Adjust the VERTICAL control until the picture is stabilized in the center of the screen.

4. Looks murky (as in Figure 1-21). Dark places fill in and everything looks very black or very white with no grays. Adjust the CONTRAST control.

The program and receiver are in color, but the picture is only black and white

1. Adjust the COLOR or COLOR INTENSITY control that governs the amount of color in the picture. If it is too low, the picture will look washed out; if it is too high, the picture will be too vivid to watch comfortably.

2. Adjust the FINE TUNING. For best results, first switch off the AUTO, or AUTO COLOR or AUTOMATIC FINE TUNING or AFT or AFC button (if your set has one) before adjusting the FINE TUNING. Then adjust the FINE TUNING control, which is probably a knob concentric with the channel selector (as in Figure 1-4). On some sets you have to push it in and then turn it to make the dial work. When adjusted in one direction, the picture becomes soft and fuzzy. When adjusted in the other direction, the sound may become raspy and cause wavy lines in the picture

Figure 1–20
Picture not centered—HORIZONTAL adjustment needed

Figure 1–21
Picture needing CONTRAST adjustment

while the picture becomes rough-edged and grainy (as in Figure 1-5). If you adjust to this second position and then back up just a little until the picture is sharp and the sound is good, you will have a well-tuned picture. The color may even come on. Now reactivate the AFT or AUTO control (or whatever), and your picture should look the best it can be.

The show is black and white, but your color receiver keeps flashing colored splotches over the picture

1. Turn the COLOR INTENSITY knob down all the way.

2. Or, turn off the AFT.

3. Or, turn off the AFT, then turn the FINE TUNING farther toward the soft, fuzzy picture position until the color flashing stops.

The receiver and show are both in color, but the color looks terrible

1. First turn off the AFT, AUTO COLOR, or whatever.

2. Then adjust HUE and COLOR to suit your tastes. Use someone's face as a guide to good color. When the flesh tones are just right, everything else generally looks good.

3. If this doesn't help, make sure the FINE TUNING is properly adjusted.

4. Finally, turn the AFT back on.

5. Do not mess with any color knobs in the back of the set. They will undoubtedly confuse the issue.

The picture has grain, snow, or ghosts (as in Figure 1-22)

1. Adjust the FINE TUNING as previously described.

2. If the set is operating on its own antenna, whether one or two VHF "rabbit ears" or a clip-on UHF bow-tie or loop, aim the antenna around in various directions to see what happens. If the antenna is a telescoping one, be sure it's all the way out to full extension.

Figure 1–22
Snowy or grainy picture

Palm-sized flat-screen television (Courtesy of Toshiba Corp.)

New Developments

What's coming next?

FLAT SCREEN TVs

Lucitron Corporation is working on its 35-inch Flatscreen, which would be 3 inches thick and made of thousands of electroluminescent elements, like tiny individual lamps. The problem is, it takes a heck of a lot of thousands of these little dots to make an entire TV picture.

Sharp manufactures a tiny screen marketed by Hycom Inc. that uses tiny thin film electroluminescent elements made of zinc sulphide and manganese. The trouble here is that the light from the screen is greenish yellow, okay for computer graphics but unsuitable for French Chef reruns.

Sinclair plans to market its $125 Microvision, which uses a normal-type TV screen that's very tiny and flattened a lot.

Perhaps the real race is between Hitachi, Matsushita, and Toshiba: all are developing pocket-sized TVs which use screens made of LCDs (Liquid Crystal Displays), the same stuff your gray-faced digital wristwatch uses for the numbers. Such TVs would consume very little power, but the black-and-gray images presently lack good contrast and the moving pictures smear somewhat. They are expected to cost about $350. Seiko has already demonstrated a 1½-inch, $400 wrist TV using an LCD screen and headphones. Perhaps they should call it the "wristwatchman."

Sony Corporation is preparing to market its 2-inch screen, wallet-sized "watchman" black-and-white TV. It will weigh about 19 ounces, measure 1.4 inches thick by 3.4 inches wide by 8 inches long and cost about $240. It will run 2.5 hours on four AA batteries. Unlike most flat-screen TVs, the watchman FD-200 uses no LEDs or LCDs. Instead, it uses a phosphor picture tube cleverly folded so that the electron gun shoots at the screen from the side rather than from directly behind (as regular TVs do).

LARGE-SCREEN TVs

The biggest color TV screen in mass production is Sony's 30-inch CVM-3000 professional MONITOR/RECEIVER. It weighs 275 pounds (mostly from the glass in the picture tube) and costs over $8000.

Some manufacturers project their TV image onto the back of a smoky screen (techniques called REAR SCREEN PROJECTION) and boast of enormous screen sizes. If you sit directly in front of one with the lights out, the picture's okay. Otherwise it's a bit dim.

Video projectors are another way to show large-screen TV images. Chapter 15 covers this subject in detail.

The largest TV screen is in Dodger Stadium in Los Angeles. The mammoth 20 × 28-foot color screen is made of 24,576 special light bulbs which individually vary in brightness and comprise a picture from these many dots of light. For $3 million, you can have one installed in your very own playroom. Imagine the Joneses keeping up with that!

Sanyo TPM 2000 with 2" screen and built-in radio (Courtesy of Sanyo Electric) Below, Panasonic TR 1010P with 1.5" screen (Courtesy of Panasonic)

OTHER TVs

Sony has introduced its KX-27 HF1 "Profeel" Video Monitor System, unusual because it's modular. You buy separate components such as TV monitor, tuner, speakers, etc., and assemble a package that fits your specific needs. It has a 26-inch screen which, because of a special COMB FILTER circuit, removes grain and snow from the picture and displays a picture 35% sharper than previous consumer models. It has a 10-watt AUDIO amplifer (that's a lot for a TV) and direct VIDEO and AUDIO inputs (which avoid converting a VIDEO signal to RF and then back again to VIDEO inside the set at a cost of 10% of the picture sharpness). It costs $1500, counting the tuner and speakers. Sears, Technika, and NEC also make high-performance component TV monitor/tuner systems. They're a few hundred dollars less than Sony's Profeel.

You can expect to see sharper pictures from TVs in the future as more manufacturers include COMB FILTERS in their more expensive models.

Three-dimensional television is still in its infancy, mostly because all systems so far require the viewers to wear special glasses, and the 3-D technique is incompatible with regular color TV broadcasts. The FCC frowns on broadcasters occupying the airwaves with programs which are unviewable to the 99.9% of the viewing population who don't have the glasses, or who don't own a special TV set.

Wide-screen TV has the same incompatibility problem with the FCC. Prototypes have been built by Sony, CBS, and NHK in Japan which prove that it's possible to have a cinema-type TV screen five units wide by three units tall rather than the 4 × 3 box we're all used to. Besides a wide-screen picture, these systems produce a much sharper picture than regular TV, comparable to projected 35mm slides. The problem with H-D (High Definition) TV is that it takes the equivalent of five TV channels at once to broadcast super-sharp pictures and special wide-screen TVs to display them. H-D TV may be a decade away.

Stereophonic TV is already in use in Germany and Japan, but the FCC has not yet decided which method should be used for making stereo audio available to the U.S. broadcast-TV public. Once they approve the standards for stereo and everybody buys stereo TV sets, we'll be stuck with that method forever. Therefore, the FCC is taking its time trying to select the *best* technique for stereo TV broadcasting. G.E., ahead of its time in this area, manufactures a stereophonic TV, but the only shows you'll hear in stereo from it (for now) are from stereo videodiscs and stereo videocassettes.

TVs in a Nutshell

The TV broadcaster electronically combines (MODULATES) the TV camera's picture (VIDEO) and microphone's sound (AUDIO) into a single TV signal (RF) which travels

through the air, or via cable TV, into the antenna input of your TV set. The TV set's tuner decodes these signals back into VIDEO and AUDIO. Guided by the VIDEO signal, an electron gun zigzags its beam across the TV screen, zapping 60 pictures a second onto the screen so that the image appears to move smoothly.

When tuning in TV channels, first turn off the AFC or other automatic controls. Next tune until the image is grainy with wavy lines and the sound is raspy. Then back up a little until the picture looks good again. Finally, switch the automatic circuits back on.

When adjusting color, first switch off the automatic circuits, fine tune to the station, and turn down the COLOR. Next adjust the CONTRAST and BRIGHTNESS. Then turn up the COLOR and adjust it for the best flesh tones. Switch the automatic circuits back on.

When you take VIDEO and AUDIO from a videocassette or disc player and run these signals directly into a TV MONITOR/RECEIVER, the picture will be sharper than if you MODULATE the VIDEO and AUDIO and send this RF signal into the TV-antenna input.

VIDEO and RF must travel through special kinds of wire, using special kinds of connectors.

Flat-screen and large-screen TVs are now coming to market, but they're expensive. Stereophonic television is on the way. Wide-screen and 3-D television are unlikely to appear soon.

Chapter 2
More About TV Antennas

HIS MASTER'S RECEPTION

Entire books are written about the many different kinds of TV antennas and methods of connecting them. Since *the antenna is the eyes and ears of the TV receiver, it is important that the signal it delivers to the set be the best possible.* Here are some different kinds of antennas and some suggestions on hooking them up.

Monopole Antenna

The simplest of the built-in antennas is the MONOPOLE (see Figure 2-1), for use only when you are close enough to the TV transmitter to receive a strong signal. To work, it must be fully extended (telescoped out), and on small receivers it receives both VHF and UHF. Point the antenna around the room, using trial and error to discover the best antenna position.

Dipole or Rabbit-Ear Antenna

This V-shaped antenna (see Figure 2-2) is a little better than the MONOPOLE because it is

Figure 2-1
MONOPOLE antenna

more directional and more sensitive. It is used under the same conditions as the MONOPOLE, but aiming the DIPOLE is a little more critical. For example, if Bugs Bunny were sniffing in the direction of a TV station, his rabbit ears, if he spread them in a V, would be oriented for the best reception of that station.

25

DIPOLE OR RABBIT-EAR ANTENNA

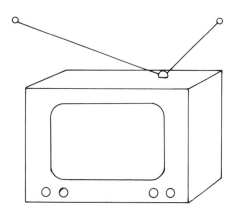

Figure 2-2
DIPOLE or rabbit-ear antenna

Loop and Bow-Tie Antennas

If your receiver has a separate antenna for UHF, it will probably look like a little loop (see Figures 2-3 and 2-4). For best results, aim it so the hole of the loop faces the direction of the signal transmitter. Some UHF antennas look like bow-ties and clip to the MONOPOLE or DIPOLE antenna (as in Figure 2-5). If you were wearing the BOW-TIE antenna, you would get the best reception while facing the direction of the TV trans-

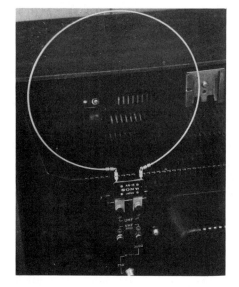

Figure 2-3
LOOP antenna for UHF

mitter. The UHF antenna must be connected to the UHF screws on the back of the set. Remember that UHF signals might not originate from the same transmitter that beams VHF. Again, trial and error will demonstrate which direction provides the best signal.

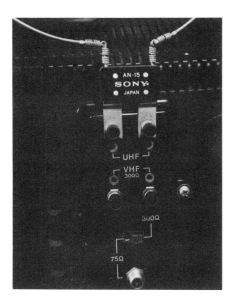

Figure 2–4
Connection for LOOP antenna

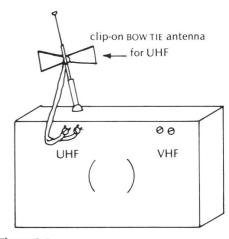

Figure 2-5
BOW TIE antenna

Portable TV Placement

Television signals do not penetrate brick or steel well, so if you are in a building with such construction, position the receiver with its MONOPOLE or RABBIT EARS accordingly, such as in front of a window. If you still have trouble, and if you happen to have some insulated (rubber-coated) wire handy, stick about three feet of the wire through the window to the outside. Strip one or two inches of the insulation off the indoor end and wrap the bare wire around the MONOPOLE or DIPOLE antenna on your set. The wire may help carry the TV signal in from the outside and transfer it to your antenna for you.

If you are in a basement or other enclosed area and have trouble getting a good picture, try touching the antenna to something metal: a curtain rod, a metal desk, a bookcase, a steam pipe, or whatever. This may improve the reception. *Caution:* Do *not* touch the antenna to any "live" electric wiring. Your TV receiver will become an instant hot plate, and you may be the first one cooked.

Portable TV Antennas

These only work well if you're pretty close to a TV station. The advantages of the portable DIPOLE (Figure 2-6) over your built-in TV antenna are: (1) It's a bit more sensitive than a MONOPOLE, and (2) It's easier to move the antenna to where the signal is best than to lug the whole TV around; perhaps balancing it on a window ledge. A television falling off a 14th floor window ledge is rough on passersby.

The power line antenna shown in Figure 2-7 is something you see ads for. "Turn your whole house wiring into a giant TV antenna," they'll say. Bullpudding! Sometimes they work; usually they don't. Don't buy one unless you can bring it back.

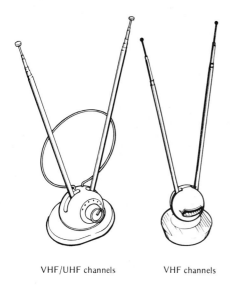

Figure 2–6
Examples of indoor portable antennas.
Courtesy of Winegard Company

VHF/UHF channels   VHF channels

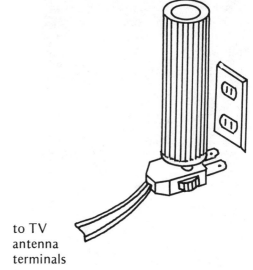

to TV antenna terminals

Figure 2–7
Power line antenna

Improvised TV Antennas

What if you're caught with no antenna at all? You've held up Sears and rushed out with a TV under each arm. When you get to your hideout you discover you forgot to steal an antenna to go with the TVs. Without an antenna, how will you be able to watch the cops chasing you on the 6 O'Clock News? Your big debut, and you'll miss it all, right? Wrong, Mugsy.

IMPROVISED ANTENNAS

With luck, perhaps you can locate a clothes hanger or, better yet, a few feet of wire. The clothes hanger you would spread open and hook to one of the antenna screws on the set. With the wire, you could hang one end out of a window (if possible) and after removing the insulation from the other end, attach it to one of the antenna terminals at the back of the set.

Experiment—who knows, you may discover a revolutionary new shape for the super antenna of the future!

If you want to be scientific about it, you could manufacture your own FOLDED DIPOLE out of inexpensive antenna wire called TWIN LEAD (shown in Figure 2-11). Such an antenna could be strung up in an attic, on the wall behind a TV receiver, under the carpet, or stuck to cardboard so you could aim it. As shown in Figure 2-8, cut a section of TWIN LEAD to a specific length for your desired channel. Although this size DIPOLE is most sensitive for only one specific channel, it will also pick up other channels to some degree. Bare the ends of insulation and twist them together. At the midpoint A, remove a ½" length of insulation from *one* of the twin wires, then cut the bare wire in half. From another piece of TWIN LEAD, bare the end and twist one of the two strands with one of the strands at A, and the other with the other. The other end of this second piece of TWIN LEAD goes to the back of your TV set. Bare this end and attach the wires to the proper terminals (UHF or VHF) as shown in Figure 1-2. Cover the bare twisted wires with tape, keeping the two wires at point A from touching each other.

Bigger Antennas

If you live well away from the television transmitter, you will need a separate, exterior antenna. Generally speaking, the bigger the antenna, the more directional and more sensitive it will be. If your neighborhood gets only one VHF channel, mount a big VHF antenna in a high place, aim it toward the transmitting station, run the antenna wire down to your receiver, and connect it to the screws marked VHF on the back of the set. To receive a UHF station, put up one of the various UHF antennas and connect the lead wire to the UHF terminals on the back of the set.

How TV Antennas Work

The broadcaster pumps his RF signal out of his antenna at a certain frequency, or channel number. This signal travels in a straight line except when it bounces off something like a mountain, a large building, or a water tower. The signal will bend slightly to reach a little ways over the horizon. As the signal

Figure 2–8
Homemade FOLDED DIPOLE antenna

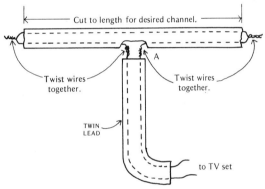

Channel	inches	Channel	inches
2	98	9	30
3	88	10	29
4	80	11	28
5	70	12	27
6	65	13	26
7	32	14	12
8	31	68	7

travels farther, it spreads out and gets weaker. If your antenna is within line-of-sight of the TV transmitter and close enough to it, the antenna will sense this tiny RF signal and funnel it down the antenna wires to your TV set.

Those long probes (called elements) that you see on rooftop antennas are a certain length in order to sense certain channels. One pair of probes, called the DIPOLE, acts much like your RABBIT EAR antenna, flattened out. It sends the signal to the antenna wire. In front of the DIPOLE are some shorter elements called DIRECTORS. Like a magnifying lens, they strengthen the signal and make the antenna more directional. Behind the DIPOLE are some longer elements, called REFLECTORS. They act like mirrors, bouncing more of the signal back for the DIPOLE to pick up. They also increase the directionality of the antenna while blocking out signals which may be coming from behind.

The more elements an antenna has, usually the more sensitive it will be (good for distant stations) and the more directional it will be (rejecting interfering signals which come from the side or back). Unfortunately, the more elements it has, the more accurately the antenna will have to be aimed to work properly; it's like a rifle, not a shotgun. Also, the heavier and more expensive it will be.

Different channels have different frequencies, and the invisible electromagnetic waves have different sizes. It takes long antenna elements to pick up channel 2, because the waves are long. Channel 13 has shorter waves, thus requiring shorter elements. UHF antennas have tiny elements to accommodate the tiny wavelengths of those ultra high frequencies.

Kinds of TV Antennas

Figure 2-9 shows various outdoor TV antennas. If you live in the boondocks with but one channel to watch, you need but a simple antenna with elements a certain length for that station. Your best bet might be a SINGLE CHANNEL YAGI, because it's relatively inexpensive and directional, giving it high sensitivity with minimal interference.

If you can receive only two stations, you may wish to buy two SINGLE CHANNEL antennas and stack them on the same pole.

If you can receive several TV stations, you have a choice of hooking up several individual antennas, or mounting a single ALL-CHANNEL antenna. ALL-CHANNEL antennas are bigger, heavier, and more expensive, and because they must contain all the different sized DIPOLES for all the different channels, there isn't much room for extra DIRECTORS and REFLECTORS, but they do the whole job in one neat package. With separate antennas, if you receive one station from the north and another from the northeast, you can aim each in the right direction. The ALL-CHANNEL antenna presents a problem, however. If both stations are very strong, and the antenna is not too directional, you might try aiming halfway between the two. If the north station is very strong and the northeast one is weak, you might aim toward the northeast. If the stations are weak or too far apart, or if your antenna is very directional, then you may need an ANTENNA ROTATOR (or ROTOR).

The ROTATOR is an electric motor which fits on the antenna mast and turns the antenna per guidance from a console near your TV. A ROTATOR needs to be pretty heavy-duty to support a large antenna in windy environs. It is likely to break when the antenna is covered with ice, so buy good quality mountain climbing gear to scale the roof.

Remember, if all your TV channels are arriving from a single direction, you do not need a ROTATOR, so don't let your dealer or TV installer talk you into one of these expensive and sometimes troublesome contraptions.

When selecting a TV antenna, one rule of thumb is to look around the neighborhood. Find out who gets good TV reception

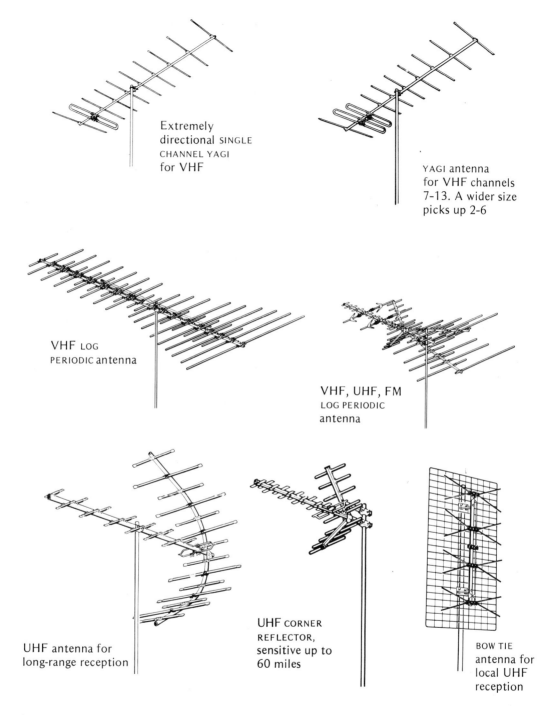

Figure 2–9
Outdoor TV antennas (Drawings courtesy of Winegard Co.)

and see what kind of antenna they used and how high they mounted it. Often, *but not always,* what worked for them will work for you. Also note: Although any antenna will pick up both color and black-and-white signals, color TVs require a better signal for clear color pictures. Thus, color generally requires bigger and better antennas than black-and-white.

Installing a TV Antenna

Buy a paperback guide on this subject for all the details. Here are some important tips they may forget to tell you:

1. Don't mount the antenna until you've marched around the rooftop (or wherever) with the antenna seeking out the best signal. Sometimes a matter of six feet can make a big difference.

2. Wear sneakers, or you'll leave long fingernail scratches down the roof and a dent in the ground.

3. To protect the shingles, stay off them when they're very hot or cold.

4. Be aware that chimneys, a popular place to mount antennas, expel acids along with the smoke. These acids will accelerate the corrosion of your antenna and connections.

5. Build your antenna mast *too strong,* especially if it must withstand much wind and ice. You don't want it coming down and shish kebabbing your neighbor's cat.

6. When you connect the antenna wire to the antenna, paint sealant over the connection to repel water and corrosion.

7. If you attach brackets or guy wires to the roof, seal the areas with roofing compound to avoid leaks.

8. As will be discussed further under Antenna Wire, if using TWIN LEAD antenna wire, use enough STANDOFFS. When running the cable past metal, such as rain gutters, use 7½" STANDOFFS.

Aiming the Antenna

There are several methods to aim your antenna if you don't have a ROTATOR. First, you could bring a small portable TV up on the roof with you and connect it to the antenna and watch the results as you aimed hither and thither. The second method requires a walkie-talkie and a helper. The helper with the brains sits in the living room deciding when the picture looks best. The one without brains teeters on the rooftop rotating the antenna. A third method replaces the walkie-talkie with a third person who relays all the messages. Experience has taught me that you do not pick 5-year-olds for this relay job, no matter how abundant or willing they may be.

Sometimes things get in the way of the signal on its journey to your antenna. Trees and wooden houses block the signal to some degree. If the signal is very strong to start with (like when you're within 30 miles of the transmitter), there's usually plenty of signal left over even after some gets trapped in the trees. Otherwise, you may need a tall mast to get the antenna over the tree line. You may also need a Piece of the Rock for when the whole works impales a Piece of the Roof in the next ice or wind storm.

Mountains are a big problem. You might have to mount your antenna on a tower to "see" over the mountain. Sometimes the TV signal will hit a mountain, building, or large metal object and bounce back and hit your antenna. This is a blessing in cases where a mountain blocked off your original signal. Instead of building a 437-foot tower to peek over the mountain at the original signal, you simply aim toward where the signal is "bouncing" from, as in Figure 2-10.

This "bounced" signal is not a blessing when you receive the original signal *and* the reflected signal together. The result is two signals, making two pictures side by side, almost superimposed. One may look like a ghost. Solution: Get a very directional antenna and aim it at the strongest signal;

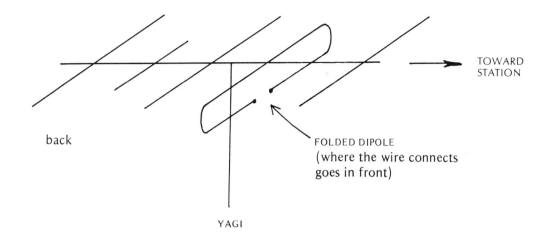

FOLDED DIPOLE
(where the wire connects goes in front)

YAGI

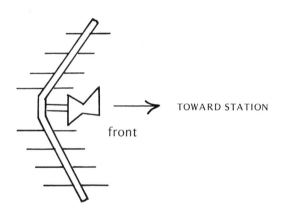

CORNER REFLECTOR

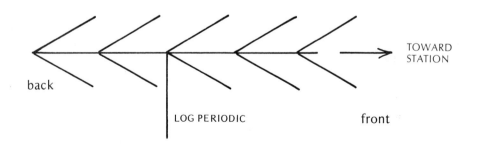

LOG PERIODIC

Antenna aiming

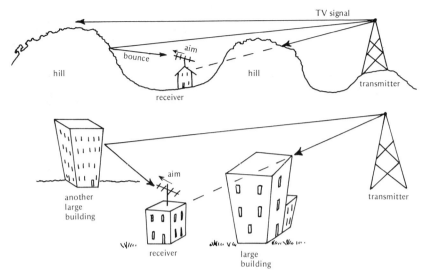

Figure 2-10
Sometimes you get a good signal aiming *away* from the TV transmitter

it will reject the weaker signal. Another solution which may become possible in a few years is called CIRCULAR POLARIZATION, or CP. This new method of transmitting TV signals, in experimental use in a few metropolitan areas, requires a special rooftop antenna to take full advantage of the signal's properties, although RABBIT EARS benefit quite a bit and regular outdoor TV antennas will still work. Your special CP antenna would "see" direct TV signals from the transmitter, but would be blind to any signals that arrive "on the bounce."

Antenna Wire

TV signals are very fussy about the wires and connectors they'll pass through. And the weaker the station, the fussier they become. A coathanger will pick up a powerful station, but to pick up a weak station you need a good antenna and proper wire. There are basically three kinds of proper wire, the flat 300Ω (the omega symbol stands for OHM, pronounced like home), TWIN LEAD; sort-of-rounded 300Ω SHIELDED TWIN LEAD; and 75Ω COAXIAL CABLE (see Figure 2-11). Each has its own advantages and disadvantages.

TWIN LEAD

Cheapest and worst of the three is simple TWIN LEAD. It looks like a narrow ribbon with thicker, rounded edges which carry the actual wire. It comes with heavy-duty insulation for outdoor applications, or with thin insulation for indoors. The indoor stuff won't last a year outdoors before it cracks and allows water in, and the outdoor TWIN LEAD is said to last only about 2 to 4 years.

The bare wire conductors can be simply wrapped around the screw terminals (clockwise for best tightness) on the TV and the antenna, and the screws tightened. Sometimes the wire comes with crimped-on SPADE LUGS (shown in Figure 1-2) which fit easily under the screws. At the antenna end it is best to paint some sealer around the bare wire ends to retard corrosion and dampness. Also anchor the wire as it leaves the antenna connection so the wire doesn't

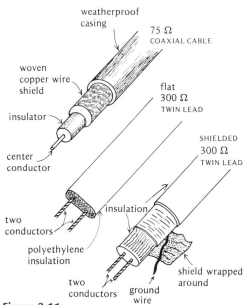

Figure 2-11
Kinds of antenna wire

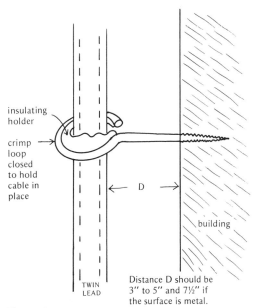

Figure 2-12
STANDOFF

flap in the breeze, fray, and eventually break. It's quite fragile.

TWIN LEAD cannot be taped to the antenna or mast, nor can it *ever* touch metal like rain gutters or aluminum windows. Doing so will attenuate (reduce) the signal drastically, and may cause "ghosts" (multiple images), flashing, loss of color, snow, and other interference. Nor can TWIN LEAD lie against a wet roof or side of a building without deleterious effects. So how do you string such wire from place to place? On telephone poles? Yup, almost. You thread it through STANDOFFS (Figure 2-12) which hold the cable away from all surfaces. There are masonry STANDOFFS for brick homes and caves, and special mast STANDOFFS for your antenna pole. Since STANDOFFS look a little ridiculous inside your home and assumedly your walls stay dry, TWIN LEAD can be tacked to the walls or trim using special *plastic* nails. Never, *never* use staples. Staples are metal, a no-no.

Being unshielded, TWIN LEAD is vulnerable to outside interference such as the tic-tic-tic of automobile ignitions, signals from CB radios, and even ghosting from reflected TV signals. One way to cancel out some of this interference is to twist the cable one turn every foot or two as you thread it along its route, like in Figure 2-13. (This will also reduce the force of wind against it.) Another trick is to run the cable far away from interferences such as the street, electric wires, neighbors' CB radio antennas, etc. Avoid coiling up excess TV cable and stuffing it somewhere like behind the TV set. This not only weakens the signal, it acts as a miniature TV antenna dragging in unwanted interfering signals. Cut the cable to the length you need, avoiding excess.

And one last indictment of TWIN LEAD. It works abominably when wet, even when new. When it gets old and cracked it may fail altogether when wet.

SHIELDED TWIN LEAD

About twice as expensive as TWIN LEAD is SHIELDED TWIN LEAD. Like its humble cousin you connect it by wrapping the two con-

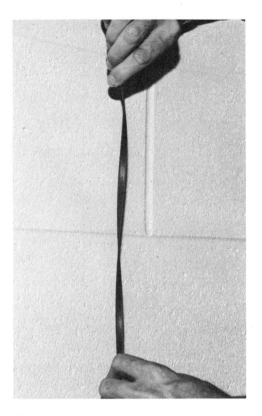

Figure 2-13
Twist TWIN LEAD along its route to cancel interference

ductors around the screws on the TV and antenna. A third wire, wrapped along the foil shield, is a ground wire which should be connected to a water pipe, a stake in the ground, or some other grounding source. Also like its cousin, its fragile wires fray, corrode, and break easily. Use sealer on the antenna connections and anchor the wire.

Unlike its cousin, SHIELDED TWIN LEAD is immune to most interference, doesn't require STANDOFFS, and can be run next to metal.

COAXIAL CABLE

A little more expensive than SHIELDED TWIN LEAD is 75 Ω COAXIAL CABLE (often abbreviated as 75 Ω COAX). COAX looks like a skinny rubber hose ¼ inch in diameter. Professionals call it by its technical names, RG-59/U for the common cable, RG-6/U for a stronger, more interference-proof cable, and RG-11/U for the twice-as-expensive heavy-duty, long-distance cable. Like SHIELDED TWIN LEAD, COAX is impervious to interference and can be attached anywhere and coiled fairly indiscriminately. Coil and tie COAX in 1½ feet diameter circles behind your equipment to hide excess. Tighter coils run the risk of kinking the cable and harming its electrical properties. COAX is not bothered by water and should last outside for ten years.

Because COAX antenna wire ends with a screw-type plug (see Figure 1-8, the F CONNECTOR) rather than bare wire strands, the connection is more solid and durable. Crimped-on F CONNECTORS are less likely to fray at the antenna end. At the set end they allow the cable to be disconnected and reconnected to the set without significant wear.

COAX conducts TV signals slightly less efficiently than TWIN LEAD, which means that if your signal is very weak, TWIN LEAD or SHIELDED TWIN LEAD will do a better job. Some of this loss is in the COAX wire itself, but much of it is due to the transformers.

Matching Transformers

Transformers? Where did they come in? Let's back up. IMPEDANCE, measured in OHMS (Ω), is an electrical attribute of electronic devices and cables. Things of the same IMPEDANCE go together; things of unlike IMPEDANCE do not—at least not efficiently. Most TV antennas are 300 Ω. Nearly all TV sets have 300 Ω antenna inputs. Connect the two with a 300 Ω TWIN LEAD and everybody's happy. There is also such a thing as a 75 Ω TV antenna. Some TVs also have 75 Ω antenna inputs. Take a look at Figure 2-4; next to the plug is a switch (in the 300 Ω position) which can switch the set to the 75 Ω mode. Switch it to 75 Ω, connect the system together with a 75 Ω cable, and you're in business. So what if your TV set

COMPARISON OF TV ANTENNA CABLES

Cable type	Efficiency dry	Efficiency wet	Rejection of interferences	Cost
300Ω TWIN LEAD	75%	6%	fair	low
300Ω SHIELDED TWIN LEAD	60%	60%	good	medium–high
75Ω COAX	40%*	40%	good	high

*Including losses by transformers.

has only 300Ω antenna inputs. What do you do with your 75Ω cable?

You plug your 75Ω COAX into a MATCHING TRANSFORMER (sometimes called a BALUN) like the one shown in Figure 2-14. This converts 75Ω to 300Ω and vice versa.

If your antenna and TV were both 300Ω and your antenna wire was 75Ω COAX, you'd install a MATCHING TRANSFORMER at both ends of the COAX to mate the COAX to the antenna and TV. Note that every time you insert something extra in your antenna line such as a TRANSFORMER, even though it's needed, you lose maybe 10% of your signal. This INSERTION LOSS can't be helped. Sometimes you can avoid unnecessary insertions. For instance, if you have a 75Ω COAX antenna wire going to the TV set in Figure 1-8, plug it straight into the 75Ω input; do not transform it to 300Ω and connect to the 300Ω terminals. Also, try to use 75Ω antennas if you're using 75Ω cable.

Antenna Joiners

You have a couple antennas on the roof, and you need to run the wire from them to your TV set. All these signals can travel over the same wire, *but you can't simply hitch the same wire to all the antennas*, at least not without losing a lot of your signal. The right way to connect antennas is through an ANTENNA JOINER or BAND SEPARATOR/JOINER like the ones shown in Figure 2-15. The wire from one antenna goes to one input of the JOINER while the wire from the other antenna goes to the other input. The JOINER efficiently combines the signals and sends them out its output. The unit is PASSIVE, which means it requires no batteries or power to make it work. ANTENNA JOINERS are designed to combine UHF and VHF antenna signals, or HIGH BAND (channels 7–13) and LOW BAND (channels 2–6) signals, or various other combinations.

75 Ω COAX
cable connects
here

300 Ω TWIN LEAD
to TV set

Figure 2-14
MATCHING TRANSFORMER (Courtesy of Winegard Co.)

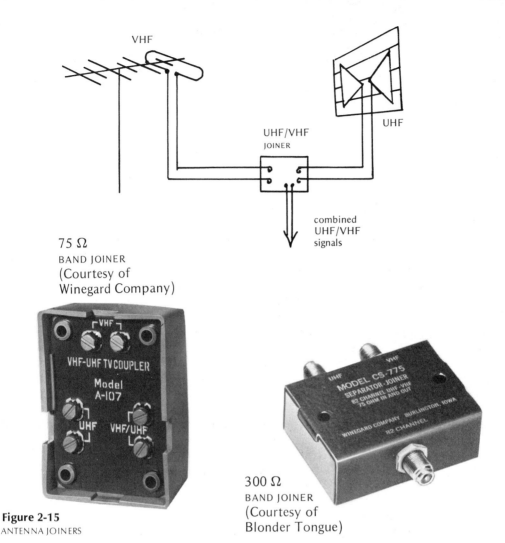

Figure 2-15
ANTENNA JOINERS

Signal Splitters

A TV SIGNAL SPLITTER is a device which takes the TV signal in a single wire and divides it into several signals. It could either divide the TV signal into equal parts for several TVs to use (in which case the SPLITTER is often called a TV COUPLER) or it could divide the signal into UHF, VHF, and FM signals to go to specific places (in which case the SPLITTER is technically called a BAND SPLITTER). Here's how the BAND SPLITTER is used:

If you live in a neighborhood that receives multiple channels, there are ALL-CHANNEL antennas that pick up UHF, VHF, and usually FM for your radio. Generally, all three signals travel through a single wire lead from the antenna. To separate the signals so that you can send each one to its proper terminals on the TV set, you use a VHF/UHF/FM SIGNAL SPLITTER. If FM is not important, you use a VHF/UHF SIGNAL SPLITTER and connect the wires as shown in Figure 2-16.

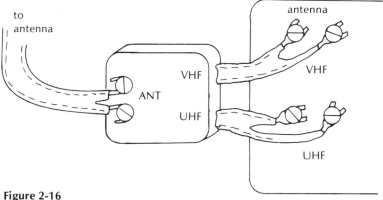

Figure 2-16
VHF-UHF SIGNAL SPLITTER

If the ALL-CHANNEL antenna is sending you its signal via 75Ω COAX, *and* you happen to have the appropriate 75Ω antenna input on your TV, there's no need to split the signal. Simply plug the COAX into the 75Ω input, switch the TV's antenna switch to the 75Ω position, and you're done.

What do you do if the antenna has a 75Ω COAX wire and your set doesn't have a 75Ω socket in the back? You buy a MATCHING TRANSFORMER/SPLITTER like the one in Figure 2-17. It does two jobs at once. It converts the signal from 75Ω to 300Ω for your TV, and it also splits the signal into UHF and VHF for you to connect to those terminals on your TV. These devices are very common and inexpensive.

SPLITTER/TRANSFORMER combinations like this are often used in institutions and in buildings serviced by Cable TV, or where subscribers share a community antenna. Sometimes you never see the cable at all, ony a socket in the wall. If the socket has screw threads on it, like the 75Ω socket on the back of your receiver, you will need a 75Ω COAX cable with an F CONNECTOR on each end to join the set and socket. No SPLITTER will be necessary. On the other hand, if the wall socket has three holes about ¼ inch apart, like in Figure 2-18, it requires a 300Ω TWIN LEAD with a matching plug. This kind of setup is usually for private homes and will require a SPLITTER if you want both UHF and VHF.

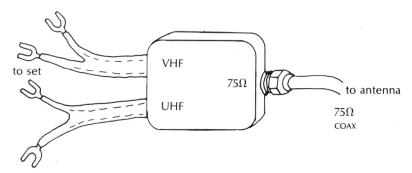

Figure 2-17
MATCHING TRANSFORMER, which converts 75Ω COAX to 300Ω TWIN LEAD. This model also acts, as most do, as a SIGNAL SPLITTER.

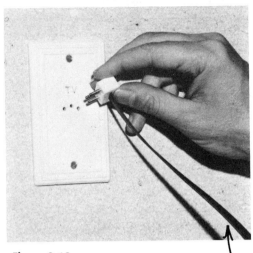

Figure 2-18
300Ω antenna wall connection

TV Couplers

What happens if you want to connect two receivers to the same antenna? In a pinch, you could attach a second piece of wire to the antenna and run this second piece to the second receiver, but this method wastes a lot of your precious TV signal. Instead, spend a few dollars on a TV COUPLER which, like the SIGNAL SPLITTER, will efficiently divide the antenna signal into two (or more) signals for two (or more) separate receivers.

Match the coupler to your needs. If you have only two TVs, buy a COUPLER with only two ports (outputs), not three or four. If, for some reason, you end up with an unused port on your COUPLER, that port must be TERMINATED with a 75Ω TERMINATOR similar to the one in Figure 1-11.

TV COUPLERS come in two varieties: PASSIVE and AMPLIFIED. The PASSIVE COUPLER (Figure 2-19) costs only a few dollars and is easy to install, but all it does is share the signal among all of the sets it serves. If there are two, each set gets half the signal strength; if three, each gets one-third, and so on. When you have more than enough signal to begin with, the degradation in picture quality that results may not be noticeable. But if your signal is weak to start with and provides a picture that is already slightly grainy, using a PASSIVE COUPLER will make that picture even worse.

Under those conditions you need an AMPLIFIED COUPLER. This accessory costs about $50, is powered by house current, and is connected in much the same way as its passive brother, but it provides much better results. Instead of simply sharing the available signal among receivers, the AMPLIFIED COUPLER boosts the signal to each receiver so that each gets full signal strength. In general, the AMPLIFIED COUPLER is a must for

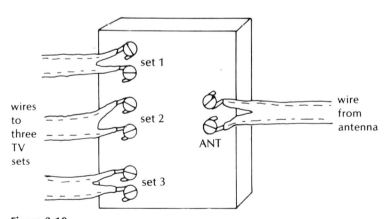

Figure 2-19
Passive multi-set TV COUPLER

institutions that connect multiple sets to a single antenna. If you buy one, make sure it has UHF capability if you want it to boost UHF channels too. Some only boost VHF.

Antenna Preamplifiers

What if you are using one antenna to serve only one receiver but you still have a grainy, snowy picture, the sign of a weak or distant signal? Perhaps you make sure the antenna is aimed right. Perhaps you buy a bigger, more sensitive antenna. If those tricks didn't work (assuming nothing is wrong with your antenna wire or with your set), you could buy an ANTENNA PREAMPLIFIER (or, as it is sometimes called, an ANTENNA AMPLIFIER or ANTENNA BOOSTER). This device usually mounts on the pole near your antenna and connects to your antenna wire. Using house current, it boosts the antenna's signal, thus improving your set's picture.

Now don't go off and buy an ANTENNA PREAMPLIFIER instead of a good antenna. The antenna itself is the place to start. Get the best signal you can first, *then* amplify it if necessary. Amplifying an inferior signal with ghosts, snow, or interference from a misaimed or too-small antenna will just give you a stronger inferior signal with ghosts, snow, or interference. For the ANTENNA PREAMPLIFIER to really help, it must have a relatively ghost-free, clean signal to start with. Once it has this "pretty good" signal, it will boost the quality to "good" or maybe even to "excellent." Also, the ANTENNA PREAMPLIFIER will help get back some of the signal that's lost as it travels down along antenna wires.

Line Equalizers

When you run a TV-antenna signal a long way through a wire (say 300 feet for RG 59/U COAX, for instance) it gets weaker. Strangely, the high channels get weaker faster than the low ones. So that all channels come out equal, you connect this little box in the antenna line and it will weaken the low channels while leaving the high ones alone, thus evening them all out.

Ghost Eliminators

Large metal structures like water towers and buildings sometimes bounce back to you a delayed TV signal which interferes with the original TV signal and causes a double image, or "ghost." Inexpensive GHOST ELIMINATORS are often advertised as the solution. Sometimes portable RABBIT-EAR antennas have GHOST ELIMINATORS built-in, with switches on the antenna bases to adjust the amount of filtering.

Don't expect much from these devices. Most models work on the simple principle of reducing the entire TV signal, thus diminishing the ghost. This is hardly an improvement since you're losing the baby along with the bathwater, so to speak. Short of hiring an exorcist or holding a seance, the best way to get rid of ghosts is to buy a good directional antenna and aim it carefully.

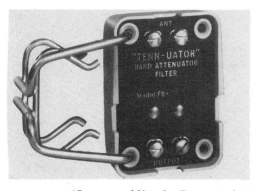

ATTENUATOR (Courtesy of Blonder Tongue Labs.)

Attenuaters

You *can* have too much of a good thing. If your TV antenna signal is too strong, your picture may bend or be contrasty, or you may hear buzzing in your sound. Some TVs

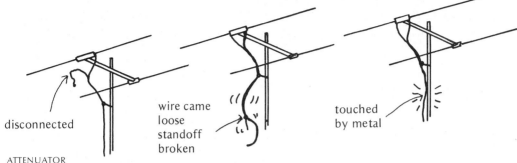

disconnected   wire came loose standoff broken   touched by metal

ATTENUATOR
TV signal problems are often caused by a faulty antenna cable

have DISTANT/LOCAL switches on the back which when switched to LOCAL will attenuate the signal coming in from a strong local station. You can buy an ATTENUATER to connect between your antenna and your TV set. They even make them selective for particular channels for times when your antenna picks up a really strong station and a really weak station in the same direction. The ATTENUATER will reduce the selected channel's signal while leaving the weak one untouched.

Common TV-Antenna Ailments and Cures

The most common TV-antenna problems don't involve the antenna at all. The trouble is generally with the antenna wire and more often involves TWIN LEAD than other types of wire. Check it to see if it is disconnected, loose, touching metal, cracked, wet, corroded at the connections, or otherwise in bad shape.

Also, you need to decide whether your problem is with the antenna signal or with your TV set. One way to do this is to disconnect your outdoor TV antenna and try a portable RABBIT-EAR antenna in its place. If the trouble disappears (taking into consideration the limited sensitivity of the little RABBIT EAR), your outdoor antenna system is probably faulty. If the symptoms are unchanged, perhaps the problem is with the TV set.

Let's start with TV interference problems:

Sporadically, you get interference in your picture and sound as follows: The interference may be in the form of a crosshatch or herringbone pattern on the screen, like in Figure 2-20. There may be bars of color floating through the picture. In extreme cases, your picture may even turn negative, like a photograph negative. The sound may buzz or sound like a garbled Donald Duck. If this is your problem, you probably have CB interference, good buddy.

Citizen's Band radios transmit on frequencies much like TV. If a strong transmitter is near your TV, its signal can sneak into your M*A*S*H reruns.

Figure 2-20
Interference from a CB transmitter

You can be more sure that the interference is from a CB if:

1. The problem is worse on channels 2, 5, and 6.

42   More About TV Antennas

2. The problem occurs only at certain times, like a certain hour of the evening (CBers are creatures of habit too).

3. You can hear the voice and you recognize CB language like "breaker, breaker," or "10-4," or at the end of every other sentence the words "for sure." The voices may seem preoccupied with someone named "Smokey." You may also hear call letters like WBS7341 or a slangy identification like "Square Dance Lady" or "Road Runner," and you may even pick up a clue to the speaker's address.

4. You take a stroll around your neighborhood and notice strange antennas looming over rooftops or out windows. You may see super-long antennas attached to cars nearby.

Whose fault is TVI (TV Interference)? If it's only on channels 2, 5, or 6 it's usually the CBer's fault. CB transmitters broadcast a strong signal at their proper frequency, and a tiny bit of signal at some multiple of that frequency called a *harmonic*. Your TV channel 2 happens to be just twice one of the CB frequencies. Channel 5 is three times another CB frequency. Your TV, when tuned to these channels, is sensitive to the CBer's harmonic and is receiving that signal in competition with your desired TV channel.

The remedies for CB-caused interference are to:

1. Locate the culprit.

2. Ask the CBer nicely to install a LOW-PASS FILTER on his or her antenna. This should filter out those nasty harmonics which bother your TV.

3. If you can't locate the CBer or you get no cooperation, try:
   (a) Checking the phone book for a local CB club. Some have committees which try to resolve problems between the CBers and TV viewers.
   (b) Contacting the FCC (listed under U.S Government, Federal Communications Commission in the phone book). They're pretty busy so writing them a letter may be faster than getting through by phone.

4. If civilized methods don't work try this:
   (a) Pray for lightning to strike the CBer's antenna.
   (b) Hire the local teenage hoodlum to rip-off the offending CB rig.

If the interference is on other channels besides 2, 5, and 6, or if the CBer was cooperative but it didn't help, the fault may be with your antenna or TV. First, let's see if it's the antenna's fault. Disconnect the antenna during the next CB transmission and see if you still pick up the interference (don't worry about your TV station at this point). If the interference remains, it's not your antenna that's picking it up, it's your TV set. It needs modification to reject the interference.

This can often be corrected by adding a special filter to circuits in your TV set. Many TV manufacturers will provide the filter free or at a nominal charge. A technician could install it but it's simple enough to do yourself. Your TV dealer may even contact the manufacturer in your behalf to order you the filter.

If, in the above experiment, the interference stopped when you disconnected your antenna, then it's your antenna that's the culprit. There are several possible remedies:

1. Install a HIGH-PASS FILTER between your antenna cable and your TV. This "passes" the TV signal to your TV while rejecting the interfering signal. Sometimes these devices are called TRAPS (because they "trap" the offending signal). Radio Shack and other TV-parts stores sell them. If one TRAP helps, but not enough, try two at the same time.

2. Another remedy for TVI is to switch from the flat TWIN LEAD you may be using to SHIELDED TWIN LEAD or to COAX. These are well shielded from outside interference.

Common TV-Antenna Ailments and Cures 43

HIGH PASS FILTER (Courtesy of Radio Shack)

Sometimes a very close AM radio station will interfere with your TV, perhaps throwing vertical, red wavy lines into your TV picture. There's not a lot you can do (other than move away or switch to Cable TV). The station's chief engineer might give you some suggestions if you call.

To read more on this subject, consult the *Consumer Electronics Service Technician Interference Handbook—TV Interference* (catchy title) available from the Electronics Industries Association, 2001 Eye Street, NW, Washington, DC 20006, or Sony's *Interference Handbook* from Sony's Technical Publications Department, 4747 Van Dam Street, Long Island City, NY 11101.

Interference in the form of black, horizontal jagged lines ripple or roll through the picture, most often on channels 2, 3, and 4. This may be caused by electrical arcing in some household appliance, most usually a three-way light bulb. Try turning off everything in the house (except the TV) and watch. If the problem stopped, switch things back on again, one at a time until you find the offender. If it's a light, try removing the bulb and scraping its base clean, then screwing it in tight before using. Then again, sometimes scraping won't help and you have to throw away the bulb.

Sporadic interference that goes tic-tic-tic, speeds up, slows down, is often accompanied by dashes of colored snow (or black-and-white snow), and sounds sort of like a car's engine. This is static from auto ignitions. Try:

1. Making sure your TWIN LEAD antenna wire is twisted about once every two feet throughout its journey from antenna to TV set.

2. Relocating your antenna and antenna wire as far as possible from the street. Try to put your house between the antenna (and wire) and the road.

3. Installing SHIELDED TWIN LEAD or COAX antenna wire.

Snowy buzzing which accompanies an electric motor running somewhere. Vacuum cleaners, electric shavers, dishwashers, dryers, power tools, and electric mixers often spit buzzing signals down their electrical cords and into the house wiring. They then pass into your TV set. Try plugging a LINE FILTER (see Figure 2-21) in between the wall socket and your TV set, or between the wall socket and the appliance.

Constant herringbone pattern on a certain channel, sometimes affected by your program's sound. The TV signal is too weak. It needs a better antenna, better wire, or a BOOSTER.

Suddenly one night, your local channel is driven off the screen by another, perhaps snowy image. This new image superimposes itself over yours so much that you can actually read words and decipher the

Figure 2-21
Installing a LINE FILTER either at the offending appliance or set itself will reduce interference (Courtesy of Radio Shack)

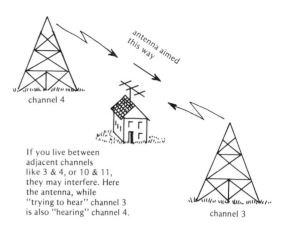

If you live between adjacent channels like 3 & 4, or 10 & 11, they may interfere. Here the antenna, while "trying to hear" channel 3 is also "hearing" channel 4.

picture. In more moderate cases, you see persistent lines or ripples (much like Figure 2-20) over your show.

This is called CO-CHANNEL INTERFERENCE. There are, for instance, lots of channel 2s. Your TV only picks up the closest one—normally. Sunspots and other atmospheric conditions sometimes ionize the air about 60 miles up, causing this air layer to reflect TV signals from distant stations. Thus another channel 2, which is over the horizon and perhaps 400 miles away, comes bouncing in to compete with your local channel 2.

There's nothing you can do. This unusual phenomenon will go away in a couple of days.

If you find picking up "Channel 2—Dublin" fascinating, you can join the community of ham radio and TV fans who seek out this phenomenon (called "TV DX-ing," where DX stands for *distance*). The club is called: Worldwide TV-FM DX Association, Box 97, Calumet City, IL 60909.

Picture displays "windshield wiper" effect (Figure 2-22) and writhing herringbone pattern. This is called ADJACENT CHANNEL INTERFERENCE and results when a strong channel's signal "leaks into" the adjacent channel that you're watching. This is most noticeable between channels 3 and 4, and channels 10 and 11.

Like people, many antennas are nearly as sensitive out their backs as they are out their fronts. In such cases, if you're in between two stations broadcasting on adjacent channels, your antenna will pick up both.

The solution is to buy a TRAP tuned to the

Figure 2-22
"Windshield wiper" kind of interference from adjacent channel

Common TV-Antenna Ailments and Cures

unwanted station (many are adjustable for various channels) and attach it between your antenna cable and your TV.

Here are some specific antenna and antenna wire problems:

Snow, ghosts, no color, color flashes on and off, picture rolls, black ridges around faces, etc.

1. Replace antenna wire if cracked, frayed, weather-beaten, or old.
2. Check for poor connections, especially between the antenna wire and the antenna where wind has a tendency to loosen the connectors.
3. Replace old or broken STANDOFFS.
4. Install more STANDOFFS, especially if the problem shows up on windy days.
5. Straighten bent antenna elements. Replace broken elements with ones the same length (you can make them yourself with a hacksaw and some ½-inch aluminum tubing).
6. If using COAX, check the screw-on connectors. Sometimes the wire twists by itself and unscrews the F connectors from their sockets (at the TV, SPLITTER, COUPLER, or elsewhere).

If the picture suddenly becomes snowy after you have disconnected and reconnected your COAX, check the F connector. Perhaps you bent that thin center wire in the plug. It should be straight, like in Figure 1-8.

Ghosts, or loss of color, or loss of detail in picture. If you're using TWIN LEAD, it's probably touching metal somewhere.

If your antenna has a BOOSTER and you suddenly get a grainy or snowy picture, or you see a wide, soft, dark, horizontal bar across the picture, perhaps sliding up the picture. Your BOOSTER's probably shot. Fix or replace it.

Antennas in a Nutshell

ANTENNA SYSTEM ACCESSORIES

Type	Where used	Function
Antenna amplifier	Near antenna	Increase strength of signal from antenna.
Balun, or matching transformer	Antenna and receiver	Match 75 Ω coaxial cable to 300 Ω twin lead.
Joiner	Antenna	Combine the outputs of two or more antennas into one signal.
Two-way coupler	Near TV sets	Enable use of same antenna by two receivers.
Four-way coupler	Near TV sets	Enable use of same antenna by four receivers.
High-pass filter	Input to TV set	Reduce or eliminate interference from radio transmitters.
Lightning arrestor	Transmission line at point of entry into building	Protect receiver against damage by lightning and heavy static charges.
VHF/FM splitter	Near TV set or FM radio receiver	Separate VHF, TV, and FM radio signals and channel them to the appropriate receivers.
VHF/UHF splitter	Near TV-antenna terminals	Separate VHF and UHF signals and channel them to the proper receiver inputs.

ANTENNA AND INTERFERENCE PROBLEMS

Symptom	Probable cause	How and what to check	Remedy
Ghosty reception	1. Antenna incorrectly pointed	1. Rotate antenna	1. Lock antenna into position where ghosts disappear.
	2. Inadequate antenna	2. Neighbor's reception	2. Replace antenna.
	3. Pickup of signals by twin lead	3. Twist twin lead and observe picture	3. If twisting twin lead works, fine. If not, replace twin lead with shielded twin lead or coax cable.
Snowy pictures	1. Inadequate antenna	1. Neighbor's reception	1. Replace antenna.
	2. Inadequate or worn out twin lead	2. Neighbor's reception	2. Replace twin lead with better type.

ANTENNA AND INTEFERENCE PROBLEMS (Cont.)

Symptom	Probable cause	How and what to check	Remedy
Picture flickers (especially when it's windy)	1. Loose antenna element	1. Antenna elements	1. Resecure loose element or replace antenna.
	2. Loose twin lead connection	2. Check connection	2. Tighten connections.
	3. Faulty connection at coax cable plug	3. Check plugs	3. Tighten, or replace plug.
Occasional streaks in picture	1. Interference from radio communications transmitter	1. Disconnect antenna	1. If disconnection worked, install high-pass filter at VHF receiver terminals.
	2. Interference from electrical devices	2. Turn electrical devices on and off while observing picture	2. Replace twin lead with shielded twin lead or coax cable, or install noise filter at device causing interference.
Poor color picture but black-and-white picture sharp	1. Inadequate antenna	1. Neighbor's reception	1. Replace antenna, or install antenna booster.
Bars and rippling on channels 2, 3, and 4	1. Interference from electrical devices	1. Shut off all appliances	1. Replace or fix problem appliance (usually a 3-way bulb).
Tic-tic-tic, or car engine sound	1. Auto ignition static	1. Watch for nearby cars when problem shows	1. Twist twin lead, or replace with shielded or coax. Relocate antenna far from street.
Another program superimposes over yours	1. Co-channel interference	1. Check for sunspots	1. No cure. Wait it out.
Windshield wiper effect	1. Adjacent channel interference	1. Try adjacent channel	1. Install tuned trap.

Chapter 3
Cable TV

HIS MASTER'S ANTENNA

How Cable TV Works

Gather round, boys and girls, and I'll tell you the story of cable TV. Once upon a time, a man lived on a mountaintop and got fabulous TV reception. The Valley Villagers below, blocked by mountains, got no TV reception. Poor Villagers—no Flintstones, no Price Is Right, no Family Feud, no Wonder Woman, no violence, no sex (well, maybe sex), no Carson reruns, zippo. The Mountain Man, whose wife was the overworked Village Librarian, hit upon an idea to lighten her workload. He offered to split his antenna signal and run a long antenna wire down the mountain to the village. He would boost the signal so that there would be enough to go around, and would run high quality coax wire on the telephone poles past all the Village homes. Since the wires, boosters, telephone pole rentals, antenna, and upkeep all cost money, the Mountain Man would charge a nominal fee to all the subscribers on his COMMUNITY ANTENNA.

The Village Fathers, after a noisy debate and a quiet payoff, granted the Mountain Man an exclusive FRANCHISE to be the only antenna in town. After all, it wouldn't make sense to have a cluster of competing antenna wires strung from pole to pole. Besides, the resulting monopoly would guarantee the Mountain Man that his large initial investment would not be endangered by competition. About half of the homes that could be reached by The Cable chose to subscribe for $5 per month. The Mountain Man sent a truck around to connect them up. Local youths, good at shinnying up telephone poles, also connected up a few homes, bypassing the $5 fee, of course.

Everyone was happy, except the Village Librarian, who now for lack of readership was cut from the payroll.

To increase subscribership, the Mountain Man added a few more antennas for more channels, pulling in some stations from 100 miles away. Viewers loved it. Local broadcasters hated it. They felt that people watching distant stations would dilute local viewership, reduce the stations' ratings, and decrease how much their sponsors paid for ads. As a consequence, the local stations would reduce their less profitable programming, such as news and local affairs. This reduction in public service from the local broadcasters would hurt those not served by Cable. The FCC (Federal Communications Commission), trying to protect the public, forbade the Mountain Man from importing distant signals. The Mountain Man could see how the FCC had a right to control the airwaves—they're for everybody to share in a controlled way—but The Cable was *his*. How could the FCC tell him what to put over *his* wire? The debate rages to this day, but the Mountain Man was allowed to import distant stations only if the local broadcasters couldn't prove significant harm to them.

While in the process of seeking out distant stations, the Mountain Man came up with a channel 4 from the next state. But he already had a channel 4 (a different one) on his system from a local station. How could he send this new station to his subscribers? Meanwhile he received no stations on channels 7, 9, and 11.

Merlin, the Village Video Wizard, came to the rescue with two solutions to the problem.

1. He invented a CONVERTER which would simply translate the channel 4 signal to, say, 7 and pump it down the wire like any other channel. Or,

2. He could buy a DEMODULATOR (a TV tuner) which would separate the channel 4 RF signal into video and audio. Then he could run the separate video and audio to a MODULATOR which combines them into a channel 7 RF signal, like any TV transmitter normally does.

Thus in goes channel 4, out comes 7, and ring goes the Mountain Man's phone as his subscribers call wondering why the channel 4 news appears on channel 7.

While the new subscribers signed on, the profits poured in. The FCC, sensing Cable's mighty impact on the population, set up regulations to guarantee "community service." One rule required the Mountain Man to build his own TV studio and make available an ACCESS CHANNEL where anybody could air their views, run video tapes of their school plays, run college courses by TV, or even sneak in a little porn on the uncensored, unrestricted channel. Not a lot of people watched (except for the porn). Eventually, the courts told the FCC they had exceeded their authority by demanding a COMMUNITY ACCESS CHANNEL.

Meanwhile, the Village Fathers, aware that the Mountain Man was making a bundle on their FRANCHISE, pressured him to keep the unprofitable-but-public-service-minded ACCESS CHANNEL open. So the previously empty channel 9 was filled with Telecourses, local sports, public affairs, and some local news produced in the Mountain Mini-Studio.

Channel 11 was still empty, so the enterprising Mountain Man aimed a camera at a drum with some cards on it and sold advertisements. Nobody watched. Improving on the idea, he bought a CHARACTER GENERATOR, which could electronically type words on the screen and roll them by. He teamed it up with mini-news and weather services so that now the viewer could watch the weather report repeatedly crawl across the bottom of the screen, hear a 24-hour radio type newscast, and see a progression of local ads on the screen.

One day the Mountain Man, while attending his local theatre, hit upon an idea: Show a popular movie over TV and have everybody pay if they wanted to see it. Better yet, show a couple of movies per week with lots of reruns, *and* no ads.

But how do you get people to pay extra for something already coming into their home? And what channel do you put it on now that all the channels are filled up?

Merlin came to the rescue again with

50   Cable TV

another CONVERTER (or DECODER) box. The box would be labeled with channels 2-13 and A, B, C, etc. Depending on where the dial was set, the incoming signal would be converted to, say, channel 3. People would connect the box between The Cable and their TV-antenna input, and would tune their TV to channel 3—forever. The box would do the tuning. Because the box had this special tuner in it, it could tune in channels a TV couldn't get. Not only did this provide a range of more channels that could be cablecast, but it made it necessary for subscribers to rent the box if they wanted to pick up those "special" channels with pay-TV movies. Merlin made a bundle manufacturing DECODER boxes for the Mountain Man to rent. Merlin's shifty nephew, Herbert, also made a bundle, copying the circuit and selling the boxes directly to the public who, for the one-time cost of the box, could now pick up pay-TV for free.

The viewers loved it. The movie theatres hated it. The broadcasters, afraid that Cable would siphon off their prime movies, lobbied against it. The FCC banned it, at least the 3- to 10-year-old movies which comprise the bulk of broadcast TV fare.

Movie rentals are pretty expensive. The Mountain Man was approached by Mr. City Slicker with the following proposal: Mr. Slicker would lease popular movies, at phenomenal expense, for cable use. Mr. Slicker would then play the movies over his own cable system, feeding the signal *to other cable systems*—for a price, of course. In cases where distribution by wire was too expensive, Mr. Slicker would beam the signals through the air using MICROWAVE transmitters (which instead of *broad*casting in all directions, *narrow*cast in a focused beam at a particular target). The Mountain Man could put a MICROWAVE receiving antenna on his tower to pick up the signals.

Mr. Slicker showed the Mountain Man a briefcase full of maps and crisscrossing lines. These lines represented a web of connections passing his movie signal along from cable company to cable company, eventually covering the map with a NETWORK. Mr. Slicker pulled out more diagrams showing how he would eventually beam his movies up to a satellite 22,300 miles in the sky. All the participating cable companies would be able to point giant "dish" antennas at the "bird" and get his movie signal without messing with the cable NETWORK. The Mountain Man couldn't wait to sign on the dotted line.

By the time the ink was dry, Herbert (remember Merlin's nephew?) had somehow acquired the plans for a satellite receiver and was selling do-it-yourself kits to crafty consumers who wanted to pick up the satellite pay-TV signals without paying.

By now, the Mountain Man was cablecasting 25 channels to his subscribers, some stations in Spanish, some specializing in sports, some religious, and some giving courses over TV. About 30% of his subscribers paid the extra $10 per month for pay-TV. The cash rolled in.

Then Merlin came up with an idea the Mountain Man didn't like: Put a high-power satellite in space that would beam pay-movies directly to the homeowner, skipping the Mountain Middleman. The villagers could simply rent a small dish-shaped antenna and converter box from Mr. Slicker and aim it at the bird to pick up Slicker's movies directly.

Suddenly one night, lightning struck Mountain Man's array of antennas, lighting up the whole sky, $50,000 worth of video gear, and the first 120 TV sets along the cable's path. The Mountain Cable switchboard also lit up and didn't quiet down until everything was fixed.

Mountain Man sold out, moved to Aruba where his wife got a job as a librarian, and they lived happily ever after, without the companionship of Fred Flintstone, Wonder Woman, Johnny Carson, or Kermit the Frog.

Decoding Cable's Signal for Your TV

To get a better picture of how cable TV packs all those channels into the same wire, and how you get them to play through your TV, let's study frequencies for a minute.

FREQUENCY EXPLAINED

When we listen to music, we are sensing sound vibrations through our ears. High notes are made of many vibrations per second (sometimes called cycles per second) and are said to have a high pitch or high frequency. Low notes have fewer vibrations per second for a low frequency.

Instead of saying vibrations per second, we'll use the more popular term, HERTZ, named after the German physicist H. R. Hertz, and abbreviated Hz.

The lowest note we humans (assuming your author is one, too) can normally hear is 20 Hz. The highest note we can hear is about 20,000 Hz or 20 KILOHERTZ (KILO means 1000), abbreviated 20 KHz. Normal speech occurs at about 1000 Hz (1 KHz).

The frequencies of the invisible electromagnetic waves used to transmit TV programs are nearly a million times the frequencies of speech. A million vibrations per second is called a MEGAHERTZ (MEGA means million), and is abbreviated 1 MHz.

Why are these TV frequencies so high? Back in Chapter 1 under How TVs Work, we saw that the electron machine gun in the picture tube zigzagged across the screen 15,575 times per second, tracing out its picture. During each of those zigs, the gun had to shoot, stop, shoot, and stop many times to create the light, dark, light, and dark parts of the picture. The TV signal had to have coded in it an electrical vibration for every one of those shoots-and-stops. It takes millions of these vibrations per second for the TV signal to carry the message to the gun to shoot-and-stop millions of times per second. There's really more to it than this, but we'll save that for the engineering books. No doubt you're preparing right now to race down to your bookstore to dive deeper into this topic with wild abandon.

FREQUENCIES USED BY CABLE COMPANIES

So that one TV channel doesn't interfere with another and so that broadcasters of other communications don't interfere with each other, the FCC has allocated certain frequencies to each. These are shown in Figure 3-1. Notice that channels 2–6 take up one set of frequencies, then there's a gap between 6 and 7, then comes channels 7–13. TV broadcasters can't use the frequencies in this gap because FM radio, police, and airplanes use these frequencies. Your average TV can't even tune these frequencies in.

But cablecasters don't have to be concerned with encroaching on the police and air traffic frequencies because the cable signal *stays in the cable*. Therefore the cable companies can use these frequencies as they please. So they fill these gaps with more TV channels. Of course, for your TV to be able to receive these frequencies, you need to rent the Cable companies' CONVERTER which translates these MID-BAND and SUPER-BAND frequencies to another frequency your TV *can* receive.

You may be wondering why the cable companies didn't just convert their TV channels to UHF (which your set *can* receive) and avoid messing with a CONVERTER box. There would seem to be plenty of UHF channels available. There are two reasons for their decision: (1) Ultra High Frequency (UHF) signals are extremely hard to transmit cheaply through long cables. Lower frequencies are easier. (2) If you could pick up the stations easily without a

Figure 3–1
Cable and regular TV frequencies

Frequency in MHz	5	54	88	108	120	174	216	318	402	470	890
Broadcast TV channels		2, 3, 4, 5, 6				7, 8, 9, 10, 11, 12, 13				14, 15, 16, ... 81, 82, 83	
Kinds of broadcasts	CB radio	VHF TV	FM radio	Maritime, weather satellite, air traffic, police, public service, two-way radio, two-meter amateur radio		VHF TV	Amateur radio, government, mobile, air traffic	Coast Guard, satellite, mobile, air navigation		UHF TV	
Names given to these frequencies		LOW-BAND TV		MID-BAND TV		HIGH-BAND TV	SUPER-BAND TV	HYPER-BAND TV		UHF BAND	
Cable TV channels	$T_1, T_2, T_3, \ldots T_{14}$	2, 3, 4, 5, 6 (unless converted to a different channel number)	FM radio or A_3, A_4, A_5	A_1, A_2	A, B, C, D, E, F, G, H, I	7, 8, 9, 10, 11, 12, 13, (unless converted to a different channel number)	J, K, L, ...V, W	W+1 thru W+17			

CONVERTER, then how could cable companies make money renting them to you?

In short, the cable company charges you to hook up to its "antenna." To discourage homeowners from climbing the phone poles and hooking themselves up for free, and for technical reasons relating to transmitting certain frequencies, the cable companies transmit the TV signals on channels your average TV set doesn't get. You then must rent a CONVERTER from them to translate these frequencies to channels you *can* get. Figure 3–2 shows how you hook up the CONVERTER.

Figure 3–2
CONVERTER hookup to TV

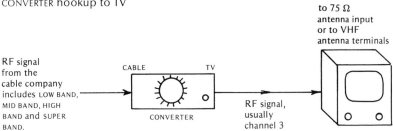

Decoding Cable's Signal for Your TV

PAY-TV CHANNELS

Besides your basic TV channels, you can subscribe (at extra charge) to a variety of other special channels offering sports, movies, children's programs, etc., depending on your cable company. Here are some of the pay-TV services generally available:

- HBO—Home Box Office, started in 1975 by Time, Inc. One to three-year-old movies, older classics, specialty programs, some sports. Many movies are "R" rated. For about $15 per month you get 300 hours of programming, including 15 new movies.
- SHO—Showtime, started in 1977 by Viacom and Teleprompter. Offers about the same programs as HBO.
- MC—Warner/Amex's Movie Channel. Like HBO and SHO, with some oldies.
- Cinemax—Like HBO, but aimed toward the more intellectual audience. Some foreign flicks.
- HTN—Like HBO, but more family oriented with G and PG movies.
- SelecTV—A little less recent combination of movies than HBO. A few foreign films, oldies, and R-rateds.
- Spotlight—Potpourri of movies for children and teens through senior citizens.
- Bravo, Escapade, PET, and EROS—Soft core adult films.
- WHT—Wometco Home Theatre. *Over-the-air* broadcasts of scrambled TV signals showing movies in the metropolitan NY area. A special DECODER (unlike the cable company's) and sometimes a special antenna are necessary to pick up these signals. WHT and similar scrambled stations may also appear on one of your cable channels, but need special DECODERS to unscramble the signals.
- USA—Madison Square Garden Sports, a joint venture of UA Columbia Satellite Services and Madison Square Garden Sports. The USA network covers 200 sports telecasts per year plus some children's programming (Calliope), live coverage of the House of Representatives (CSPAN), and programming specializing in entertainment of interest to black audiences.
- ESPN—Entertainment and Sports Programming Network, started in 1979, and owned primarily by Getty Oil and Anheuser-Busch. Sports programming 24 hours per day. Unlike the movie channels, ESPN takes advertising.

So, if the cable is connected to your home, and you're already renting the CONVERTER box, what stops you from watching pay-TV channels without paying? (Aha, we get to the real reason why you're reading this chapter.) The answer is a simple FILTER (or TRAP), like those described in Chapter 2 for reducing ADJACENT CHANNEL INTERFERENCE. When the cable company rents you their CONVERTER *without* the pay-TV option, it contains a FILTER which stops the pay-TV channel from getting through. What keeps a subscriber from renting the CONVERTER, opening it up, and removing the FILTER?

1. You've got to understand some electronics to know which part to remove.

2. It's illegal. You're *renting* the box; it's not yours to tamper with. Such electronic piracy is considered theft-of-service, a crime.

3. Crafty cable companies sometimes don't put the FILTER in the CONVERTER, but instead, put it in the junction box of the telephone pole where the cable branch splits off for your house. Equally crafty subscribers sometimes climb the pole, open the box, and remove the tiny plug-in FILTER, but again, they're tampering with the cable company's property in order to steal service.

Broadcast pay-TV decoder (Courtesy Wometco Home Theatre)

One discovery some subscribers have made is that if their CONVERTER needs repair, and the cable company gives them another one, the sloppy companies sometimes will exchange a filtered CONVERTER for a non-filtered one. The subscriber gets home, discovers he's got free HBO, and *never* says a word. People generally perceive this more as finding money on the sidewalk than theft-of-service.

Over-the-air pay-TV broadcasters, like Wometco Home Theatre, use another method for keeping their signal from being intercepted for free. They SCRAMBLE their signal, then rent you a DESCRAMBLER (or DECODER) to make the program watchable. They SCRAMBLE the picture by messing up the SYNC (remember SYNC?) so your picture bends and tears across the screen. Also, they code the AUDIO signal into the RF transmission in such a way that your TV set can't play the sound. The rented DESCRAMBLER has a speaker in it, which gives you the sound while your TV gives you the picture. Figure 3-3 shows how it's hooked up.

Most of these CONVERTERS and DECODERS have one common fault. If you have a remote control on your TV, forget about it. The channel switching is done manually on the CONVERTER box (unless *it* happens to be remote controlled). All programs once switched at the box are viewed on the same channel on your TV, rendering your TV's remote control useless. This problem also bugs you if you're a home video recording enthusiast. You may want to program your recorder to record one show on one channel, then another show later on another channel. Your recorder will switch channels as programmed, but your CONVERTER won't be switched, and it's the CONVERTER that tunes in the stations. A

Figure 3–3
Over-the-air pay-TV hookup

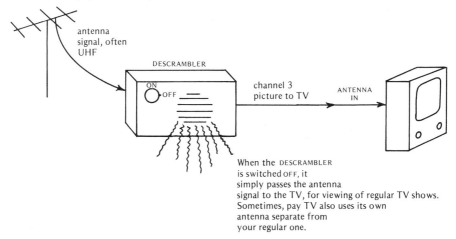

Decoding Cable's Signal for Your TV

further irritation to the home recordist involves one of the reasons you bought your machine in the first place—to watch one show while recording another. The CONVERTER translates only one station at a time, so unless you have two CONVERTERS, both your TV and your videocassette recorder will be receiving the same channel. What to do? Read on, McDuff.

UP-CONVERTER
(Courtesy of Comprehensive Video Supply Corp.)

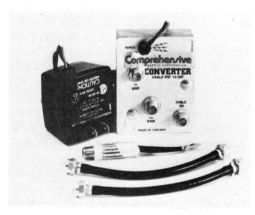

BYPASSING
THE CONVERTER

You could buy an UP-CONVERTER (or a BLOCK CONVERTER, as they're sometimes called). This little box takes all the incoming cable frequencies and converts them to UHF frequencies. Figure 3-4 shows how it's hooked up. Once you've figured out where to tune your UHF dial to pick up your familiar stations, you're all set. You can now use your TV's remote control to select the various converted-to-UHF stations, and your programmable videocassette recorder can simply select the UHF channel it's supposed to record. UP-CONVERTERS are sold by Vidcor, mainly through electronics parts outlets, or through RCA's Special Products Division under the name Channeltrak. Energy Video sells its $80 Channelizer. Sigma Sound sells its $45 Channel Plus. Etco sells its $40 thirty-channel Cable TV converter #187AE047. All of these UP-CONVERTERS will translate the cable frequencies to UHF frequencies for your TV set to tune to.

Some new TV sets coming out now, like the Sony KV-2645RS and the RCA Color-Trak, tune in the MID-BAND and the SUPER-BAND frequencies. They are regular TV receivers but with 105 or more channel capacity. RCA's ColorTrak 2000 picks up 127 channels. Simply connect The Cable into the antenna input of your TV and dial up the channel you want. This extra channel feature on TVs and videocassette recorders is called CABLE-READY.

Some TV sets, like the Sony KV1945R, aren't quite CABLE-READY but because they can be variably tuned through the frequencies, they can be "tricked" into pulling in several MID-BAND channels.

Figure 3-4
Connecting an UP-CONVERTER to your TV

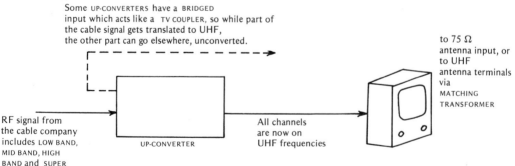

56 Cable TV

Having a CABLE-READY TV or an UP-CONVERTER can save you money. Many cable TV companies will reduce your subscription fee by about $3.50 per month if they don't have to provide a CONVERTER box to you.

The cable companies will have my kneecaps for printing this, but guess what little bonus you get with an UP-CONVERTER or CABLE READY TV? Yup. Pay-TV is in there somewhere. If they didn't TRAP that frequency before it came into your house, your UP-CONVERTER will give it to you now.

The devious consumer who moves into a house and finds The Cable still connected may consider buying a CONVERTER box in the alley from a guy named Scarface. Or maybe the local kid genius will offer to build him one for $30 in parts. Neither of these are brand new ideas.

The schematic (electronic parts) diagram for the DECODER for a large Southern California *broadcast* pay-TV system, ON TV, appeared in *Electronic Engineering News*, October 27, 1980. The same circuit is used in Chicago, Ft. Lauderdale, and Los Angeles. The January 1981 issue of *Radio Electronics* also shows how to build a pay-TV DECODER. You can buy half-assembled kits to finish yourself for under $150 from small "pirate" shops.

Is this legal? The FCC and several state courts say no. The California Legislature has passed an anti-pirate law with fines up to $2500 and/or 90 days in jail. The home user is unlikely to be caught; it's the manufacturers and sellers of these DECODERS who are feeling the pressure from the pay-TV companies. On the other side of the legal question are The Americans for Free Airways, people who cite the 1934 Federal Communications Act, saying all airwaves are free. They maintain that no law prohibits them from tuning any kind of receiver to any station they can pick up. Electronic parts stores, under pressure not to sell pay-TV kits and circuit parts, also feel put upon. They maintain that they are selling "parts" not CONVERTERS. It's none of their business how people assemble the parts and for what application. You don't blame the gunpowder and lead industries for assassinations, or the liquor companies for auto accidents.

The battle rages on. Perhaps the Supreme Court will settle it.

And what about "pirate" cable CONVERTERS? Are they legal? This may have to be sorted out by Congress. There are an estimated 35,000 pirate Z channel CONVERTERS in Los Angeles. What do the cable companies do about it? Their detectives mostly search for missing FILTERS on telephone pole junction boxes and apartment basements. When they find a violator, they simply cut off his cable service. Ouch!

WAYS OF CONNECTING YOUR CABLE

See Figure 3-5 for some possible cable configurations. Not everything may work; you may have to experiment. Everything you learned in the last chapter applies to cable as well as antennas. You use MATCHING TRANSFORMERS, SPLITTERS, COUPLERS, and BOOSTERS the same way as before. Essentially, the cable is like your antenna wire, only you're renting the signal on it from someone else.

Standard 1 set, no CONVERTER (for systems which use standard TV channels for much of their programming):

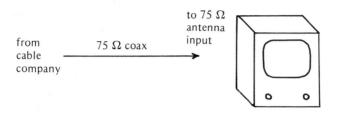

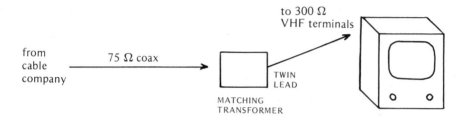

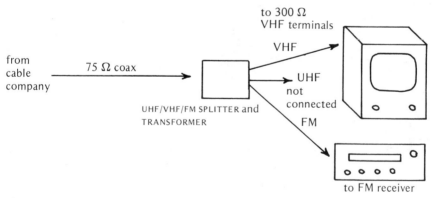

Figure 3-5
Various cable connections

For two sets, no CONVERTER:

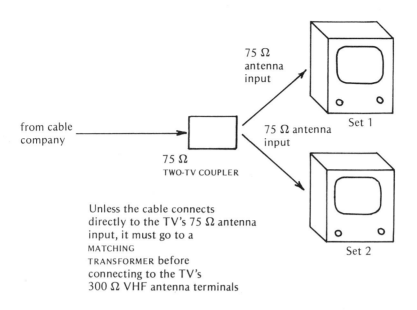

For other combinations, no CONVERTER:

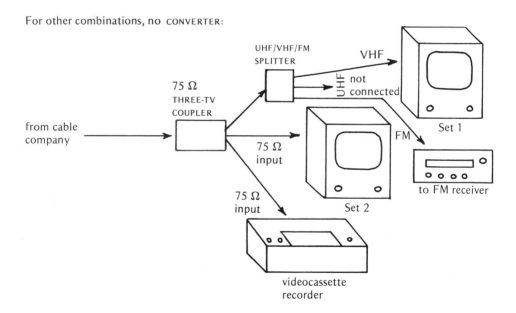

Figure 3–5 (*Cont.*)
Various cable connections

One set with CONVERTER:

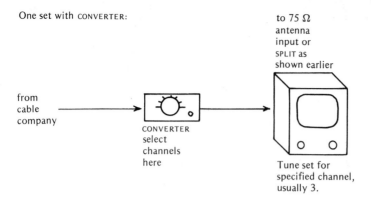

Two sets, one CONVERTER. Both sets always get same program:

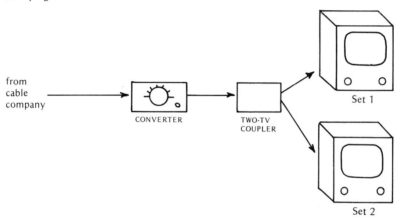

Two sets, one CONVERTER. One set can tune separately to normal channels if they can be received without a CONVERTER, while other set can get all channels on the system:

Figure 3–5 (*Cont.*)
Various cable connections

Two sets, two CONVERTERS:

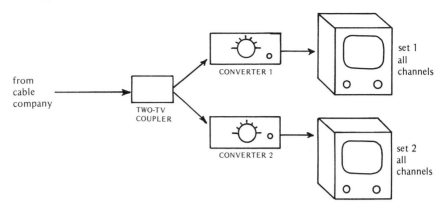

One set, UP-CONVERTER:

To 75 Ω antenna input or through MATCHING TRANSFORMER, send the UHF signal to the 300 Ω UHF terminals

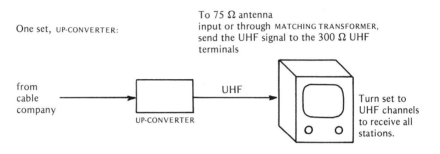

One set, UP-CONVERTER (for systems which use standard TV channels for much of their programming):

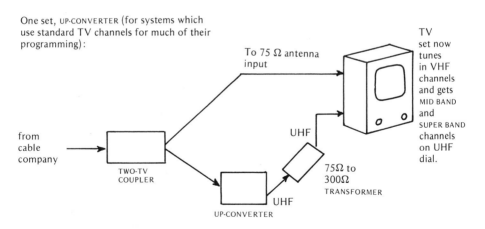

Figure 3–5 (*Cont.*)
Various cable connections

One set, UP-CONVERTER with built-in COUPLER (for systems which use standard TV channels for much of their programming):

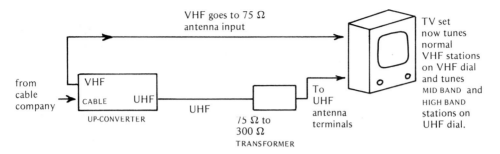

One set, one FM radio, UP-CONVERTER:

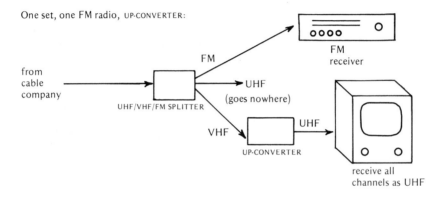

Two sets, videocassette recorder, UP-CONVERTER:

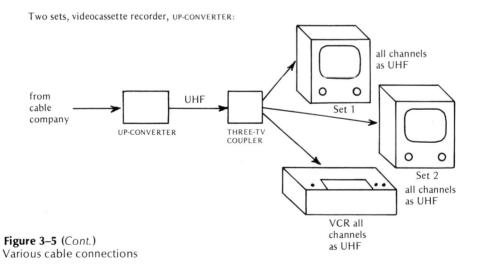

Figure 3–5 (*Cont.*)
Various cable connections

UP-CONVERTER and PAY-TV DECODER (for systems which use standard TV channels for much of their programming including SCRAMBLED TV):

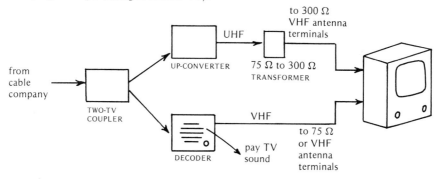

Use of CABLE READY TV with cable TV:

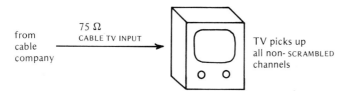

Use of CABLE READY TV with cable TV having some SCRAMBLED channels:

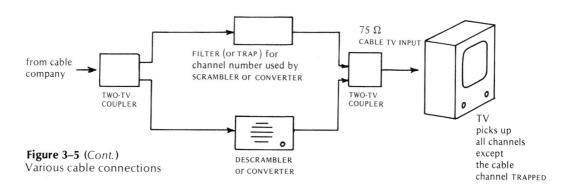

Figure 3-5 (Cont.)
Various cable connections

Use of CABLE READY TV with cable TV having some SCRAMBLED channels

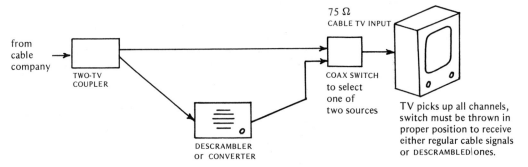

Simultaneous use of UP-CONVERTER and CABLE TV CONVERTER or DESCRAMBLER. Permits remote controlled TV (or programmable VCR) to receive any channel, including SCRAMBLED ones

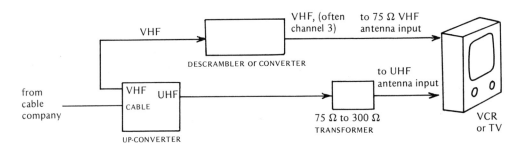

Figure 3–5 (*Cont.*)
Various cable connections

64

Two-Way Cable

Most cable systems are designed to send TV signals in one direction—to your home. The source from which all these TV signals emanate is called the HEAD END of the cable system. The viewer, I guess, would be at the tail end. At substantial expense, the cable equipment can be made to pass TV signals *upstream* too, so that a college, for instance, could play instructional tapes over the cable company's cable and into homes.

More can travel through a cable than just TV programs. Computer data, alarm signals, all kinds of coded information can traverse the wire—in both directions—if the system is so designed.

The subject of INTERACTIVE TV (where the viewer, through a keypad, can send signals *from* his or her home *to* the cable HEAD END, or to a computer, or to other viewers) is a complex one, and only tangentially relates to Cable TV. TELETEXT, VIEWDATA, VIDEOTEX and other systems which allow the home viewer to select and display printed data on a home TV screen will be discussed in detail in Chapter 17 (paperback Volume II), Video and the Computer. Here, we'll just stick with cable INTERACTIVE TV.

AT&T, CBS, and other wealthy companies are dropping big bucks into the INTERACTIVE TV field and research is going on in Salt Lake City, in Washington, DC, in Coral Gables, as well as in Great Britain, France, Canada, and Japan. Sammons Communications is sending news, community notices, and sale bulletins to 200 cable subscribers in Dallas. In Columbus, Ohio, Warner Cable is making it possible for viewers with an Atari home computer to access an information retrieval service called Compuserve.

QUBE

The most well known and ambitious two-way cable project started in Columbus, Ohio in 1977. Warner Cable got 27,000

Newer, smaller Warner Amex QUBE III home console (Courtesy of Warner Amex Cable Communications)

people to sign up for QUBE (for an extra $3.45 per month on top of their regular service at $8.50 per month). QUBE subscribers received a button-covered box the size of a telephone. These buttons allow the viewer to: (1) choose one of 30 channels offered on the system, and (2) respond to questions that the television might ask. The coded responses of all the QUBE viewers travel *upstream* to the cable offices to be tallied by a computer.

Some of the features of a TV system that "listens" to the viewers are:

1. Special programs with panels of experts debating controversial topics could query the viewers for their viewpoints. The question might be posed, "How should card cheats be punished?" and on the screen would appear:

    1. Life in prison
    2. Electric chair
       TOUCH NOW

The viewers would press their selections on their boxes. The computer would tally the votes, and the viewers would know how other respondents felt on that issue, and the panelists could discuss the results. Conceivably, mayors could make policy in light of viewer responses during a talk show. For more frivolous examples, an advertiser could test-market several packages to see what had the most viewer appeal. A "Gong Show" could be set up where contestants would get the hook when too many viewers disapproved of the performance. Viewers could vote to select the best tanned beauty at the local pool. Viewers could vote to select whether a drama should end happily or tragically; the appropriate tape would then be fed into the video tape player to satisfy the most viewers.

2. Signalling the computer with a certain code would start an up-to-date news read-out rolling across the screen.

3. Items on the screen could be ordered by typing in a certain sequence on the keyboard. Billing would be automatic (the computer knows who you are when you "talk" to it and charges could be simply added to your cable bill). Shopping by TV would allow you to remain in your comfy chair while you ordered groceries, appliances, or exercise equipment.

4. Instead of paying for pay-TV movies whether you watched them or not, you could watch—and be billed for—just the ones you wanted. The computer, after giving you a two-minute grace period to decide whether you wanted to watch a show, would sense which channel you were on and would bill you accordingly. To avoid unauthorized use of this Premium channel, a special key must be inserted into the selector console to activate the service. QUBE offers a selection of ten pay-per-view channels at once. A College Board coaching course costs $1.50 per lesson. Soft-core porn is $3.50 per show. Some football games are $7.00, and boxing matches cost up to $10.00 each.

5. If a fire/burglary alarm system were connected, an alarm signal could be passed through to the police or fire station automatically via QUBE.

6. A special code, known only by the adults in the household, could energize a certain channel, bathing the room in an evening of porn, for instance, billed automatically on a pay-as-you-lust basis.

7. Ralph Nader could ask how many people were ripped-off buying a certain product. When the buttons are pressed, the computer prints out a list of viewers who could be contacted, or approached as potential volunteers in a crusade against the offender.

INTERACTIVE systems like QUBE have some disadvantages too:

1. A Scripps-Howard news organization study found QUBE viewers ambivalent about INTERACTIVE TV. They were renting QUBE because of features other than interactivity, such as:

   (a) Twenty more channels than "basic" cable service.
   (b) More movies and sports on QUBE.

2. The QUBE survey technique is bad. If a question is put to the viewers and 80% of the respondents press the button indicating disapproval of something, does that mean 8 people responded that way out of a paltry total of 10, or a substantial 800 out of 1000 disapproved. Since QUBE doesn't release actual figures, only percentages, there's no way to tell how meaningful these statistics are. How many 5-year-olds pushed the button and were counted in the survey? What kinds of people happened to be watching at that moment—teenagers, executives, homemakers? Without better control of the data, the survey results cannot be taken seriously.

3. What keeps someone from peeking into the computer's mind and seeing how many times *you* watched "Captain Lust," and how you indicated *you* felt about abortion?

4. Pay-as-you-view bills can add up unexpectedly quickly. Many bills run $40 per month and a few exceed $150 per month.

5. Shop-at-home orders seem to have a high return rate. There seems to be a lot of impulse buying followed by second thoughts.

Future Developments

How about more channels? Present cable TV systems are stretching when they deliver 50 channels at once. Unless there's a surprising new circuit breakthrough, that's about all the channels a coax cable can hold. Why? Because each channel needs a FREQUENCY of its own. As we keep adding new FREQUENCIES, the FREQUENCY number keeps getting higher. You may remember from the EQUALIZER discussion in the previous chapter (of course you remember) that higher FREQUENCIES have greater difficulty traversing the wire. So we reach a point where we can't add new FREQUENCIES without having trouble pumping them through the wire.

No law says we have to transmit TV via a wire. How about shooting the signals out of a laser and into a long glass rod that reaches from the cable company to your house. Crazy? No, FIBER OPTICS.

Here TV signals MODULATE (vibrate at certain frequencies) the intensity of a light from a laser. The light is funneled down a long, thin glass fiber, and it comes out the other end relatively unchanged. At the other end is a light sensor, sort of a photocell, which converts the incoming light into electrical impulses. This electrical signal is DEMODULATED (decoded) and tuned in by your TV. Light can be MODULATED at extraordinarily high FREQUENCIES, and these light signals have little trouble passing through long glass fibers. It's coming.

In Higashi-Ikoma, Japan and in Arnhem, the Netherlands, they have another way to get more channels into their homes. Consider your telephone for a moment. You can handle only one call at a time on the phone, one "channel," if you will. If you made your TV system like a phone system, you'd receive only one channel at once. Lousy system, right? But with your phone, you can dial up millions of other phones, putting millions of other conversations at your ear. Similarly, a dialable TV cable system, using but one puny wire, can call up potentially millions of different channels.

The Higashi's and Arnhem's don't get 50 channels at once, but they can get as many channels as there are to dial up, one at a time. The Arnhem's presently have 36 channels at their fingertips, but that number is easily expandable. The Higashi's can dial regular TV programs, videocassette libraries where automated tape players will play their desired selection (like a juke box), and they can even dial *each other*.

Potentially with this "switched optics" system, you'll someday be able to "call up" the TV channel you want to view, and have thousands of channels to choose from.

There's another way to get 50 or more channels on your TV *and* avoid paying a fee for the pay-movie channels (although the legality of this process is under debate). Remember how Mr. Slicker sold the Mountain Man TV programs beamed by satellite from the movie distributor directly to the cable companies? You could get a "dish" and pick up scads of channels just like the cable companies do. In fact, if you aim your antenna at the RCA Satcom III-R bird, you'll get nearly all the stations that the cable operators get, plus some more they don't subscribe to. Such ventures are not off in the distant future somewhere, they're happening now. Over 1500 satellite-TV receivers are already being installed each month. Check out Chapter 18 (paperback Volume II) and see what you're missing.

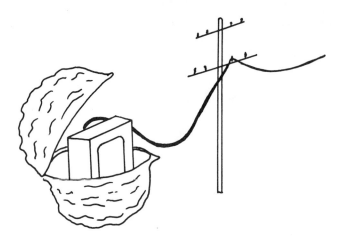

Cable TV in a Nutshell

Cable TV is like a shared antenna. The cable company maintains several antennas at its HEAD END and sends the TV signals down a coax wire past people's homes. For a fee, they can connect to the wire and receive the shows.

The cable company may provide a studio for local origination of programs, and may have 24-hour news, weather, and ads running on a special channel. The company may connect with a NETWORK to receive some other cable company's signal to add even more channels. If the company buys a special satellite-receiving antenna (called a "dish" because of its shape), then even more programming from across the country can be available.

Each channel has its own FREQUENCY, and UHF channel FREQUENCIES are hard to transmit through The Cable. That leaves only VHF channels, but there are only 12 of those, 5 in the LOW-BAND FREQUENCIES and 7 in the HIGH-BAND. To make more channels available, the cable companies take advantage of the fact that some frequencies, set aside for *broadcasting* police, air traffic, and amateur radio communications, don't have to be set aside for *cablecasting*. So they put TV channels in these gaps where other signals used to be. To tune in these MID-BAND and SUPER-BAND FREQUENCIES, one needs a CABLE READY TV or a CONVERTER.

The CONVERTER that cable companies use translates the cable FREQUENCIES into a single channel that the standard TV set can use, such as channel 3. One problem with such a system is that only one channel is translated at a time, and this is done by twisting a dial on the CONVERTER. Thus if you have two TVs and one CONVERTER, both must receive the same channel at once. If you have a videocassette recorder and a TV, both get the same show at once. If you have a programmable videocassette recorder, it will only pick up the station to which the CONVERTER is switched. The same is true for your remote-controlled TV; the remote is rendered useless. To change channels, you must jump up and physically turn the CONVERTER's dial.

One solution is to buy an UP-CONVERTER. This box connects to the cable and converts all the TV channels to UHF FREQUENCIES which your TV can pick up. Remote controls, programmable videocassette recorders, multiple TVs, and recorders all can work freely again.

Pay-TV is usually cablecasted on one of the FREQUENCIES which must be CONVERTED to be watched. Often, CONVERTERS without the pay-TV option have FILTERS in them to block the pay-TV channel. Sometimes these FIL-

TERS are at the connection box between the main cable and your house. Some people purchase CONVERTERS on the "black market" in order to avoid paying for cable service, or to receive pay-TV for free.

Broadcast and cablecast pay-TV can also be SCRAMBLED. To make the TV signal usable, one must rent a DECODER. Circuits for these can also be purchased on the "black market." Circuit diagrams and parts lists appear in electronics magazines.

Two-way cable allows the viewer to send signals to the cable company's HEAD END as well as receive them. QUBE is one such system which features a voting/survey capacity, teleshopping, and selection of various data such as news, weather, advertisements, and public affairs. A large selection of pay-TV movies and sports are available on a pay-per-view basis.

If you don't get cable or many broadcast TV channels, try a satellite receiver.

Chapter 4
Videocassette Recorders

HIS MASTER'S TAPE

If you're thinking of buying a videocassette recorder (over 4 million people did this year), then this chapter will show you what you're getting into. You'll see how complicated this gadget is (and isn't) to operate and the kinds of things it can do. And if you're *still* interested in buying one, have a glance at Chapter 19 (paperback Volume II), Buying Home Video Equipment, before taking a plunge with your hard earned wampum.

If you already have a videocassette machine, this chapter will repeat a lot of things you should have read in your instruction manual. Beyond that, you'll see some tips the instruction manuals miss. And finally you'll get a dose of Everything You Ever Wanted to Know About Video Tape But Never Quite Felt Compelled to Ask.

If you're the kind of person who never trusts a machine with more than one button, that's okay. This chapter is designed to waltz you gently into a crowd of video buttons and features without your ever realizing you were dancing to the beat of the Video Revolution.

Words to know

VIDEO TAPE
A long plastic ribbon impregnated with a substance on which you can record video signals. It's usually threaded up on one reel and winds off onto another when played or recorded.

CASSETTE, VIDEOCASSETTE
A small case that holds a full reel of video tape and an empty take-up reel. After the cassette has been inserted into the VCR, the machine automatically draws the tape from the cassette,

> threads it, plays it, and winds it back onto the take-up reel. You can remove a cassette in the middle of a program (the machine automatically unthreads the tape before it ejects the cassette to you) and come back later to pick up where you left off. Rewinding, though recommended, is not required unless you want the tape to start at the beginning again.
>
> VCR
> Videocassette recorder, a machine that can record a VIDEO-CASSETTE or play it back.
>
> VTR
> Video tape recorder, a machine that can record a video tape or play it back. Technically, VTR means only reel-to-reel (as opposed to cassette) video tape recorders, but many use the term more generally to mean any video recorder, even VCRs.
>
> FORMAT
> An attribute of any VCR which makes it able to play tapes made on someone else's VCR. Two VCRs must have the same FORMAT (tape size, tape speed, cassette shape, and electronic properties) to be compatible. VHS is one popular FORMAT and BETA is another, but the two are incompatible with each other.

Basic Operation of a Videocassette Recorder

PLAYING A VIDEOCASSETTE RECORDER AFTER SOMEONE ELSE HAS SET IT UP

Let's start easy. Your wife had to leave the country in a hurry—something about a phony stock deal. The evening before she left, she video tape recorded instructions on how to run the company in her absence. When you awaken the next morning, the videocassette is on her pillow with a note saying the fridge is stocked with TV dinners (how appropriate) and the tape machine ready to play. What do you do next?

The safest thing to do at this point is to turn State's Evidence. The second-safest alternative is to refer to the machine's instruction booklet. Let's assume your wife shredded the instructions along with all her personal files before she left. This leaves only the third-safest thing: Read this book and try your luck with the tape machine. Here goes:

1. *Turn on the TV* you'll use to watch this presentation. While it is warming up, *turn on the power for the tape player.* A little pilot light will probably come on to tell you the machine is getting power. If it isn't, look for a TIMER switch and flip it (some timers must be ON for the VCR to work, others must be OFF).

2. *Press the* EJECT *button* on the tape machine and its mouth will open.

3. Remove the videocassette from its cardboard (or whatever) box. Holding the cassette so that its label is right side up (readable), *insert the cassette* into the machine's mouth. The little trap door on the cassette goes in first. (See Figure 4-1).

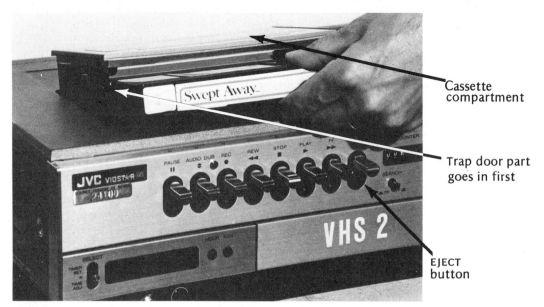

Figure 4–1
Inserting videocassette

4. *Press down the cassette compartment (closing the mouth) until it clicks, locking into place.* (Gulp!)

5. *Tune the TV set* to correspond to the preselected videocassette recorder channel, usually channel 3 or 4. At this point you should be viewing a blank TV screen. If you see a regular TV show playing:

(a) Maybe you're on the wrong channel.

(b) Maybe the antenna signal, instead of the VCR's signal, is going to your TV.

Look for an OUTPUT SELECT (or TV/VIDEO, or TV/CASSETTE or TV/VTR) switch on your VCR and flip it to VCR (*away* from TV).

6. *Press the* REWIND *button* to back up the tape to the beginning, if necessary.

7. *Press the* PLAY (or FORWARD or FWD) *button* on the VCR to set things into motion.

8. Adjust loudness, brightness, color, and so forth on your TV receiver. Stopping, playing, rewinding, and fast forwarding (winding ahead) are all functions of the VCR.

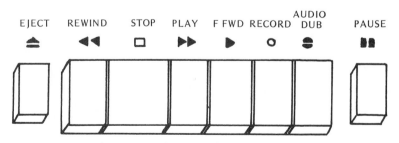

Function controls

> **Basic functions
> of the video tape machine**
>
> PLAY or FORWARD or FWD
> Plays the tape so that the program can be watched normally.
>
> FAST FORWARD or F FWD or FF
> Quickly winds the tape forward in time toward the end of the show.
>
> REWIND or RWD
> Quickly winds the tape backward in time toward the beginning of the show.
>
> STOP
> Makes the tape stop moving. The picture on the TV screen also disappears.
>
> PAUSE or STILL
> Makes the tape stop moving, but the picture on the TV screen holds still. (The image is called a "still frame," or "freeze frame.") On older machines, the picture may disappear in this mode. During PAUSE, the motor (you can hear it) in the tape machine keeps going. *Do not* stay in this mode for too long (over two minutes) as it can wear out that part of the tape. There is a spinning "head" in the VCR that reads the picture off the tape while rubbing against the tape. Too much rubbing in one place results in damaged tape.

NORMAL TV VIEWING

VCRs, with the exception of some portables, are designed to connect up *between* your TV antenna and your TV set. While watching normal TV, the antenna signals travel from the antenna, *pass through the VCR untouched* (the VCR could even be OFF), then into your TV, which you can watch normally. When watching the VCR's program, you flip a switch which electrically disconnects the antenna and substitutes the VCR's signal on channel 3 or 4. Then, you simply tune your TV to one of those channels. Some machines make this substitution automatically, (flipping their own switch) when you play a tape.

Here are the steps you take to watch normal TV with a VCR hooked up:

1. *Turn on the TV set.* Make sure the VCR timer is switched OFF (or ON on some models; more on timers later).

2. *Switch the TV/VTR SELECTOR to TV.* (This switch could also be called OUTPUT SELECT or PROGRAM SELECT on some models). The VCR does *not* have to be turned on.

3. *Select the desired TV channel* on your TV set.

That's all.

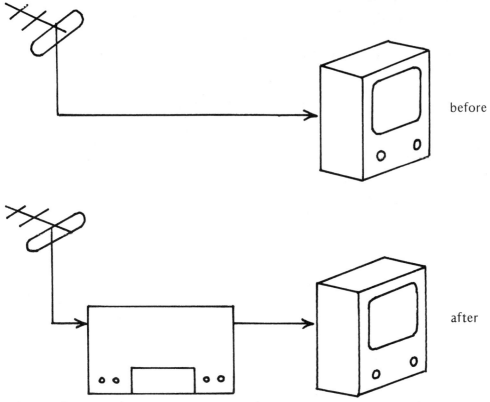

VCRs usually connect between your antenna and your TV

RECORDING TV PROGRAMS OFF-THE-AIR

Let's assume the VCR has already been properly hooked up to the antenna and TV (you'll see how to do this later). Essentially, you have to tell the VCR what *source* to listen to (camera, another VCR, or the TV antenna), and what *output* to send to your TV (the show *it's* recording or *some other* shows coming in on the antenna).

You also have to tell the VCR which channel to record and at what speed it should record it.

One thing the recorder *will* do by itself is ERASE the tape automatically as it records. So if you have the two-hour "Lynda Carter Special" recorded from last week, and this week you record a half hour of morning test pattern, you'll lose the first quarter of Lynda Carter. She'll become . . . da Carter. Do you realize that if you recorded "Gilligan's Island" over the first 20 minutes of Lynda Carter and 15 minutes of the "Muppets" over the end of Lynda Carter, you could label your tape "Gilligada Carpets." The right montage of "Eight Is Enough," "Archie Bunker's Place," "CHiPs", and "Little House on the Prairie" could result in a show called: "Eight Is Enough Bunk CHiPs on the Prairie." There's no telling what you'd get overlapping "The Incredible Hulk," "Dolly Parton," and "I'm a Big Girl Now." . . . Just a thought.

Back to business, here are the steps you follow to record shows off-air:

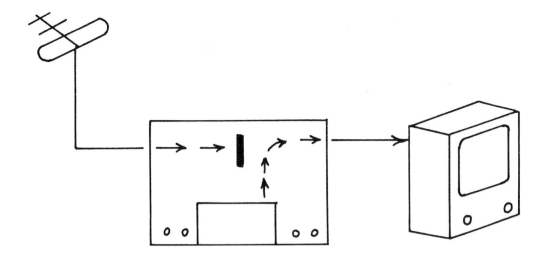

To *play* a tape, switch the VCR's *output selector* to VTR or VCR or VIDEO, or CASSETTE or whatever, *away* from TV. This feeds the cassette program to your TV and inhibits the antenna signal. Tune the TV to channel 3 or 4.

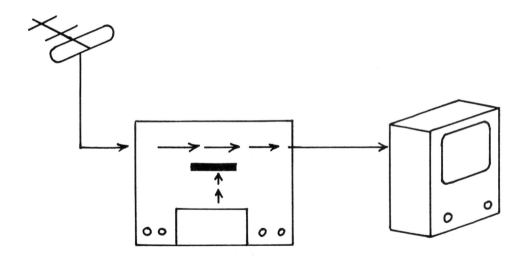

To watch regular TV, switch the VCR's *output selector* to TV. This feeds the antenna signal to your TV and inhibits the cassette and RF generator. Tune the TV to regular channels.

Playing a tape versus watching normal TV

1. *Turn on the VCR and the TV set.* Tune the TV to the proper channel, usually 3 or 4. Turn OFF the VCR's timer (or turn it ON depending on manufacturer).

2. *Load the cassette into the VCR* as described earlier. Make sure you have the proper length tape for the desired recording time.

3. To *tell the VCR which source to listen to,* flip a switch, depending on the type of machine you have:

- VHS—Flip the INPUT SELECTOR switch to TUNER. Switch the timer to OFF.
- JVC—Flip the REC SELECT to TV.
- BETA—Flip the INPUT SELECT switch to TV. The timer should be turned to ON.
- Others—There are only 1437 other combinations of switch names and positions for the 20 or so other brands of VCRs. Read your instructions.

4. *Tell your VCR what signal to send to your TV.* It's best to monitor the beginning of your recording to make sure everything is okay, so you'll want your TV to receive the VCR's signal that's being recorded. Depending on the machine, you do this by:

- VHS—Switching the TV/VTR SELECT to the VTR position.
- JVC—Flipping the VIDEO/TV switch to the VIDEO mode.
- BETA—Flipping the ANTENNA or OUTPUT SELECT switch to CASSETTE.

5. *Set the recorder's* SPEED. For highest quality sound and pictures use the fastest speed (shortest play time). On VHS machines, this may be labeled SP (Standard Play—two hours on a standard T120 tape). On newer BETA machines, this may be X2 (two-hour recording on a standard L500 tape). Older or industrial BETAS can record at the X1, or one-hour speed, which is the preferable speed if you have such a machine. If the program is too long to fit on the tape at that speed, or if you're a cheapskate who likes to cram as much on a tape as possible regardless of picture quality, switch to a slower speed (longer play time). On VHS machines this may be labeled LP (Long-Play—four hours on a T120) or SLP or ELP (Super-Long Play or Extra-Long Play—six hours on a T120). On a BETA this may be labeled X2 or X3 (two- or three-hour modes on an L500 tape).

More on tape lengths and recording times later.

6. *Switch the VCR's tuner to the channel you want to record.* For best tuning, disengage the AFT button and fine tune the station manually, using the procedures described in Chapter 1 for tuning your TV. Then switch the AFT back ON. *Take note:* If your antenna signal is poor or your VCR is poorly tuned, your recording will look bad. And once you've recorded a show poorly, it will remain that way—forever or until you erase the tape and record something new.

If your TV set isn't perfectly tuned to start with, you will have difficulty knowing when you've properly tuned your VCR. So make sure your TV is tuned well, perhaps by first playing a tape which you know is good and viewing it on your TV while fine tuning the TV set for the best picture. Now the TV is well-tuned-to-the-VCR. Incidentally, the tuning of the TV set *will not* affect your recording. Your TV is just a monitor, showing results. *Your VCR's tuner makes the results.*

7. *Reset the tape* INDEX COUNTER *to "000"* by pressing the button next to it. Later this can help you find things on your tape.

8. If the TV says your picture and sound are okay, *start recording.*

- Most VHS—Press the PLAY and RECORD buttons simultaneously.
- Many BETA—Press the single wide RECORD button.

76 Videocassette Recorders

9. *Check your recording.* Would you rather know something was wrong right away, or after two hours of recording *nothing*? Before the show starts, do steps 1–8 above, recording just 15 seconds of show. Then play it back. If it's okay, rewind and get ready to record for real. This precaution may sound silly now. I guarantee it won't after you've lost a few shows.

10. *Rewind the tape* to the beginning when you're finished recording.

RECORDING ONE PROGRAM
WHILE VIEWING ANOTHER

Your wife's stock scandal will appear on "Eyewitness News," but that's the same time your favorite "Tic Tac Dough" is on. Now, with unbridled anticipation, you prepare to watch the important program, "Tic Tac Dough," while simultaneously recording the news for later review—if you have time.

1. *Carry out steps 1–9 from the previous section.* Recording TV Programs Off the Air.

2. Once satisfied that your recording is successfully underway, *send your antenna signal straight to your TV* by:
- VHS—Switching the VTR/TV SELECT to TV.
- JVC—Flipping the VIDEO/TV switch to TV.
- BETA—Flipping the ANTENNA or OUTPUT SELECT switch to TV.

3. *Select your desired broadcast TV channel on your TV.* This will not affect your recording. The VCR is recording the channel *it* is tuned to, while the TV is showing the show *it* is tuned to.

If you want to check up on your recording from time to time, simply flip that VTR/TV SELECT (or whatever) switch to VTR and tune your TV channel to 3. This will not effect your recording.

RECORDING
WITH AN AUTOMATIC TIMER

These timers all work differently. Some are so complicated you need a doctorate in electrical engineering just to program them. Nevertheless, here are the basics.

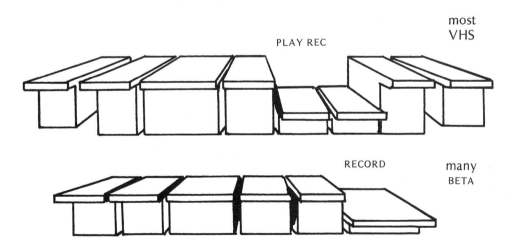

Press RECORD-and-PLAY on most VCRs, or just RECORD on some

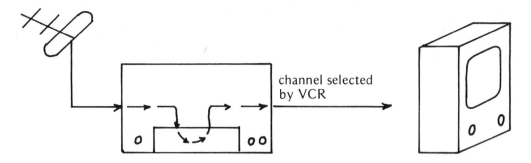

To see what your VCR is recording off-the-air:

1. The VCR must be listening to the right source. Switch its INPUT SELECT to TUNER or TV.
2. Your TV must be listening to the VCR. Switch its OUTPUT SELECT (or VTR/TV SELECT or ANTENNA switch or whatever) to VTR (or VCR or CASSETTE or VIDEO, or whatever). Tune the TV to channel 3 or 4.

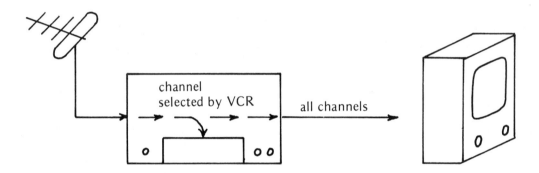

To watch other TV broadcasts while your VCR records one of them:

1. The VCR must *still* listen to the antenna. Switch its INPUT SELECT to TV.
2. Your TV must listen directly to the antenna. Switch the VCR's OUTPUT SELECT to TV. Tune the TV to the regular channels.

Watching your recording while you record versus watching normal TV

1. For VCRs with separate timers, *connect the timer* to the VCR following the instruction manual (where have you heard this before?).

2. *Prepare the VCR to record*, following steps 1–9 earlier, Recording TV Programs Off the Air. Once you know it's tuned and working okay, have the VCR rewound and ready to go. You may turn off your TV now, it's not needed. Besides, if your prim Aunt Prudence is visiting, with the TV off she'll never know you're recording "Prisoners of Passion." In fact, she can use the TV (if you've switched the OUTPUT SELECTOR switch to TV) and watch Masterpiece Theatre, oblivious to what's going on tape in the same room.

3. If you haven't already, switch the timer ON, and *set the timer turn-on time*. Most are adjusted like your typical alarm clock or digital LED timers.

4. *Switch the timer to* TIMER SET *or* AUTO SET, *or* TIMER/SLEEP, depending on the unit used.

5. *Switch the VCR to* RECORD. It shouldn't start recording yet, but the RECORD INDICATOR light may light up to indicate it's ready to. Some newer machines don't require this step.

6. All VCRs stop at the end of the tape, but only some have timers which can be programmed to stop at the end of a show.

These are the basics of operating your VCR. From here on things will get more complicated, but taken one step at a time, the mystery will unravel.

Connecting Up Your VCR

First, read your instru . . . oh, you've heard this before?

TV CONNECTIONS

Generally, you will connect the TV antenna to the VCR as if it were a TV set.

1. *Disconnect the 75Ω coax Cable TV cable from the TV* (if it has it), or disconnect the 300Ω antenna wire from the VHF and UHF terminals of your TV set.

2. *Reconnect these cables to the corresponding* ANTENNA *inputs on the back of the VCR*. As with a TV, if the antenna wires are 75Ω coax, you either have to connect to the 75Ω ANTENNA input or use a MATCHING TRANSFORMER/SPLITTER, which will give you 300Ω UHF and VHF wires to connect to those inputs on the VCR.

3. Hook your TV set up to the VCR's TELEVISION OUT (or VHF OUT and UHF OUT, or RF OUT) sockets. These wires may be provided with the machine, or you may have to buy them in a radio/TV store.

Do not use CB cable for any VCR or TV connections. It looks like TV coax, but it's 50Ω. Use only 75Ω RG-59U cable or 300Ω twin lead to connect the VCR to the TV.

4. *Plug in the TV and VCR's power cords.*

5. Pick an unused channel in your area and *switch the RF* OUT *or* RF CONVERTER *to the vacant channel*, usually channel 3 or 4. This switch is usually in the back of your VCR.

6. *Tune your TV to this RF channel.* You may wish to play a good prerecorded tape while making this adjustment so you'll have a picture to evaluate as you tune.

VIDEO CONNECTIONS

See Figures 4-2 and 4-3 for a look at some of your VCR's inputs and outputs.

Video in Most VCRs are designed to accept straight video signals as well as RF. This video could come from a camera, another VCR's VIDEO OUT, a SWITCHER/FADER or some other special effects device, the TV OUT from a TV monitor/receiver, or the VIDEO OUT from a separate TV tuner.

Video sources are very standardized, which means any VIDEO OUT will work okay with any VIDEO IN. Sometimes the plugs and

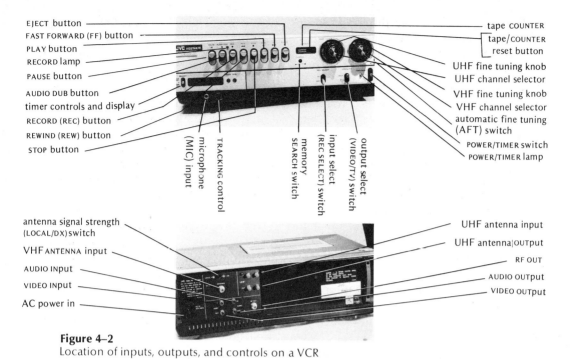

Figure 4–2
Location of inputs, outputs, and controls on a VCR

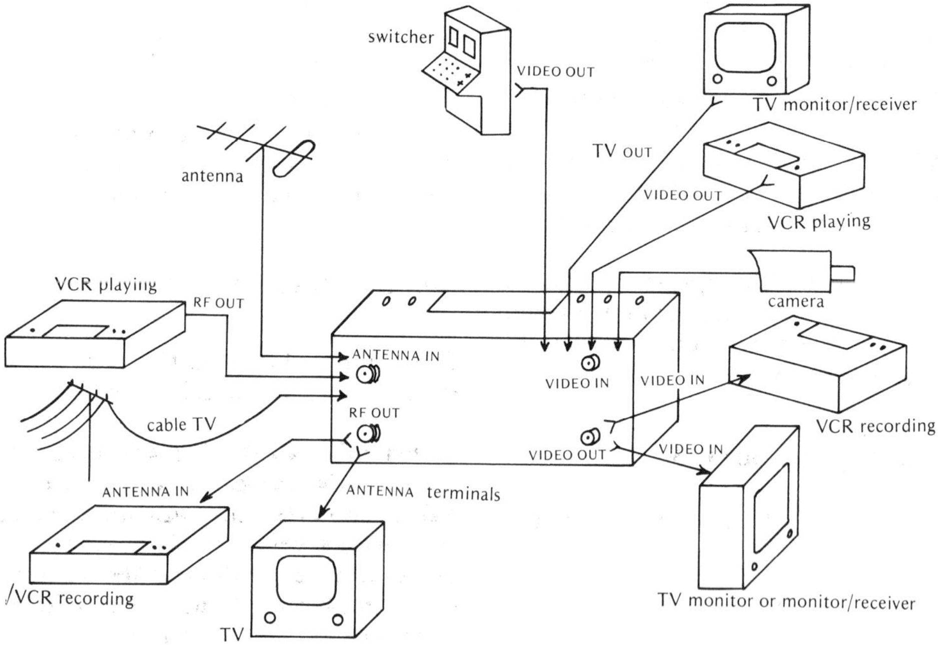

Figure 4–3
The ins and outs of TV and video signals

sockets don't match, requiring you to use an ADAPTER like those in Figure 1-13.

In the back of most VCRs, you'll see a VIDEO IN (or CAMERA IN) jack, usually with a PHONO (RCA) socket. On some VCRs, simply plugging something into the VIDEO IN socket will tell the machine to disregard the antenna signals it's getting and listen only to whatever you plugged in. On other models, the socket is activated manually by flipping a switch:

- VHS—Switch INPUT SELECT to CAMERA.
- JVC—Switch the REC SELECT to AUX.
- BETA—Switch the INPUT SELECT switch to VCR.

Direct video input gives sharper pictures than their RF counterparts, so use straight video whenever possible. When the video signal is converted to RF and sent into the ANTENNA input of the VCR, about 10% of the sharpness is lost in the conversion.

Camera in Industrial and studio TV cameras are usually designed for indoor use. They have AC power cords and standard video outputs, perfect for connecting to your VCR's VIDEO IN socket.

Portable VCRs and TV cameras (discussed in detail later) connect together with a fatter camera cable which runs power from the VCR to the camera, audio from the camera's built-in mike to the VCR, video from the camera to the VCR, REMOTE PAUSE from the camera to the VCR, and sometimes video from the VCR to the camera's viewfinder. All these wires fit into a single umbilical. To accommodate this connection, portable VCRs have a multipin CAMERA IN socket through which this blizzard of signals travel.

A few console (nonportable) VCRs like the JVC 6700U Vidstar have convenient multipin CAMERA sockets for direct connection to portable cameras. Most decks do not, however, and require a separate POWER SUPPLY or CAMERA ADAPTER to provide power to the camera and proper AUDIO, VIDEO and REMOTE PAUSE signals to the VCR. Its connection is shown in Figure 4-4. The camera plugs into the ADAPTER; the ADAPTER powers the camera while splitting up the AUDIO, VIDEO, and REMOTE signals for the VCR to use.

Video out Among the goodies in the back of most VCRs you'll find another phono-type socket labeled VIDEO OUT. This signal can be sent to another VCR, to a TV monitor, to a modulator which converts the signal into RF, or to specialized equipment which can change or improve the signal.

Its greatest use is when sending your VCR's signals to another VCR for copying. Again, the direct VIDEO OUT is a "cleaner" signal than the RF OUT, better for copying.

Serious videophiles may purchase video MONITORS (which cost about $300 more than regular TVs) and send the signal there directly for the "cleanest" of pictures. The difference between an RF picture and a video picture hardly warrants the effort on a small TV screen, but if you use a video projector to watch your tapes, the large screen will blow up this degradation. So go straight video when blowing up the image.

AUDIO CONNECTIONS

Sound is recorded automatically when you record the picture from a TV show off the antenna. The sound part of the program is also automatically included in the RF signal you send to your TV set from your VCR.

What if you want to make your own sound, to sing along with the "Lawrence Welk" show, or to capture Baby Snookems' first words while your camera catches the accompanying grimace?

The separate audio inputs are activated either by inserting a plug into them, or by flipping a switch—like the INPUT SELECT switch—to the CAMERA position.

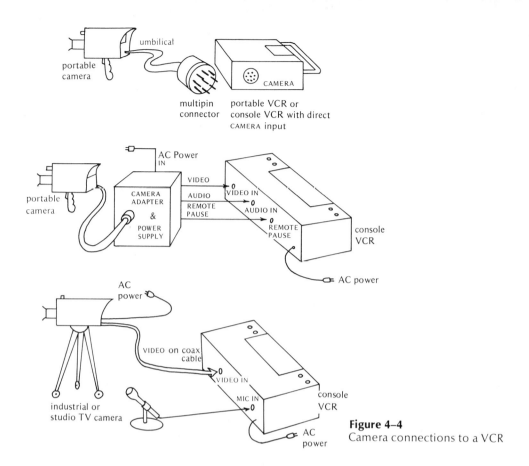

Figure 4-4
Camera connections to a VCR

Mic in (Microphone in) This is where you can plug a microphone in and record your own audio track along with the TV's or the camera's picture. An adapter may be necessary if the mike plug doesn't fit the VCR's socket. This input can also be used while DUBBING new audio in the place of audio already recorded on the tape. More on this in the Audio chapter.

The MIC socket is appropriate for *weak* audio signals, like the ones you get from microphones, most record turntables, telephone pickup coils, electric guitar pickups, or the MICROPHONE OUTPUT of an AUDIO MIXER. In general, if your audio source has no power supply of its own, doesn't use batteries, and doesn't need to be plugged into the wall to work, it should be plugged into the MICROPHONE INPUT or MIC IN of the VCR.

Audio In VCRs are likely to have another input labeled AUDIO IN or LINE IN, or AUX IN or HI LEVEL IN. These inputs are not as sensitive as the MIC inputs and can take the stronger audio signals you would get from an AM or FM tuner, an audio tape deck, any preamplifiers, the LINE OUT, or HI LEVEL OUT, or TAPE OUT, or AUX OUT, or PREAMP OUT from any MIKE MIXER, VCR, a hi-fi or a radio, an audio cassette deck, a monitor/receiver, a movie projector, or a turntable that has built-in preamplifiers. In many cases, if the audio source needs electricity to operate, the audio signal should go to the AUDIO IN of the VCR rather than the MIC IN. See Figure 4-5 to review.

A word about preamplifiers: A microphone turns sound into a tiny electrical signal. A preamplifier changes the tiny

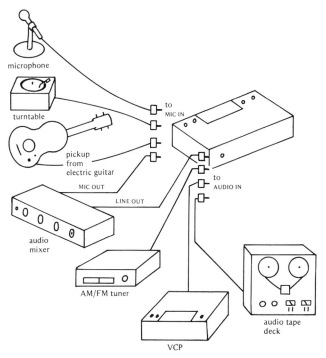

Figure 4-5
High- and low-level audio sources

signal into a medium-sized electrical signal. An amplifier turns a medium-sized signal into a big electrical signal. A speaker changes a big signal into sound. Medium-sized signals are easiest for electronic equipment to handle, so most audio devices have PREAMP outputs for sending medium-strength signals to other devices. They also have LINE and AUX inputs to receive medium-strength signals from other devices.

Some audio devices have earphone, headphone, or speaker outputs. In most cases, signals from these outputs are too loud even for the AUDIO IN of a VCR: The recorded sound comes out very raspy and distorted. There is a cure for this and it's called an ATTENUATOR. It contains resistors which "throw away" most of the audio signal so only a small amount gets through to the audio input of the VCR. The ATTENUATOR has a socket at one end that you plug your source into, and a plug at the other end that goes into your audio input. If you have to plug a strong source into your VCR and you *don't* have an ATTENUATOR, make sure that you at least use the AUDIO IN (not MIC) and be sure the volume control on the source is turned very low.

Some VCRs don't have AUDIO IN sockets. To make it possible for you to send HI LEVEL signals into your VCR through its MIC input, the manufacturers include an ATTENUATOR with the VCR. Again, plug it in between your source and the MIC input and try to keep the source's volume low to prevent distortion.

Audio out This socket sends a HI-LEVEL audio signal out for use by other equipment as shown in Figure 4-6. The signal is too weak to power an earphone or a speaker, and is too strong for a MIC input on most devices. It's handy for sending sound to your hi-fi or to another VCR while copying tapes.

Connecting Up Your VCR 83

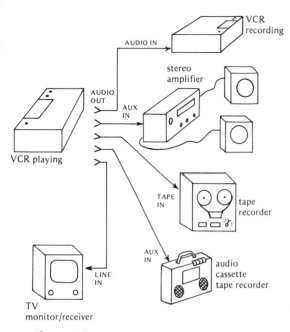

Figure 4–6
AUDIO OUT from the VCR to other equipment

OTHER CONNECTIONS

Remote REMOTE PAUSE is a function that you'll find on nearly every late model VCR. Normally, you can PAUSE your VCR during playback, which stops the tape from advancing, but leaves everything else in the machine the same. Some (mostly older) machines make the screen go black on PAUSE. Most VCRs now display a "freeze frame" or still picture in this mode. While in the RECORD mode, you can temporarily interrupt the recording process to leave something out (like a commercial) by hitting the PAUSE button on the machine. REMOTE PAUSE allows you to do the same thing at some distance from the machine by plugging in this simple pushbutton device, as shown in Figure 4-7.

Remember that REMOTE PAUSE, like regular PAUSE, should not be used for more than a couple minutes at a time to avoid damaging the tape.

More expensive and advanced VCRs permit other VCR functions to be remotely controlled, like REWIND, STOP, PLAY, SCAN, and even CHANNEL CHANGE. This feature is often called FULL-FUNCTION REMOTE CONTROL, however some brands are "fuller" than others. For instance, you can't remotely hit the RECORD button on the Sony SL5800, Zenith VR9750, Toshiba V8000, Akai VP7350, and Sanyo 5050, whereas *you can* on the Panasonic PV1750 and NV8200, JVC HR2200, Hitachi VT8500, and the Mitsubishi HS300U. These Hitachi and Mitsubishi models, as well as the newer Matsushita decks also permit REMOTE CHANNEL CHANGE. This feature can make your primitive TV set act like a remote controlled unit when it's connected with such a VCR. One European VCR sports 44 remote controlled functions. That's "full."

Even more recent, advanced, and expensive is WIRELESS REMOTE CONTROL, requiring no electrical connection between your keypad and the VCR. These systems transmit the control commands via infrared light signals or by radio waves.

One down-to-earth thought about remote controls: No law says your VCR has to be near the TV. You can usually place the TV and VCR 20 feet apart with 20 feet of antenna wire between the two without any problem. Thus, the VCR could be next to

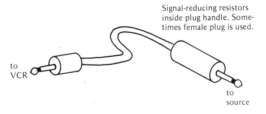

Audio ATTENUATOR

84 Videocassette Recorders

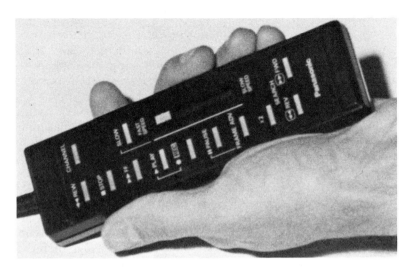

Remote control

your easy chair putting you within comfortable arm's reach of *all* the VCR controls. This set up not only saves a buck, but is also reduces the number of components in your system that could break down.

Convenience Outlet In the back of some VCRs you'll find an electrical outlet. This AC power socket is handy for plugging in small accessories such as separate timers, converters or decoders, or a small lamp. If the outlet says UNSWITCHED, it provides power whether the VCR is turned ON or not (handy for timers which must keep going after the VCR is turned off). If the outlet is labeled SWITCHED, then it turns on and off with the VCR (handy for activating decoders during the time a show is being recorded).

These outlets are usually labeled something like "300 W MAX" which means they can feed up to 300 watts of power to an appliance (like three 100-watt light bulbs). Plug your 1000-watt steam iron into it and you risk zapping your circuits or blowing a fuse inside the machine. Depending on your TV's wattage, it may or may not be a good idea to plug the TV into this outlet.

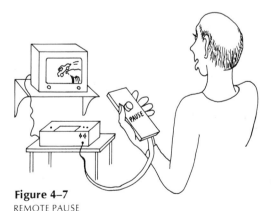

Figure 4–7
REMOTE PAUSE

Other Controls and Features Found on VCRs

There are over 25 brands and 200 models of VCRs. Before this page leaves the typewriter, this number will be 201. Some features and controls are found on nearly all VCRs while others appear on just a few specialized ones. So, for simplicity (in a field of unyielding complexity) I will break this list into two groups: Common controls and features, likely to be found on most VCRs, and Less Common controls and features, found on advanced or specialized VCRs.

COMMON CONTROLS AND FEATURES

Dew indicator For the VCR to operate properly, the tape has to slide smoothly through its internal mechanism. When it's damp in there, (like when the machine is cold and the air is humid) the surfaces become "sticky" and the tape risks getting hung up and damaged. The DEW INDICATOR senses this dampness (lighting to appraise you of this) and makes you wait until it's safe. If the DEW INDICATOR brings your TV production to a standstill, just leave the machine ON and wait for the light to go out. Sitting and staring at the machine will not shorten this process. When the light goes out, your VCR will function normally again.

Standby lamp It takes a moment for a VCR to withdraw the tape from the cassette and thread it into the machine. During this time, the VCR won't accept any commands like PLAY, EJECT, etc. This indicator is most common on BETA VCRs because they thread up the tape as soon as you put the cassette in and are thus "busy" for a moment before they are willing to accept commands.

Sometimes, if the tape fouls up inside the machine, STANDBY lights up and the VCR switches itself to STOP. Try winding or rewinding the tape a ways. This may straighten things out.

Counter/Reset/Memory This is helpful when trying to locate things quickly on a tape (described in detail later). Working like the odometer on your car, it keeps track of how much tape you've used. Press RESET to make the counter read "000" at the beginning of a tape and thereafter write down the numbers which correspond to different events on the tape.

Most VCRs also have a MEMORY button next to the COUNTER. Say you're playing or recording "Sheriff Lobo" and you think you'd like to come back to the exciting scene where he wrecks the entire police motor squad (again). First, you switch MEMORY ON. Next press the COUNTER RESET to "000." Later, when you want to find that spot, simply press REWIND. When the counter reaches "999", it will STOP. You're there.

One disadvantage of this system is that by pressing RESET mid-tape, you no longer are keeping chronological track of elapsed tape. This may or may not be a problem for you depending on whether you're trying to keep a log of what's where on the tape.

Speed select VHS recorders are equipped with one, two, or three speeds: SP (2-hour Standard Play), LP (4-hour Long Play), and SLP (6-hour Super Long Play). BETA recorders similarly can have up to three speeds: X1 (1 hour), X2 (2 hour), and X3 (3 hour). Of course, different length tapes will change how long the VCR can record at a given speed.

You select the tape speed when recording. Most machines automatically sense and adjust themselves to the correct speed when playing back.

Generally, select the fastest speed (shortest recording time) for highest quality recording. More on this later.

Tracking Generally, you shouldn't have to adjust this control, leaving it in the FIX position. When playing a tape, if your picture shows a band of hash across it like in Figure 4-8, jiggles a lot (and you're not watching "Charlie's Angels") it's time to adjust TRACKING. Turn it until the picture clears up.

Figure 4-8
TRACKING maladjustment

This control compensates for minor differences between similar FORMAT recorders and tapes. You are most likely to have to use it when playing a friend's tape, or even when playing your own tapes when your machine gets worn and out-of-adjustment. This control does nothing when you're recording, only when you're playing tapes.

Audio dub When you record a program, you automatically record both picture and sound. With AUDIO DUB, you can go back and rerecord *new* sound, erasing the old sound as you go, and leaving the picture untouched.

To activate the AUDIO DUB feature:

1. Plug a microphone into MIC IN, or some other audio signal source into AUDIO IN as described earlier.

2. You may wish to check your sound before you start recording by listening to it through your system. You could:
    (a) Plug an earphone into the EARPHONE jack if your VCR has one.
    (b) Run RF to your TV and monitor your sound there.
    (c) Connect your hi-fi to the AUDIO OUT jack on the VCR.

On some decks, you will automatically be able to hear your audio source with the VCR on STOP. On others, you must press AUDIO DUB to send the sound through to the monitoring outputs. On a few, the machine may actually have to be recording before you can hear your source.

3. Find where you wish to start dubbing new sound.

4. Press AUDIO DUB and PLAY together.

5. You will only hear the new sound being recorded. The old sound is being erased as you go.

6. When done, press STOP.

Dropout Compensator (or DOC) All video tapes have occasional tiny defects in them which result in specks of snow appearing on the TV screen whenever your video signal from the tape "drops out." An example appears later in Figure 4-23 where DROPOUTS are discussed in detail. A DROPOUT COMPENSATOR is a circuit built into many VCRs to "cover up" these unsightly blemishes. The DOC will successfully hide tiny DROPOUTS from your eye, but large ones (perhaps caused by creases, scratches, or dirt) cannot be hidden and still appear in all their glory.

LESS COMMON CONTROLS
AND FEATURES

Noiseless still frame Remember how PAUSE could stop the tape amid playing, and on recent machines would display a still picture? Sometimes that picture would stop with a horizontal band of hash across it (which looks exactly like the TRACKING problem shown back in Figure 4-8). This hash is called NOISE or a NOISE BAR. (Just as in audio where "noise" is unwanted sound, in video, "noise" can be unwanted stuff in the picture, like hash, grain, snow, specks, whatever). Without the NOISELESS STILL FRAME feature, you would have to adjust your TRACKING control each time you PAUSED the tape in order to remove the hash.

With NOISELESS STILL FRAME (as it's called on the JVC HR6700 Vidstar) the VCR's circuits push that hash off the screen to display a clear picture. On the Panasonic PV1650, this feature is called CLEAR STILL.

If your VCR has this feature and it seems to be displaying the NOISE anyway, you can correct it by turning the V LOCK ADJUST control on the back or bottom of your VCR.

Frame advance This is a STILL FRAME subfeature. Once in the STILL FRAME mode, a second control allows you to advance to the next video picture and stay there; or on some models, by holding the button down you can continuously advance the tape slowly, picture by picture, like viewing a movie one frame at a time.

Usually, those familiar NOISE BARS will roll across the screen as the pictures progress.

Noiseless frame advance All of the above, but without the NOISE BAR. The result is like watching a slide show.

Variable-speed slow motion Sometimes combined with STILL FRAME or FRAME ADVANCE features, this one allows the tape to proceed slowly forward. There will usually be a NOISE BAR rolling through the pictures as the tape moves.

Variable-speed noiseless slow motion The above without the NOISE BAR. This feature works best on tapes you've recorded and are playing on the *same machine.* Tapes acquired elsewhere may still display some NOISE.

Double, triple-speed play The tape passes through the machine at twice or three times normal speed while it's playing. Although the picture moves at the fastest pace, funny things happen to the audio. The industrial Panasonic NV8200 and 8310 units play the sound like a speeded-up record. The Donald Duck sound is nearly incomprehensible. The RCA VDT 625, Sony 5000 series, and Hitachi VT8500 simply mute the audio during speeded play. JVC's HR3600 and its Vidstar HR6700 include a special SPEECH COMPRESSOR circuit which renders the sound much more intelligible. Another name for this circuit is LIPLOC. Is there someone in your family who you wish had LIPLOC?

High-speed picture search This feature shuttles the tape *forward or reverse* at 5-, 10-, or maybe 20-times normal speed with its picture still showing. Various brand names for this include BETASCAN, OMNISEARCH, SUPERSCAN and SHUTTLE SEARCH. Presently, a NOISE BAR intrudes into the picture during the SEARCH mode, but on some brands it's worse than others.

Variactor tuning On some TVs and VCRs you change channels by turning a knob, "clunk-clunk." On more expensive ones, you push a button in a row of buttons in order to change channels. This latter type is VARIACTOR tuning. It allows you to go from channel 2 to 8 directly, skipping the channels in between, a feature called EXPRESS TUNING. It allows REMOTE CONTROL and PROGRAMMABLE TIMERS to switch channels for you. It cuts down on wear and trouble from the "clunk-clunk" part of the tuner. And often, the VARIACTORS can be tuned to many MID-BAND and SUPER-BAND cable-TV frequencies, freeing you from the restrictions of the CABLE CONVERTER BOX as described in Chapter 3.

Solenoid controls Older or cheaper VCRs have FUNCTION CONTROLS that look like piano keys or levers, are fairly hard to press (like the keys on a manual typewriter), lock into place when pressed, and spring back with a *snap.* Institutional and reel-to-reel VTRs often sport a single large control which you twist with a "clunk-clunk" from mode to mode. In both cases, the machines are manually operated, and their mechanical controls have no brains. *You,* the user, supply the brains. *You* must press STOP, then wait a moment before changing modes, from REWIND to PLAY for instance. *You* must press STOP, then wait a moment, before pressing EJECT to get your tape out of the machine. *You* must *remember* not to PAUSE in the same spot on the tape too long. If you don't STOP mechanical decks between modes, (1) they may eat your tape, or (2) their buttons simply won't go down *unless* you press STOP.

Conversely, the newer, more expensive VCRs have SOLENOID CONTROLS, feather-touch pushbuttons, much like your pocket calculator. These machines have some brains of their own. If you're in PLAY and you push REWIND, the machine will (1) Disregard your command unless you first press STOP and then wait a moment, or (2) will press its own STOP and then will REWIND. Such machines won't allow you to EJECT a tape until *it* has finished unthreading it back into the cassette. Recent SOLENOID VCRs, if left on PAUSE too long, will switch themselves to STOP, thus protecting the tape. In short, automatic VCRs have built-in safety features to protect themselves from clutzes and impatient operators.

Damped cassette eject Without this, when you press EJECT the cassette compartment snaps open with a cheap-sounding "clunk." With it the VCR opens its mouth silently, or with a faint woosh. No big deal.

Electronic program indexing Earlier you saw how by pushing the MEMORY button and setting your COUNTER to "000" you could prepare your VCR to later stop at that point when rewinding. That only indexed one spot on a tape. ELECTRONIC PROGRAM INDEXING aids in locating a number of spots on a tape.

Sharp's VCR allows you to assign an index number from 1 to 99 to various segments of the tape. You later recall them by number.

Akai uses a system they call Inter-Program Location Search (IPLS) which searches for snowy sections of unrecorded tape between recorded segments. This system only works if you leave a little blank space between your recordings; butt them together and you're out of luck.

Most advanced is the system which records a brief index signal on the tape whenever you press RECORD. Record 20 segments on your tape and you leave 20 index markers on it. When in FAST FORWARD or REWIND, a sensor in the VCR stops the tape when you pass a marker. If you wish to wind the tape all the way without stopping, this feature can be disabled.

One problem with this feature: The index signal puts a "glitch" (a momentary interruption of your smooth picture) into your recording and inhibits "clean" editing of your tapes. On the JVC HR6700 and Sony SL-5600 and 5800 you can't stop this reference marker from being recorded. But in the Toshiba V8000 and Hatachi VT8500 you *can* override this feature if you prefer glitch-free edits to indexed ones.

Scene-to-scene stabilization In the Editing chapter, you will learn how to PAUSE your recording at crucial places to edit out commercials, change scenes, or to change shots or angles. However, every time you interrupt the smooth motion of the tape, such as stopping it or pausing it, you cause a glitch. Different manufacturers have attempted to minimize these unsightly blemishes from your backyard epics. Sony does it with a Timing Phase Circuit on their SL 3000 portable. JVC calls their method Edit Start Control. Both systems endeavor to match the new incoming SYNC signal with the old recorded SYNC signal and thus maintain a smooth flow of SYNC pulses recorded on your tape. When played back, the smooth SYNC flow yields a steady picture on your TV.

Hitachi and Matsushita use a different system, which they call Scene Transition Stabilization. Every time you hit PAUSE while recording, the tape backs up a little. When you UNPAUSE, the circuits start the tape moving, line up the SYNC signals, and record the new picture over the tail of the old one. Disadvantages of this method include:

1. Precise editing is difficult. You have to allow a two-second "tail" at the end of each scene so the VCR doesn't start the next scene on top of something you wanted to keep.

2. Since there's an overlap of video signals, the color gets messed up for a couple seconds. This shimmering spectrum of colors is called VIDEO MOIRE.

Programmable timer The simplest timers turn your VCR on at a certain time during a 24-hour interval to start a recording in your absence. You would have to make sure the proper channel was selected and that the VCR was in the RECORD mode and switched to the TV input in order for your recording to occur in your absence. The VCR would continue recording until it ran out of tape and then stopped.

Improved models turn your VCR OFF as well as ON, and can do so over a several-day period.

A further-improved type is the SERIAL TIMER which turns your VCR ON and OFF at the same time each day until the tape runs out. This would be good for faithfully recording your favorite serials each day while you're away enjoying a 444-day sightseeing tour in Iran. (You may need an extra-long tape though.)

The above timers have one thing in common: The channel being recorded is always the same. To be truly PROGRAMMABLE, the timer must store commands to change channels as well as record at certain times.

While the little microcomputers which make your VCR PROGRAMMABLE add flexibility to your deck, they can also make it more difficult to operate. Depending on the complexity of the unit, your attitude and persistance, and how often you use this feature, you *may* find programming your VCR to be too difficult to master or remember. The Sharp VC-6800 VCR, for instance, requires you to punch an exact sequence of 18-odd keystrokes into its calculator-like panel *for each channel* you wish to program in advance. The machine will record up to seven different channels for numerous recordings up to one week ahead—if you can figure out how to program it to do so.

Stereo Dolby New top-of-the-line VCRs such as the JVC HR-7650U, Quasar VH-5623UW, Akai 7350 VHS decks and the Marantz VR 100 and 200 BETA machines will record and play back in stereo. Splitting the tiny audio track into two tracks is how they give you stereo, but the narrower tracks degrade the fidelity of the sound. By adding DOLBY (more on DOLBY in Chapter 7) to the machines, the manufacturers get back some of the sound quality they lost. So except for the stereo aspect, stereo DOLBY sounds about the same as monaural non-DOLBY. Incidentally for tape traders, stereo DOLBY tapes will play on standard machines and vice versa (but the sound will not have the stereo or DOLBY effect).

Framus Most advanced among the VCR features is the FRAMUS, or INTEGRATED SQUARTIN-HABIT. This clever device regulates how the HIGH-BAND RF signal is sent to the VCR's internally clocked CROBOSTAT to provide inertial reference for the servo motors and thus deliver a rock-solid picture regardless of TRACKING interference. This feature is usually found on VCRs having their own built-in KITCHEN SYNC, but it may appear on semiprofessional models equipped with everything but the KITCHEN SYNC.

Enough about the bells and whistles on VCRs. Let's learn a bit more about what they do, how they differ, and how to handle them.

Player/Recorder Compatibility

You get a phone call from this "business acquaintance" of your boss. He has this "very revealing" video tape of your boss which he wants you to see. Can you play it on your machine?

The first question you should ask (besides how much does he want for this tape) is:

What kind of machine was the tape recorded on? Nowadays, *a tape recorded on one machine will almost always play back on another machine of the same make and model.* Any healthy Technicolor 212 should play back a tape made on any other healthy Technicolor 212. Why the word *"almost"* in the above statement? If the gentleman's recorder wasn't running exactly right when making his tape, the tape won't play back correctly on your machine. (If your player isn't working right, of course, it won't play that tape properly either.) Like the mad genius who understands what he's talking about although nobody else can, your friend's tape recorder may be able to play the tapes *it* makes, but nobody else's VTR can play them. Barring machine problems, however, tapes can be interchanged between machines of the same make and model.

What if this mysterious gentleman describes a tape recorder different from any you own? Or, more likely, he knows nothing about video equipment; all he can do is describe the tape clutched in his sweaty hand. Now how do you tell if your machine can play it?

For one video tape or videocassette machine to play tapes made on another machine, the following criteria must be met:

1. Both machines must have the same FORMAT.

2. Both must use compatible reels or cartridges or cassettes.

3. Both recorder and player must be color in order to display color.

4. Both must use the same television standards (one can't be a foreign country's standard).

Let's attack these criteria one-by-one:

VIDEO TAPE FORMATS

Tape comes in various widths to fit various video tape machines. Every tape machine can work with only one width of tape. Generally, the wider the tape, the higher the picture quality and the more expensive the video tape and the tape equipment will be. There's ¼-inch, ½-inch, ¾-inch, 1-inch, and 2-inch tape. But there's more to FORMAT than just size. There's tape speed and other electronic differences which make not all ½-inch tapes playable on all ½-inch machines.

To simplify matters, various manufacturers and video associations have agreed to standardize on several FORMATS. Thus *if a tape is recorded on a certain* FORMAT *machine, it should play back on the same* FORMAT *machine,* regardless of manufacture (with a couple minor exceptions to be discussed shortly). The more popular FORMATS are listed below:

Format	Tape size	Notes
CVC	¼ inch cassette	CVC (Compact Video Cassette) is presently used by Technicolor and Elbex. Portable VCR costs about $1000. Normal length cassette plays 30 minutes.
EIAJ	½ inch reel-to-reel	Stands for Electronic Industries Association of Japan. Most common FORMAT in schools in the '70s. New VTRs cost about $800. Used ones are a bargain for $200. A 7-inch reel of tape plays 1 hour.
VHS	½ inch cassette	Comprising almost three fourths of the home VCR market, VHS (Video Home System) is also spreading among industry and schools. Actually three formats, VHS-2, VHS-4, and VHS-6 (or SP, LP, and SLP, as some call it) represent three speeds yielding 2, 4, or 6 hours of playing time (depending on tape length—these playing times are for a common T120

Format	Tape size	Notes (Cont.)
		videocassette). A tape recorded at the 6-hour speed must be played at the 6-hour speed. A VHS-2,4,6 machine will play all three speeds (it's switchable), but a VHS-2 machine can't play a tape recorded at the VHS-4 or VHS-6 speed. The VCRs cost about $700–$1200.
VHS-C	½ inch minicassette	Stands for VHS Compact. Recent ultraminiature (5.3 lbs) VCR recording in the VHS format, but using smaller 20-minute cassette, about the size of a deck of cards. The mouth of a VHS-C machine is too small to hold a normal VHS tape, but a VHS-C cassette can be played in a normal-sized VHS player if you stick the tiny cassette in a special adapter first.
BETA	½ inch cassette	BETA, introduced by Sony, comprises about one quarter of the home VCR market. Actually three formats, BETA-1, BETA-2, and BETA-3 (or X1, X2, and X3, as some call it) represent three speeds yielding playing times of 1, 2, and 3 hours (depending on tape length—these figures are for the common L500 cassette). A tape recorded at the 3-hour speed must be played at the 3-hour speed. A BETA-1,2,3 machine is switchable to play all three speeds, but a BETA-1 machine can't play a tape recorded at the BETA-2 or BETA-3 speed. The VCRs cost about $800–$1200.
¾U	¾ inch cassette	Widely used in schools and industry. VCRs cost about $2500. A cassette plays up to 1 hour. Picture is about 20% sharper than that from ½-inch home VCRs.
A	1 inch reel-to-reel	Older Ampex FORMAT introduced in the '60s. Ampex also refers to this as the VR-7900 FORMAT. VTRs cost about $20,000 and up. Tape plays 1½ hours.
B	1 inch reel-to-reel	Introduced by Bosch-Fernseh (they refer to it as their BCN series), also used by IVC, Phillips, and RCA, this FORMAT is most popular for professional VTRs overseas. VTRs cost $37,000 and up. Tapes play for 1½ hours.
C	1 inch reel-to-reel	Used on professional equipment by Ampex, Sony, Marconi, and RCA. VTRs cost $24,000+. Tapes play 1½ hours.
QUAD	2 inch reel-to-reel	Stands for "quadruplex" (the other recording configurations are called *helical*). It's an older FORMAT used by professionals. VTRs cost about $90,000. Tape plays 1–3 hours.

There are other, less popular VTR FORMATS still hanging around in school basements and rich uncles' attics. Some are defunct FORMATS like Sanyo's 2-hour V-Cord II, Quasar's VX (precursor to present ½-inch home video), and some are granddaddies to the ½ inch reel-to-reel decks still used today.

Incidentally, video discs also come in different FORMATS which are incompatible with each other. See Chapter 16 (paperback Volume II) for more on disc interchangeability.

COMPATIBILITY AMONG REELS, CARTRIDGES, AND CASSETTES

Your boss's "associate" may hand you a tape recorded in the same FORMAT as your VTR, but if his tape is on a 7-inch reel and your VTR is a portable which uses only 5-inch reels, you're going to have trouble playing the tape. If somehow you can get his tape on a 5-inch reel, you'll be all set.

If he hands you a CARTRIDGE, loaded with

½-inch tape, the *tape* is likely to be standard EIAJ. The trouble is, one manufacturer's CARTRIDGE won't necessarily fit into another machine. You'll have to wind the tape into a compatible CARTRIDGE or onto open reels. Put another way, the tape will play all right, but you have to package it properly. What exactly is a CARTRIDGE, you say? A CARTRIDGE, or VIDEOCARTRIDGE (sometimes abbreviated as CART) is a box that holds a full reel of tape but *no* empty reel. When the CARTRIDGE is inserted into the machine and played, the machine automatically draws the tape from the box, threads it, plays it, and winds it around an empty reel built into the machine. Before you can remove a cartridge, whether in the midst or at the end of a program, you must first rewind all the tape back onto the reel inside the CARTRIDGE. These were the first home video recorders, introduced in the '70s. Anyone who has one probably has long whiskers.

As for cassettes, all VHS cassettes are compatible and will play interchangeably if played at the proper speeds; all BETA cassettes are compatible and will play interchangeably if played at the right speeds. This means that a VHS-2,4,6 machine can play any VHS recording. A BETA-1,2,3 machine can also play any BETA tape. Incidentally, all BETA VCRs regardless of what *other* speeds they play (with the exception of the Sony SL-8600), will play BETA-1 (1-hour) tapes.

All 3/4U videocassettes are interchangeable with one tiny (literally) exception: Portable 3/4U VCRs take smaller cassettes (like the Sony KCS-20). These mini-cassettes will play okay in standard sized 3/4U VCRs, but standard sized cassettes won't fit in a portable VCR's tiny mouth.

TELEVISION STANDARDS

A tape recorded in Europe on an EIAJ recorder may not play on an EIAJ VTR here in the United States. European electricity is different and European TVs make the picture in a slightly different fashion. Even though the FORMAT is the same, tapes made using different TV standards are not interchangeable. Color is even more complicated.

The United States uses the NTSC system of recording the color. Europe and Asia use other systems such as PAL and SECAM. Some specialized VCRs and TVs are designed to be multistandard and will play tapes in these three standards. See Chapter 20 (paperback Volume II) for more on exchanging tapes with people in foreign countries.

Back to your associate and his "revealing" tapes: Ask him what kind of recorder his tape was made on and from this try to deduce its FORMAT. If your VTR uses that FORMAT and the other criteria listed here are met, you can tell him to bring over his tape for a looksee. What do you mean, he hung up?

One last note on VCR compatibility: All VCRs, both home type and industrial, are *electronically* compatible. The video and audio signals are pretty much standardized among *all* machines. This means that although I can't play your VHS tape on my BETA machine, I *can* copy your tape onto my machine. You just bring over your VHS player and some wires, we'll connect my machine to yours, and while yours plays, my BETA will record your show on my blank BETA tape. Now your show will be in *my* FORMAT.

Getting More Out of Your VCR

Earlier, you learned the basics. Now we'll cement some of these sketchy descriptions together with more detail on how to operate your VCR and cover some handy points which were left out for simplicity sake.

FINDING THINGS
QUICKLY ON A TAPE

You have started playing your wife's tape, "How-to-run-the-business-while-I'm-gone" and as expected, her introductory comments

Counter	Contents
000	Start of tape
010	Introduction
159	How to run the office
271	Where the safe deposit box key is kept
282	Where the executive washroom key is kept
366	A lesson on creative accounting
405	Farewell
410	End of recording

cackle on and on, so you decide to skip ahead to the juicy stuff, like where she keeps the second set of account books.

You press FFWD (FAST FORWARD) and wait while the tape zips ahead. Although machines differ, a 2-hour videocassette recorded at the fastest speed will skip through ½ hour in about 1 minute. So, figuring Her Ladyship for *10 minutes* of introduction, you FAST FORWARD *20 seconds,* then press STOP.

If your VCR happens to feature HIGH SPEED PICTURE SEARCH, you could take some of the guesswork out of this process by using that mode. SEARCH, however, is usually not as fast as FAST FORWARD and may be more efficiently used *after* you get close to your desired section of the program. Just to be tough on ourselves for this example, let's pretend we don't have a SEARCH feature.

After winding ahead 20 seconds, press STOP, then PLAY, and begin watching again.

You discover that the introduction was shorter than you thought, and you've gone past the "Accounting" part of the program. So to locate the end of the introduction, you switch to STOP, wait a moment, then to REWIND for a few seconds, then to STOP—wait a moment—then to PLAY. And so it goes until you find where you want to start viewing.

If your wife had indexed the tape by COUNTER NUMBER, this search would have been easier. She could have written something like the chart above on a sheet of paper and put it with the tape or in the tape box.

To use this index, you would:

1. Before starting the tape, press the RESET button, making the COUNTER numbers 000 for the tape's beginning.

2. To find the end of the introduction, FAST FORWARD to, say, 155 on the counter. If you play the tape from this point, you should be watching the end of the introduction and the beginning of the "How to run the office sequence."

The index allows you to jump ahead to 405 to watch the blithering au revoir and then back to 271 for the safe deposit box information.

Perhaps 366 is where the "Second set-of-books" sequence starts. (What's this? After watching it you plan to pull a "Rosemary Woods" and erase this section before the IRS subpoenas your tape?)

The counters on most inexpensive video tape equipment aren't too accurate and may err by 10%. And if you are playing on a JVC machine but the tape was recorded on a Panasonic machine, the difference may be as much as 30%. So don't put *too much* trust in the counter. Learn your machine; experiment to see how accurate it is, and try a friend's tape too.

Say you had recorded four half-hour shows on a tape at the SP mode (fast speed), had never bothered to index them, and now wanted to see just the end of show #2. You essentially want to get yourself maybe 50 minutes into the tape. To do this, you could FAST FORWARD the tape about 1½ minutes. But this method is pretty inaccurate. What would be nice is a COUNTER which told directly how much *time* had elapsed. (Incidentally, the Sony SL-2000 does just this; it tells elapsed time on recorded tapes.)

Say you had recorded several shows on a tape and your COUNTER read 1430 at the end of the last show. Do you dare to try and squeeze one more 30-minute show onto the tape, with the risk of running out just before the climax? What would be nice is a COUNTER which told directly how much *time* was left on a tape.

To convert your COUNTER numbers into *time* numbers, you need a conversion chart.

Video Information Systems, Inc. (Box 145, Department VM3, Deep River, CT 06417) sells for $17 what they call their "Dial-a-Time" calculator. You would rotate a cardboard disc until its pointer indicated your COUNTER number, and then you would read the recording time used and remaining from various windows in the disc. One problem with such calculators is that they *may* not be accurate for your particular machine. No one can construct a chart that works for everyone—there are so many models of VCRs, many with quirks of their own.

Figure 4–9
Relation of counter numbers to time

Your counter forward	ELAPSED TIME AND TAPE			Counter forward	Counter backward	TAPE AND TIME LEFT			Your counter backward
	T120 VHS-2	T120 VHS-4	T120 VHS-6			T120 VHS-2	T120 VHS-4	T120 VHS-6	
	Hr min	Hr min	Hr min			Hr min	Hr min	Hr min	
000	0 00	0 00	0 00	0000	8155	2 00	4 00	6 00	_____
_____	02	04	06	0062	8217	1 58	3 56	5 54	_____
_____	04	08	12	0119	8274	56	52	48	_____
_____	06	12	18	0173	8328	54	48	42	_____
_____	08	16	24	0225	8380	52	44	36	_____
_____	10	20	30	0274	8429	50	40	30	_____
_____	12	24	36	0322	8477	48	36	24	_____
_____	14	28	42	0367	8522	46	32	18	_____
_____	16	32	48	0411	8566	44	28	12	_____
_____	18	36	54	0453	8608	42	24	06	_____
_____	20	40	1 00	0494	8649	40	20	5 00	_____
_____	22	44	06	0534	8689	38	16	4 54	_____
_____	24	48	12	0573	8728	36	12	48	_____
_____	26	52	18	0611	8766	34	08	42	_____
_____	28	56	24	0648	8803	32	04	36	_____
_____	30	1 00	30	0684	8839	30	3 00	30	_____
_____	32	04	36	0719	8874	28	2 56	24	_____
_____	34	08	42	0754	8909	26	52	18	_____

FIGURE 4–9 (*Cont.*)
Relation of counter numbers to time

Your counter forward	ELAPSED TIME AND TAPE			Counter forward	Counter backward	TAPE AND TIME LEFT			Your counter backward
	T120 VHS-2	T120 VHS-4	T120 VHS-6			T120 VHS-2	T120 VHS-4	T120 VHS-6	
	Hr min	Hr min	Hr min			Hr min	Hr min	Hr min	
_____	36	12	48	0788	8943	24	48	12	_____
_____	38	16	54	0821	8976	22	44	06	_____
_____	40	20	2 00	0853	9008	20	40	4 00	_____
_____	42	24	06	0885	9040	18	36	3 54	_____
_____	44	28	12	0916	9071	16	32	48	_____
_____	46	32	18	0947	9102	14	28	42	_____
_____	48	36	24	0977	9132	12	24	36	_____
_____	50	40	30	1007	9162	10	20	30	_____
_____	52	44	36	1036	9191	08	16	24	_____
_____	54	48	42	1065	9220	06	12	18	_____
_____	56	52	48	1093	9248	04	08	12	_____
_____	58	56	54	1121	9276	02	04	06	_____
_____	1 00	2 00	3 00	1149	9304	1 00	2 00	3 00	_____
_____	02	04	06	1176	9331	0 58	1 56	2 54	_____
_____	04	08	12	1203	9358	56	52	48	_____
_____	06	12	18	1229	9384	54	48	42	_____
_____	08	16	24	1255	9410	52	44	36	_____
_____	10	20	30	1281	9436	50	40	30	
_____	12	24	36	1306	9461	48	36	24	_____
_____	14	28	42	1331	9486	46	32	18	_____
_____	16	32	48	1356	9511	44	28	12	_____
_____	18	36	54	1381	9536	42	24	06	_____
_____	20	40	4 00	1405	9560	40	20	2 00	_____
_____	22	44	06	1429	9584	38	16	54	_____
_____	24	48	12	1453	9608	36	12	48	_____
_____	26	52	18	1477	9632	34	08	42	_____
_____	28	56	24	1500	9655	32	04	36	_____
_____	30	3 00	30	1523	9678	30	1 00	30	_____

FIGURE 4–9 (*Cont.*)
Relation of counter numbers to time

Your counter forward	ELAPSED TIME AND TAPE			Counter forward	Counter backward	TAPE AND TIME LEFT			Your counter backward
	T120 VHS-2	T120 VHS-4	T120 VHS-6			T120 VHS-2	T120 VHS-4	T120 VHS-6	
	Hr min	Hr min	Hr min			Hr min	Hr min	Hr min	
____	32	04	36	1546	9701	28	0 56	24	____
____	34	08	42	1568	9723	26	52	18	____
____	36	12	48	1591	9746	24	48	12	____
____	38	16	54	1613	9768	22	44	06	____
____	40	20	5 00	1635	9790	20	40	1 00	____
____	42	24	06	1657	9812	18	36	0 54	____
____	44	28	12	1678	9833	16	32	48	____
____	46	32	18	1700	9855	14	28	42	____
____	48	36	24	1721	9876	12	24	36	____
____	50	40	30	1742	9897	10	20	30	____
____	52	44	36	1763	9918	08	16	24	____
____	54	48	42	1784	9939	06	12	18	____
____	56	52	48	1805	9960	04	08	12	____
____	58	56	54	1825	9980	02	04	06	____
____	2 00	4 00	6 00	1845	0000	0 00	0 00	0 00	____

Perhaps as good as "Dial a Time" (and saving you some money) is the conversion chart in Figure 4–9. It lists the times and the COUNTER numbers for a standard T120 cassette played on a VHS deck at the 2-, 4-, and 6-hour modes. It also tells the time remaining on a tape for any speed you wish to use. To use the chart, simply "000" your tape at the beginning. To FAST FORWARD your tape to the 30-minute mark at the 2-hour mode, wind ahead until your COUNTER reads 684. One hour and 45 minutes into the tape at the 4-hour mode would yield a COUNTER number of about 1044 (interpolating a little to estimate numbers not listed on the chart). If your COUNTER reads 1430 and you want to know whether there's time left for a half-hour recording, look across from 1429 under "Time Left" and you'll see that about 38 minutes remain at the 2-hour mode.

If you're near the end of the tape, and for some reason you didn't zero your COUNTER at the beginning or had RESET it somewhere midway, you needn't rewind to the beginning of the tape to get straightened out. Wind to the end, then RESET your counter to 000, and use the "Counter backward" column to find the segments you want. For instance, to find the beginning of the last half-hour show on your T120 tape in the 2-hour mode, you'd:

(a) FAST FORWARD to the end of the tape.
(b) RESET the counter to 000.
(c) REWIND to 9678,

... and take a look.

Unfortunately, because cassettes generally contain 5-20 minutes more tape than they say they do, this backward method may be off by just about that amount. This is at least a rough measure, anyway.

Incidentally, it takes about 3 minutes and 40 seconds to rewind a T120 cassette. A round trip to the tape's beginning to zero your COUNTER, then back to near the end to find out if you have enough tape left to record a show, would take you nearly 7 minutes of rewinding and winding. By that time you may miss the *start* of your show.

Because COUNTERS vary so much from machine to machine, the COUNTER numbers in Figure 4-9 may not be useful to you. For this reason, there are blanks for you to write in your own COUNTER numbers for your particular deck. This process may take a while; here are a few suggestions:

Easy method #1

(a) Switch to the 6-hour mode, select a channel with lots of half-hour shows for the next 6 hours, press RECORD/PLAY, and let the machine record to the end of a standard-length tape (like a T120).

(b) Rewind the tape, zero the COUNTER, then skim through the tape looking for show-starts/show-ends. Use your TV listings to determine the show lengths and fill in as many COUNTER numbers as you can.

Better method #2

(a) Aim a TV camera at a clock. Record the clock using the 2-hour mode to the tape's end.

(b) Rewind, zero the COUNTER, then play the tape, taking down the COUNTER numbers, perhaps SCANNING or FAST FORWARDING between intervals to speed up the process. This method works nicely for the backward count too. You can calculate the times for the 4- and 6-hour modes from your 2-hour list.

Better method #3

(a) Tune to a cable TV station that gives the time and record that at the 2-hour mode.

(b) Play back the results, copying down the COUNTER numbers.

Better method #4

(a) Tune your short wave radio to WWV, the 24-hour time announcements at 5, 10, 15, or 20 MHz on the dial. Dominion Observatory Canada at 7.335 Mhz is also good. Record using the AUDIO IN jack, or just a microphone in front of the radio's speaker. Use the 2-hour mode.

(b) Play back the results and correlate the times you hear with the COUNTER numbers.

If you use a tape other than T120, the chart still works. Shorter tapes will just end sooner, but their COUNTER numbers should correlate with the times up until then. Longer tapes (like the T180) are thinner and will yield a whole new set of COUNTER numbers.

BETA users can use this chart, after some modification. If you use the standard L500 or the L750 tape, mark that type of tape at the top of the chart. In place of VHS-1,2,3 you'd mark BETA-1,2,3. Then you'd set out correlating times to COUNTER numbers as described before. The L500 and L750 tapes are different thicknesses, so the chart will work for only one or the other. If you use both kinds of tapes you'll need two charts.

SETTING UP YOUR TIMER FOR A FOOLPROOF RECORDING

It's going to happen to you—I guarantee it—and more than once. You'll get home from that dull cocktail party, thirsting for that movie you had the machine record in your absence, only to find—*nothing*. Aaaugh! What went wrong?

Lots of things can foul up your simple or PROGRAMMABLE timer. Most of these are *things you forgot to do*. Murphy's Law ("Anything that *can* go wrong, *will* go wrong") applies exquisitely to VCR timers. Here are some of the reasons for go-wrongs:

1. SAFETY TAB *missing from tape*—Later in this chapter, under All About Tape, you'll see that you can remove a tab from the back of any cassette and render it unerasable (and unrecordable). This protects your tape from being accidentally erased. It's easy to forget and try to use a "protected" tape later for recording—it won't work. Without the tab, your VCR won't record when the timer tells it to do so.

2. PAUSE *still on*—If the PAUSE button is down, or you have a remote control with *its* PAUSE down, your VCR won't budge.

3. SOURCE, *or* INPUT SELECTOR *is on* CAMERA—The timer will start your VCR, all right, and will record the *camera's* signal in your absence, not the VCR tuner's signal. Switch it to TUNER (or TV or whatever), so the VCR gets the right signal.

4. *AM/PM snafu*—You thought you set your timer to record "Wall Street Week" at 10:30. What you got was Daffy Duck cartoons. Surprise! Could it be that you set your timer for AM instead of PM?

5. *Shut-off time not reset*—Your recording of "The Shining" began all right, but it went blank just as Jack Nicholson chased the boy into the maze. Just as you were really getting hooked! "What happened?" you shriek as you search for an axe!

You forgot to enter *the new* shut-off time for *this* program. The VCR still remembered the *old* shut-off time from your *last* recording.

6. *Timer disengaged*—This go-wrong relates to VCRs with electronic tuning. You set up your timer to record a show in your absence. This time you just happen to be home for the show, so you disengage the timer, press RECORD/PLAY, and watch the show pressing PAUSE during the commercials to edit them out. Later you play a little of the tape to see how it's coming out. Surprise again! While *you* were watching channel 5, your VCR was recording channel 9. When you disengaged the timer, you canceled the channel you had programmed in for recording. The VCR then automatically reverted to the channel you had watched last, which was 9.

To catch such errors, set your VCR/TV or ANTENNA switch to VCR so your *TV is showing what the VCR is getting*. If the VCR is screwed up, you'll see it immediately on your TV and can fix the problem expeditiously.

7. *Program schedule change*—When you doublecheck your *TV Guide* for the correct program time (you *were* going to do that, weren't you?) check the schedule for that channel for the previous couple hours. Is there a baseball game or presidential address lurking in there somewhere? If there is and it runs over, your favorite "Hogan's Heroes" rerun may turn out to be 10 minutes of postgame interviews and a snippet of Hogan. So check, and if necessary, select a shutoff time perhaps an hour later than scheduled.

The way to beat Murphy is to be careful. Do like airline pilots do before a takeoff: Go through a checklist. Glue one to your VCR's timer. It should include:

- POWER—ON
- SOURCE or INPUT SELECT switch—TV
- Tape speed—(pick one)
- Enough tape—(check program length)
- SAFETY TAB—not missing
- PAUSE—off
- Timer—Right day. AM or PM. Shutoff time set. Timer engaged.
- Record button—down (on older VCRs)

Perhaps you'll want to do a sample recording, just to make sure all is working.

One thing you can't safeguard against is power outages. Your VCR needs power to run. Your timer needs *constant* power to keep the correct time. A 10-minute outage can result in your VCR recording the tail end of Charlie's Angels and the front of Dolly Parton (perhaps the best parts of both anyway). Your PROGRAMMABLE TIMER, unless it has a backup battery power supply, will completely run amok if power to it stops even briefly. Its memory forgets everything—even which channels are which if your unit has electronic tuning. If you live in an area plagued with power outages, lightning strikes, or giant machines (like air conditioners or big motors) which dim or flicker the lights from time to time, don't expect much reliability from your timer or your VCR.

SETTING YOUR VCR'S TIMER TO THE NEAREST SECOND

Most network shows begin right on the button. If your recorder starts taping "Superman" early, you become the proud owner of two minutes of hair spray and feminine hygiene ads followed by the story of how Lois Lane gets herself in a mess. Will Ol' Supe come to the rescue in time? You'll never know. The machine stopped two minutes before the conclusion. On the other hand, if your recorder starts taping late, you may end up with a who-done-it without the part showing what was done.

There *are* ways to set your timer accurate to the second and avoid these little irritations.

The nondigital timers are easy to start. Just turn them to one minute later than the right time and then wait for the minute hand to hit 12. Three seconds before it hits the 12, pull your VCR's plug. The hand will stop at 12. When the correct time comes along (using your quartz watch, or dial-a-time, or WWV on the short-wave radio as a guide), plug the VCR in and voilà, your timer is synchronized to the rest of the universe.

This method doesn't work with digital electronic timers. However you can press *both* the hour- and minute-setting buttons, which sets the clock to 12:00:00. Hold them down until you want the clock to start counting, then release. Do it so *your* seconds are at zero when the world's seconds hit zero. Setting the hours and minutes now to their correct times will not affect the seconds readout. This same method also works for many time displays which appear on the TV screen of some newer receivers.

DRYING OUT YOUR VCR WHEN THE DEW LAMP GOES ON

When you move your VCR from a cool place to a warm or humid place, water condenses inside it making the mechanism sticky. This could damage your tape if the DEW sensor inside your VCR didn't shut the machine down, forcing you to wait until it dried out.

So what do you do, twiddle your thumbs while half the neighborhood mixes themselves drinks in your kitchen as they wait for your Laurel and Hardy anthology to begin? No way! Just dig out your handy hair drier, switch it to its *lowest heat,* open the cassette compartment, and remove the cassette. Then holding the drier about 6 inches from the hole, blow the mechanism dry. One minute may do the trick, and just in time—Cousin Scooter just found the Chivas Regal and Aunt Mabel is pouring her second cup of Bristol Cream.

Note: I did not say *high* heat with the nozzle *stuck into* the cassette hole. You don't want to melt anything, just dry it.

WATCHING YOUR COLORS WHEN USING A CAMERA

You will learn later how color TV cameras tend to change their color balance with changes in lighting. If your camera has a viewfinder in it, the viewfinder will be black-and-white, fine for determining sharpness and contrast, but useless for displaying color alignment. When possible, connect a color TV to the VCR when taping and observe *there* what color renditions your camera is giving you. Then adjust the camera as necessary.

SQUEEZING OUT THE HIGHEST QUALITY PICTURE AND SOUND

Start with a good picture and sound while recording. Use a strong enough antenna, properly aimed, good antenna wires, and tight connections. FINE TUNE your stations carefully. To really be sure your VCR is receiving a good signal, monitor the beginning of the recording through your TV.

Use the fastest speed (shortest recording time) for highest picture and sound fidelity.

If playing back through the antenna input of your TV, FINE TUNE your TV carefully.

For better sound, run the VCR's AUDIO OUT to your hi-fi's AUX IN or TAPE IN and listen to your sound through your $300 sound system rather than the TV's $6 sound system. Since you're likely to be feeding only a single, monaural signal to a stereo (two-input) amplifier, the sound will come out only half of the system. To use both speakers, or both channels, either (a) look for a STEREO/MONO switch on the hi-fi and flip it to MONO, or (b) buy a Y ADAPTER (Figure 4–10) which splits your audio for feeding two inputs.

Figure 4–10
Y ADAPTER

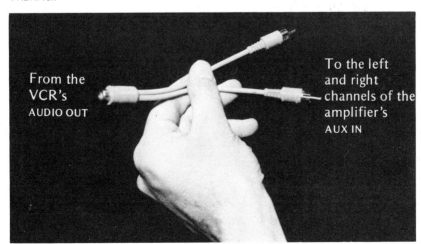

From the VCR's AUDIO OUT

To the left and right channels of the amplifier's AUX IN

Speaking of Y ADAPTERS, some alert readers may have noticed that it would be possible to plug a Y ADAPTER to their VCR's VIDEO OUT and feed the signal to two places, or conversely use the adapter to feed two sources into one VIDEO IN. Feeding the signal to two places with a Y ADAPTER will work, but it will weaken your signal badly, creating a faint or snowy picture. Feeding two sources (like two VCRs) into your VCR's VIDEO IN via the Y ADAPTER won't work at all. You'll just get mush. Chapter 14 (paperback Volume II) tells the right equipment to use for splitting or combining video signals.

For a 10% sharper picture, use a monitor or monitor/receiver and use the VCR's VIDEO OUT directly. Definitely use straight video when playing your VCR through a VIDEO PROJECTOR.

Before starting a recording *at the very beginning* of a tape, first PLAY about 10 seconds of blank tape. The first few seconds of every tape receives the greatest wear from being threaded and unthreaded, and will display snowy specks (DROPOUTS) in your recording. About 10 seconds in, the tape should remain fairly "clean."

If you plan to make *an important* recording, first make a sample recording and play it back to make sure everything is working okay. If you are recording long stretches of material with short breaks (perhaps commercials) here and there, use these respites to rewind and play back a little of your recording, just to see how it's coming. *Simply viewing your TV screen as the recording is being made doesn't completely guarantee that the recording is going onto the tape.* A good picture and sound on the TV just tells you that a good signal is *going into the VCR*. It doesn't tell you that the VCR is getting the show *onto the tape.* Only playing the tape back can definitely prove that the recording is coming out okay.

If a show is going to run past the end of one tape and onto another, don't bother rewinding the first tape just then. EJECT it, pop in the new cassette, and start recording immediately. This will save you 3 minutes and 40 seconds of missed program while you rewind the tape. You can rewind later. It won't hurt the tape to wait.

If you plan to swap tapes with others, *again* use the fastest (SP or BETA-1, or possibly BETA-2) speed for recording. The faster speeds make more stable tapes. Because tapes recorded on one home VCR will play less solidly on another VCR, it helps to start with the most stable signal you can get.

RECORDING
A SERIES OF SEGMENTS

BETA VCRs When you finish recording and press STOP on a BETA VCR, the tape stays threaded in the machine. If you press RECORD again for another show, your tape will begin recording right where the last show left off. If, however, you press EJECT, the VCR *will unthread itself*, losing your place. When you reinsert the cassette, the tape immediately rethreads. Now where is the end of your last show? It's about a half minute removed from where it used to be.

Here's how to line up your recordings without losing anything:

1. If you are just temporarily stopping the tape, say for a couple minutes, use PAUSE or STOP. This will disturb nothing and you'll start up right where you left off. If you're stopping longer, hit STOP. The tape will stay put.

2. If you hit EJECT, REWIND, or anything else that might upset your place, simply back up a little ways on the tape and hit PLAY. When you come to the end of the previous segment, hit PAUSE. Then hit RECORD while still PAUSED. UNPAUSE to commence recording from where you left off. In fact even if you didn't leave off here but wished you had, you can UNPAUSE at this point; the new recording will automatically erase the unwanted material.

And here's another suggestion for ensuring you don't miss the beginning of a show: Press RECORD about 1 second before the show actually starts. It takes the machine about that long to stabilize its recording circuits and yield a smooth picture.

VHS VCRs When you press STOP on a VHS deck, the tape unthreads itself immediately, losing your place. Early in the tape it may only cut a couple seconds off the tail of your last recording. Toward the end, this "unthreading factor" may cut off as much as 13 seconds. FAST FORWARD and REWIND also introduce this "unthreading factor."

To line up your recordings without losing anything:

1. Record 13 seconds past the end of every show before hitting STOP. When you hit RECORD/PLAY again, it will either start up exactly at the end of the previous show, or start a little thereafter. You may end up with a few seconds of between-show garbage but at least you don't lose anything.

2. This method is a tiny bit more trouble, but very accurate. About a minute before making *any* recording, back up a tiny bit and hit PLAY. Watch for the end of the previous show. When it arrives, hit PAUSE. Then hit RECORD/PLAY. When ready, UNPAUSE.

RECORDING
FROM MULTIPLE SOURCES

You may have noticed that your VCR will only accept one video and one audio signal at a time. If you have two cameras or two microphones you're out of luck. If you bought gold back when it was $35 an ounce, you can afford a VIDEO SWITCHER and an AUDIO MIXER to complete your burgeoning studio. Figure 4-11 shows how they connect to your VCR. In Chapters 7 and 14 (paperback Volume II) you'll see more

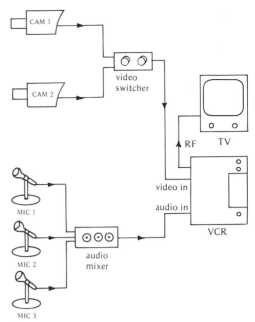

Figure 4-11
Multiple sources going to VCR inputs

about mixing several sources together for recording.

For what it's worth, there is one way to use your VCR to switch between sources. Your VCR has both an ANTENNA input and a VIDEO (or CAMERA) input. On some units, simply plugging in the CAMERA or VIDEO plug cuts out the ANTENNA signal, but on others you can connect both and switch from one to the other at will by flipping the INPUT SELECT (VHS) or REC SELECT (JVC) or VIDEO/TV switch (BETA). Just imagine recording the buildup for the Johnny Carson show and then switching to Uncle Victor prancing out from behind the full length window curtains and taking a bow to thunderous applause.

Whether you can switch audio from the ANTENNA signal to your microphone also depends on the type of VCR you have. On some, when you plug the microphone into the VCR, the mike automatically overrides any other audio signals. On other decks you automatically switch audio when you switch the video from TV to VIDEO or CAMERA.

Figure 4–12
Glitch on screen caused by switching from one source to another

COPING WITH CABLES

As your den accumulates video machinery, a jungle of wires will grow behind the equipment, complete with exotic birds and distant drumbeats. There are several ways to make a path through the cable jungle. One way is with a machete. Quick, but expensive. A slower but more civilized way is to *label* the cables with a masking tape tag telling where each goes (or comes from). Then coil up any excess wire (except twin lead) and use tape to hold the loops together. Unravel any kinks or knots.

When you do make the video switches described above, you will record a little glitch in the picture (see Figure 4–12).

Some VCRs have INPUT SELECT switches with three positions, TV/CAMERA/LINE where LINE means straight video connected to the VIDEO IN socket. With such a switch, you could conceivably plug one TV camera into the CAMERA input of the VCR, plug *another* camera (or a VCR's VIDEO OUT) into the VIDEO IN socket, and switch between cameras one and two using the INPUT SELECT switch.

The most terrifying sound in the video jungle is "oops—CRASH." It's the two-often-heard sound of a shoe snagging a cable and bringing a camera or light terminally to the floor. If you're running long extension cables from cameras, mikes, or lights, slip the cables under a carpet or tape them to the floor using wide duct tape. Run them *over* doorways, if possible.

If the cable you tripped on happens to be a mike cable, the predominant sound in the jungle will become "bzzz" or "crackle, crackle," or "(silence)." You have partially or wholly disjoined the wire's connection to its plug.

Another sure way to damage a wire is to disconnect a plug by pulling on the wire itself. Instead, *always* grasp the plug by the plug body. This avoids strain on the wire in the one place where it's weakest anyway, the place where it flexes the most, where the cable enters the plug.

Portable VCRs

Who said you can't take it with you? Today's home *console* VCRs are so small already that they would have qualified as "portables" only a few years ago. All they needed was battery power and a carrying handle.

Tape cables down. Tripups are dangerous and expensive

RECORDING WITH A PORTABLE VCR ONCE SOMEONE ELSE HAS SET IT UP

There's really nothing to it (now observe how the author stretches "nothing" into five pages of incomprehensible tedium):

1. Find something to shoot, preferably not something in the dark.

2. Remove the lens cap. Thereafter, don't aim the camera at anything too bright, like the sun. If your brother-in-law, Harold, isn't too bright, aim it at him.

3. About a minute before you're ready to begin taping, switch the VCR's power to ON. Since either the camera can PAUSE the VCR (with its trigger) or the VCR's PAUSE button can PAUSE it, let's fix things so the camera does the PAUSING; it's more convenient that way. To do this, you will first want the VCR's PAUSE *off* and the camera's PAUSE *off*. A methodical way to coordinate this is to press PLAY and look to see if the tape is moving. If it is, you are all set; the camera trigger will PAUSE you. If it *isn't*, pull the camera trigger. If this UNPAUSES you, you're all set. If it doesn't, UNPAUSE the VCR's button. If that doesn't get things started pull the camera trigger again. Eventually, you get the machine UNPAUSED and playing.

4. Now pull the trigger to PAUSE everything (after all that). Next press RECORD.

The camera's viewfinder will light up either when you switch the VCR's POWER to ON or when you press PLAY or RECORD. Either way, the camera will take about 30 seconds to "warm up" and give you a picture.

5. While in the PAUSE mode, uncap and focus the lens and adjust the IRIS (f stops). More on this in the Camera chapter.

6. When ready to record, pull the camera trigger and the VCR will switch from PAUSE to FORWARD and will begin recording.

7. To stop recording temporarily, pull the trigger again and the VCR will switch to PAUSE.

8. To start recording once more, pull the trigger again.

9. To *finally* stop recording, switch the VCR to STOP. (If you wish, you may PAUSE the recording first by pulling the camera trigger.) In the STOP mode, some VCRs and cameras keep consuming electricity; others don't. To save your batteries you may wish to switch the POWER to OFF if the machine will be idle for a while.

10. To play back a sample of what you recorded, first switch to REWIND for a ways. When the tape is rewound, switch it to STOP.

11. To play, switch it to PLAY and look in the tiny viewfinder on the camera. The image will appear there *if* your camera happens to have an electronic viewfinder (a tiny black-and-white TV set) in it. To hear sound, find the earphone. (It may be in a little pocket in the carrying case.) Stick the plug into the EARPHONE socket on the VCR or camera; the other end sticks into your ear (or is it the other way around?).

SETTING UP A PORTABLE VCR FOR USE

Recording from the camera The camera connects to the VCR via a multipin plug. Just line it up, push it in, and screw the tightening collar to hold the plug in.

Next, check the VCR's CAMERA/TV switch. When operating with a camera, this switch must be in the CAMERA position.

On most portable TV cameras, sound is picked up by a sensitive mike built into the front of the camera and recorded automatically.

When using color cameras and VCR equipment, setting up takes a little longer. Color-TV cameras have to be "warmed up" first and then adjusted to give good color.

Follow the manufacturer's directions on how to shade, register, and get a proper *white balance*.

Recording audio Instead of using the mike built into the camera, you can substitute your own, say a LAVALIER or a SHOTGUN mike. These mikes plug into the MIC input of the VCR. Since VCRs usually take a mini-plug for their mikes, an adapter may be needed to match your mike's plug to the machine's socket.

If the sound source is prerecorded and requires a LINE LEVEL INPUT, look for an AUX or AUDIO IN socket on the VCR for this.

In all cases, the audio level is automatically adjusted.

AUDIO DUBS are possible by pushing the DUB button on the deck while switching the VCR to FORWARD. The mike in the camera may be used during this process, or you may plug in a separate mike.

Recording off-air broadcasts The big difference between console and portable VCRs is that the consoles *include* the TV tuner, while the portables, for weight reasons, don't. The tuners are separate and connect to the VCR through a multipin umbilical cable or through several cables to the VIDEO IN, AUDIO IN, and REMOTE inputs. To pull in distant stations, tuners have ANTENNA inputs which connect to your rooftop antenna or The Cable.

Sometimes these tuners have PROGRAMMABLE TIMERS with all the features the console models have. Also included in most units is a power supply that runs your VCR without discharging your battery. In fact, most units *will charge* your VCR's battery. Of course, they must be plugged into a wall outlet, connected to the VCR, and turned *on* to do this.

To really be sure you've accurately tuned the tuner for a recording, you may wish to view the results on a TV set beforehand. Here's how:

Attaching a TV set Most portable VCRs have a socket labeled RF OUT or TV OUT or ANTENNA OUT. Connect this to the TV's ANTENNA terminals. Switch the VCR's PROGRAM SELECTOR (if it has one) to VCR to send the VCR's signal (rather than the antenna's) to the TV. Switch the TV to channel 3 (if that's what the VCR's RF GENERATOR puts out). That's all. The setup is a lot like your console VCR/TV setup. Figure 4-13 diagrams it.

Most portable VCRs (and console models too) have detachable RF GENERATORS. Needless to say, you have to have one before your VCR will pump anything out its RF socket. This handy option costs about $35 if yours is missing, and it plugs into a cavity in the back of your VCR as shown in Figure 4-14.

As with the console VCRs, it's possible to run direct audio and video to an audio amplifier and speaker, and a TV monitor. Or you could run both signals to a TV monitor/receiver for viewing.

Sometimes the portable VCR isn't used with a tuner. It's used to make tapes out in the field (or the woods) and then it comes home to connect to the TV for viewing. Industrial VCRs, and some older or cheaper VCRs were designed to work this way and use an external ANTENNA/VTR switch when feeding a TV. Figure 4-15 shows the hookup. In fact this kind of hookup is used for video games and many other devices which connect to your TV. With the ANTENNA SWITCH, you can select whether your TV will "see" the antenna signal (ANT position) or "see" the VCR's signal (VCR position).

Why bother with the switch, you're wondering? Why not simply connect the VCR's RF cable to the TV's antenna terminals and then connect the TV antenna to the same terminals? Such a connection *would work,* although not awfully well. You could get a grainy, ghosty signal on your TV. Here's why: RF doesn't care where it goes when it comes through the RF cable from the VCR. If the TV is still connected to its antenna,

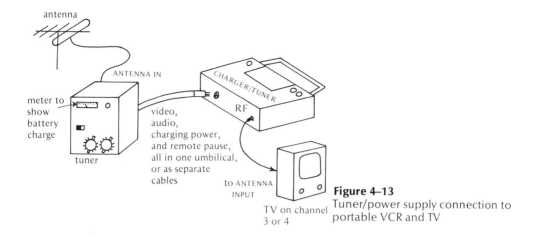

Figure 4–13
Tuner/power supply connection to portable VCR and TV

Figure 4–14
Installing RF GENERATOR. RF GENERATOR feeds TV through channel 3 or 4, depending on unused channel in your area

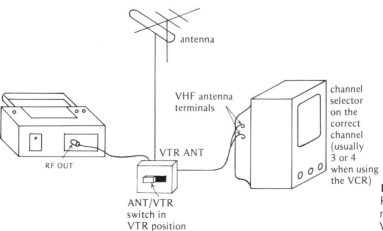

Figure 4-15
Receiving RF on a TV receiver from a portable VCR

some of the RF signal will go into the TV as it should, and the rest will detour out the antenna wires to the rooftop antenna. In fact, the RF going up the antenna will actually broadcast out the antenna a little and may interfere with other people's TVs. The Federal Communications Commission frowns on renegade TV broadcasters' scattering signals willy-nilly over the airwaves.

Cable TV systems face the same problem. If your TV set is still connected to your building's antenna system when you pump RF into your set, everyone on the system will get interference from your signal.

Time for a true story. A young man and his wife bought a portable VCR and camera. After taking it back to their apartment, they decided to set up the camera to record themselves making love. When finished, they eagerly played the tape back on their home receiver. How exciting!

The next morning the couple noticed funny stares from other tenants in the lobby. Someone in the elevator asked, "Haven't I seen you somewhere before?" In their exuberance, the couple had neglected to disconnect the master antenna cable from their TV before attaching their VCR to it. It is not known exactly how many tenants had watched this X-rated gem on their TV sets that evening or had recognized their neighbors as the main characters.

The same danger exists (but not to the same degree of embarrassment) with video games that you connect to your TV antenna terminals. The games send out an RF signal that must go into the TV set *only* and not detour up the antenna wires for all to see.

In short, when sending RF to a TV set, make sure:

1. The set is disconnected from its rooftop antenna.

2. The set is disconnected from any master antenna or cable system.

OR

3. If you have an ANT/VCR junction box with a switch, be sure the switch is in the VCR position.

POWERING THE VCR

For short on-location shootings, a rechargeable battery which slides into the VCR will power the VCR for up to 1 hour. See Figure 4–16. If you press RECORD and FORWARD, but do not pull the camera trigger, you start using power, even though the tape may be PAUSED and not moving. Add this "standby" time to the time you spend actually shooting and you can estimate that your battery will probably power you through only a half hour of tape.

Figure 4–16
Powering the VCR

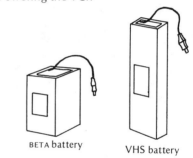

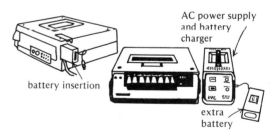

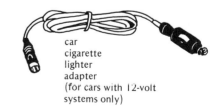

As the battery gets older, it will serve shorter and shorter duty cycles. To check your battery power (and to estimate how much longer you can shoot before this battery dies), glance at the battery meter somewhere on the VCR. *The meter may register only when the VCR is in* RECORD, *not when it is in* STOP. In these cases, if the VCR is just sitting somewhere and you wish to check the battery, merely push the RECORD button and look at the meter. The meter will give the answer in about a second. You may then release the lever, and the machine will return to STOP. If the meter reads way in the white, the battery has a lot of life left. If the meter reads near the red, the battery may have 5 minutes or so of life left. When the meter reads in the red, the battery is not sufficient to power the VCR. *Even though the motor runs, if the battery meter reads in the red, don't record.* You'll be wasting your time. The VCR motor speed starts to drop (imperceptibly at first) when the battery is nearly expended, rendering the recorded passage unplayable. If you are recording and suddenly notice that your viewfinder picture begins to jiggle or roll, that may also be a sign that your battery is weak. Check the meter. Many portable TV cameras are equipped with LOW BATTERY POWER lights inside the viewfinders to warn you of impending power loss.

To operate the VCR on location for up to 3 hours, there are optional, external, rechargeable battery packs that can connect to the VCR.

To operate near vehicles using their battery power, there is a "cigarette lighter" plug and cable which can suck power from your car while you're shooting. If your car battery is old, it may be wise to start your car every so often, unless you enjoy being stranded on-location.

For extended shooting, bring along your AC adapter, which can power the VCR from a wall outlet when your battery is used up. The adapter can also recharge the VCR's internal battery.

None of the above power supplies is yet designed to power anything *other than* the VCR and its camera. TV monitor/receivers and other equipment must be powered separately, either by batteries or by AC.

If you wish to play back some of your recording "in the field" (a good idea, just to make sure everything is being recorded okay), nearly all cameras which have electronic viewfinders will allow you to view your tape through the viewfinder. Plug in an earphone for sound, and you can see and hear your recordings without dragging along a separate TV. Take note that this playback also takes electricity and drains batteries.

You'll find more about VCR batteries and how to charge them in the Maintenance chapter. One interesting note on how to squeeze more life out of a battery bears stressing. When you "power up" your VCR and camera, they start draining your battery—even if you're not actually taping. "Power down" the system whenever possible to conserve juice. Many VCRs will permit you to be recording, hit PAUSE, then shut off all power, and later turn the power on, *with* PAUSE *still on,* and pick up exactly where you left off. Most solenoid-operated (feather-touch pushbuttons) VCRs won't let you do this. They switch themselves to STOP when you kill their power. As you learned earlier, PAUSING allows you to assemble scenes together neatly, while STOPPING the VHS VCRs withdraws the tape into the cassette and loses your place on the tape. Some VCRs have a POWER SAVER switch which essentially allows the VCR and camera to "power down" while PAUSED. This is a nice feature that really can save a lot of battery power if you have to deal with long delays between recording scenes.

VIDEOTAPING
IN THE WINTER

Portable video equipment, if it could talk, would beg to stay inside where it's dry and warm. But ice skating, skiing, snowball

VIDEOTAPING IN THE WINTER

fights, snowmobiling, and traffic-snarling blizzards occur outdoors (you needed this book to tell you that). What problems will you and your equipment face in the cold?

Shortened battery life Above all others, this may be your biggest problem. Up to 50% of your battery's life is lost when the temperature reaches freezing. Expect your 1-hour batteries to last about 30 minutes. This goes for NiCads (used by Akai, Technicolor, and newer lightweight VHS and BETAs) as well as Gel Cells (older VHS and BETA).

To avoid this problem, keep the battery warm, like in your pocket or wrapped in a blanket, until you're ready to use it. Then chuck it into the machine.

You might even keep the whole VCR (and thus the battery) warm with a blanket while it operates. By all means, don't set the recorder down on the ice or frozen ground; this will cool it and its battery very quickly.

Try reducing the power demands on the battery. Don't stand around in PAUSE or STANDBY for any length of time. Either shoot, or kill the power. Remove the plug-in RF modulator if it is not being used. Save-a-watt.

Bring two batteries for more shooting time. Keep the second battery warm until used.

Condensation Ever notice how moist the outside of a glass of iced tea becomes on a humid day? VCRs and cameras suffer the same problem. Whenever they are cooler than their environment, dew collects on them. This isn't such a big problem when you move outdoors to shoot in the cold, but when you come back in with frigid gear, it starts collecting humidity from its warm surroundings. The dampness is devastating to the electronics, makes the tape stick (instead of slide) inside the VCR, and will most likely trip the DEW SENSOR, a safety device in the VCR which turns it off (wisely) when there's excessive condensation.

When bringing cold equipment inside, immediately cover it with a plastic bag and let it warm up for an hour or more. The bag will seal out the humidity-laden air so that water can't collect on the goodies. When the equipment reaches ambient temperature, it's ready for use. Take it out of the bag first.

One way to hasten the drying-out process is to open the cassette tray (removing the cassette), and blow warm (not hot) air from a hair dryer into the VCR. In a few minutes the machine should be defrosted. Let the inside temperature stabilize for a couple more minutes before chucking in your cassette.

Don't try to play a cold tape in a warm machine and vice versa. The tape will act "sticky" if it's not the same temperature as the machine.

None of this applies to batteries. It's perfectly fine to use warm batteries in a cold VCR.

Icing Go out in a blizzard and your lens may ice up. Incidentally, shoot in a sea breeze and salt spray will cloud your lens. Salt's not so great for the VCR either.

Sticky fingers No, this has nothing to do with the theft of cold video equipment (making it "hot" equipment) but rather reminds us that slightly damp skin will freeze to very cold metal. If it's really cold, wear gloves.

Glare Snow and ice reflect a lot of sunlight. Shoot at a high f-stop. Attach a neutral density filter to your lens. Use a polarizing filter to reduce snow and ice surface reflections. Keep the sun to your back, lest your subjects are backlit so strongly that they look like silhouettes.

Video equipment (except batteries), works quite well when it's cold. It works poorly while it's changing temperature. Your camera may give poor color. The tape may stick. Your lens may fail to zoom smoothly. When you bring it outdoors, give your equipment a few hours to become acclimated to the cold. When you come in, do the same.

If it gets *really* cold (below −40°F) don't use your VCR. Its circuits could become damaged.

Now that you've got your equipment ready, *go out and shoot* those winter sports. The movement and frolic of ice and snow games makes fascinating TV viewing. Watch Uncle Vic fall off his skis, brother Rob take that toboggan jump . . . tail first. Catch Mom, with her skates creeping in opposite directions, and Dad covering three guys with gravel and slush as they try to push his car out of a snowbank. Great fun awaits the winter videographer.

VIDEOTAPING
AT THE BEACH

Video is an action medium, and what better place for action than at the beach? The beach, with its games, kids, and bathing beauties, can be a recording gold mine—if you know how to insulate your equipment from the harsh environment. Here are your enemies and how to avoid them:

Salt air Ocean beaches are bathed in an invisible spray of salt air. The salt corrodes switches, gums up VCR mechanisms, and clouds lenses.

Wrap your equipment in plastic bags and tie them closed. White or clear plastic is

VIDEOTAPING AT THE BEACH

better than black because black will get hot in the sun. Since your wrapped equipment can't "breathe" and cool itself, on hot days, run it for only about 15 minutes at a time. Then let it cool a while. Keep it shaded for best cooling. Wrap the camera so the lens sticks out (it has to "see") and make a little hole and tape it tight to your viewfinder opening so you can "see" too. Later, you wipe the salt spray off the lens and viewfinder glass with a wet swab. Use *fresh* water for this.

Dampness Strangely, your greatest danger isn't a tidal wave or your boat capsizing. It's the "dry land" danger of dripping swimsuits, kids with water buckets, and half-empty drinks spilling into your gear.

Next in line comes the sea dampness which permeates your equipment and collects. Then comes the total dunking—a disaster at best.

Like before, wrap everything while shooting. Store your gear in waterproof cases. A watertight ice chest makes a fine temporary home for your goodies (and it may even float in a capsize). Suitcases or other containers will work too. Avoid cardboard; it gets soggy when wet. Pack towels or whatever around the equipment to cushion it during travel.

For long-term storage in damp environments, buy some packets of SILICA GEL from your photo store and pack them in with your equipment. *It* will absorb the humidity rather than your machinery. When the packets get damp, simply bake them under low heat in the oven to dry them out. (If they're cloth-covered, wrap them in perforated tin foil to keep the cloth sacks from scorching).

Sand You know how it feels in your bathing suit. And you know how it tracks into your car, into your home, somehow into your bed, and magically into your breakfast cereal the next day. Similarly, that abrasive, destructive grit will find its way into your machinery unless you wrap it and store it—tightly. Also, keep your cables off the ground—they'll transfer sand to your storage box.

Heat Store everything out of the sun. Heat damages tape and makes VCRs "sticky" inside.

Above all, don't move your equipment quickly from a cool, air-conditioned environment to a sunbaked beach. It will collect water like a cold drink can, and the salty condensate will eat battery contacts, switches, everything.

If you are changing from cold to hot places, wrap the equipment in plastic (keeping out the moisture) and let it "warm up" to outside temperatures.

Although camera care and techniques will be covered in the following two chapters, here are a few special tips on beach and water shooting:

Sun and reflections Aim at the sun and you'll "burn" the camera tube, blinding it. Period. Reflections from water, chrome, glass, and wet beaches can be dangerous too. Try using a POLAROID or NEUTRAL DENSITY FILTER on your lens, and/or shooting at high f-numbers. Cloudy or hazy days provide excellent light for shooting. Perhaps wait for the sun to fall lower in the sky for reduced brightness and glare.

Rocky boat scenes Boats rock. When shooting from a boat, your scene will rock. And so will your viewers later as seasickness sets in. If you'd prefer to feed your audience popcorn instead of Dramimine, stabilize your scene by:

(a) Planting your legs and rolling your hips with the boat so that the horizon stays still in your viewfinder.

(b) Keeping the horizon out of the picture so people don't notice the rocking as much.

After the damage is done Even after wrapping the equipment, you may come home and find sand or saltwater film on your equipment. Here's what to do:

1. With a freshwater-moist sponge, wipe down your VCR and camera casings.

2. Decake dried salt with undiluted alcohol on a soft, lint-free cloth. Avoid scented cleaners; they leave a film.

3. Take off any removable covers and open any trap doors or compartments on your VCR and camera, and using an alcohol or freshwater-dipped swab, clean the seals, gaskets, and edges of the covers.

4. Clean around switches and movable parts.

5. If dampness gets into your camera lens, fungus will grow, causing a fine, spidery, light-colored buildup on the glass. You can minimize the likelihood of this by "drying out" the lens in a warm, dry place (like under a lamp), after shooting. Otherwise, if you notice the fungus growing as you hold the lens up to the light, bring it to a camera store for cleaning. If the fungus builds up too much, the camera store may be unable to remove it.

6. If your boat capsizes and "deep sixes" your gear:

    (a) Turn it off.

    (b) Remove the battery

    (c) Swim—don't paddle—to your nearest bucket of fresh water and dunk it. Slosh it around to get out all the salt—immediately. Repeat this rinse five more times, each time using a new batch of fresh water.

    (d) Take it in for service or, if you're handy, open it up and dry it out in a warm place. Blow a fan on it for a day. Still, it will be wet *somewhere*. Check carefully for moisture. Don't try to use it until you're *sure* it's dry. With luck, it may work. Then again, your drowned machine may still need service.

    (e) Lastly, go back to the scene of the accident and look for survivors, your family maybe.

Ways to Wire Your Video System

Figures 4-11, 4-13, and 4-15 showed several ways to connect up your VCR. If you have several TVs, the wiring gets a little more complicated. If you have several VCRs, the wiring starts looking like spaghetti. In fact, in honor of this nest of video wires, an organization called Videofreex published an excellent little book titled *Spaghetti City Video Manual* (Praeger, 1973). If you get cable TV, then you add a whole new tangle to the ball of wires. Let's save the multiple VCR wiring diagrams for the chapter Copying a Video Tape. As for the cable connections, look back at the examples in Figure 3-5 and try to imagine a VCR in place of any of those TVs, with the VCR's TV hooked directly to the VCR. Any of these configurations, except the one with the SCRAMBLED TV DECODER with its own built-in speaker will work fine with a VCR in the place of a TV.

For the person with cable TV and a SCRAMBLED pay-TV channel, the best setup may be the last one in Figure 3-5 (substituting a VCR for the TV set there). Essentially, the cable signal goes to the UP-CONVERTER. The UP-CONVERTER's VHF signal goes to the DESCRAMBLER. The DESCRAMBLER's signal goes to the VCR's VHF input. The VCR's VHF output then goes to your TV's VHF input. Meanwhile, the UP-CONVERTER's UHF output goes to the VCR's UHF input and the VCR's UHF output goes to your TV set's UHF input. Believe me, this all makes sense.

Figure 4-17 diagrams some ways to connect your VCR to several TVs and lists the advantages and drawbacks of each setup. If you have cable TV, it would be advantageous

Method #1

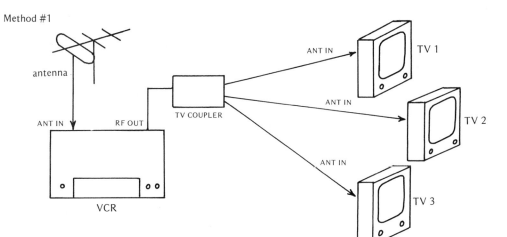

Capabilities:
1. With VCR's output switched to TV, all TVs can view any channel. The VCR can simultaneously record any channel. The recording cannot be viewed simultaneously.
2. With the VCR output switched to VCR, all TVs see only what the VCR is recording or playing. They must also be tuned to the VCR's channel (usually 3 or 4).

Note: If the antenna signal is weak, there may not be enough signal to feed three TVs, especially if their wires are long. In such cases, instead of using a PASSIVE TV COUPLER (uses no power), buy an ACTIVE COUPLER (which amplifies the signal while splitting it). Or, insert an RF AMPLIFIER or ANTENNA AMPLIFIER or ANTENNA BOOSTER *between the VCR and the* COUPLER.

Method #2

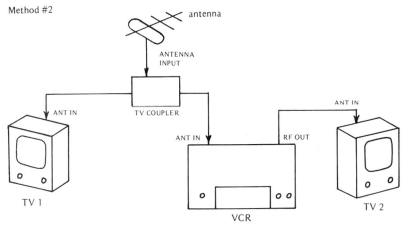

Capabilities:
1. TV 1 is isolated from VCR and can be used for normal TV viewing only (great for the kid's room while Mom and Dad view "Captain Lust.")
2. TV 2 displays regular TV channels if VCR's output is switched to TV. If output is switched to VCR, then TV views only what the VCR is recording or playing. TV must be tuned to VCR's channel to display the VCR's signal.

Figure 4–17
Connecting your VCR to several TVs

Method #3

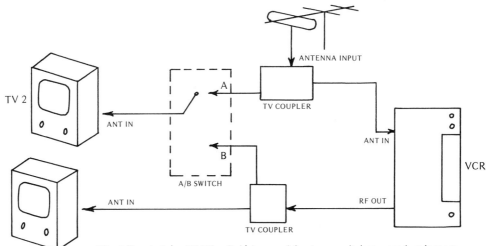

The A/B switch (or COAX switch) is a special antenna switch you can buy in most electronics stores. It allows you to select one of two sources.

Capabilities:

1. With A/B switch on A, TV 2 is isolated from VCR and views regular broadcasts.
2. With A/B switch on B, TV 2 watches what the VCR is putting out, which could be (a) regular TV channels if the VCR output is switched to TV, or (b) whatever the VCR is recording or playing if the output is switched to VCR.
3. TV 1 can view regular broadcasts or taped program depending on position of VTR's output switch.

In short, anything can be viewed on either TV with this setup, depending on where the switches are set.

Figure 4-17 (*Cont.*)

to install an UP CONVERTER (described in Chapter 3). Without it, all your TVs and VCRs are limited to receiving the one channel the cable company's DECODER BOX is tuned to. With it, all the cable channels are translated to UHF frequencies and this assortment of channels is available to you just like with a regular antenna. In short, these diagrams work the same whether you receive antenna signals or UP CONVERTED cable TV signals.

Try to keep cable runs as short as possible to avoid TV signal interference and to maintain a lusty signal.

All About Tape

WHAT IS VIDEOTAPE?

Video tape is a thin plastic film impregnated with a fine powder (cobalt doped gamma-ferric oxide) which can be magnetized. As with audio tape, the HEADS inside the recorder generate an oscillating magnetic field. The tape, sliding over the HEADS, gets magnetized. You can't see the magnetism; recorded tape looks no different physically than blank tape. The magnetic vibrations on the tape are transformed electronically into a picture and sounds by the magnetism-sensing playback HEADS of your VCR, in conjunction with various electronic circuits.

Video tape is also very *unlike* audio tape. For one thing, it must capture and retain over 200 times the amount of information (magnetic vibrations) that audio tape needs to hold. Secondly, it must be manufactured to such exacting standards that it's incredible there's so much of it around. Imagine manufacturing a ribbon of tape, about 800 feet long, *exactly* ½-inch wide, and .75 mil (.75 thousandths of an inch or 20 microns) thick. The plastic ribbon must neither stretch nor shrink significantly, yet be strong enough to withstand the wear-and-tear of threading and winding. The tape must be smooth enough to slip easily through the mechanism and not abrade the delicate spinning VIDEO HEADS inside the machine. The magnetic powder on the tape is specially formulated to hold the kinds of magnetism used in video recorders, and must be very pure, uniform, and resistant to flaking or shedding from its base. For these reasons, ½-inch wide audio tape and discarded computer tape are totally unsuitable as substitutes for video applications. Even the videocassettes themselves have to be built to exacting tolerances with 36 parts to meter the flow of tape from one hub to the other without sticking or jamming.

Like audio tape, video tape can be erased (demagnetized) and used over.

As shown in Figure 4-18, video tape comes in a variety of sizes. Professional quad video tape is 2 inches wide and comes on large reels. Professional/industrial tape is 1 inch wide and comes on 7-inch reels. Tape used in industrial and educational application is ¾ inch wide and comes in U-Matic cassettes. The home user buys cassettes of tapes ½ inch wide and occasionally ¼ inch wide.

Packing all that information onto a tape isn't easy. And the less tape used per hour, the tighter the signal gets squeezed and the lower the quality of the picture and sound. Notice in Figure 4-19 how many square feet of tape are used per hour by the professional and industrial recorders. By comparison, home machines use very little tape.

Figure 4-18
Kinds of video tape

¾" U-Matic videocassette ½" EIAJ video tape 1" videotape

½" BETA videocassette ¼" CVC videocassette ½" VHS videocassette

Figure 4–19
Tape speeds and consumption

Format	Width (inches)	Speed (inches/sec)	Sq ft/hr
BETA-1	½	1.57	19.6
BETA-2	½	.79	9.9
BETA-3	½	.53	6.6
VHS-SP	½	1.31	16.4
VHS-LP	½	.66	8.2
VHS-SLP	½	.44	5.5
3/4U (industrial, educational)	¾	3.75	70
EIAJ (older industrial, educational)	½	7.5	94
QUAD (broadcast professional)	2	15	750

HOW THE TAPE GETS RECORDED

When you chuck a videocassette into your machine and start recording, strange things begin to happen (inside the machine, too). Little mechanical fingers pluck a loop of tape from the cassette and wind it around the VCR's HEAD DRUM, as shown in Figure 4-20. The spinning drum holds a pair (or two pair) of VIDEO HEADS which swipe across the tape 60 times per second. Each swipe records a TV picture on the tape. Figure 4-21 shows the path of the VIDEO HEAD and also where audio, etc., get recorded on the tape. Since the tape is moving too, each swipe of the VIDEO HEAD hits a different spot, covering the tape with diagonal magnetic traces as diagrammed in Figure 4-22.

TAPE SPEEDS AND PLAYING TIME

The slower the speed, the more you get on a tape and the fewer tapes you have to buy. Sounds like a good deal, eh? Not exactly. You sacrifice audio and video quality as you

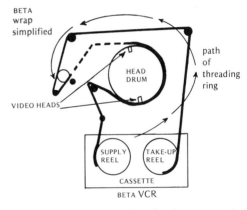

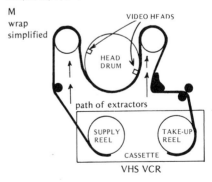

The BETA format uses a tape threading ring to extract the tape from the cassette. The VHS format uses the so-called "M" wrap which requires two extractors but utilizes a simpler tape path.

Figure 4–20
Threading patterns

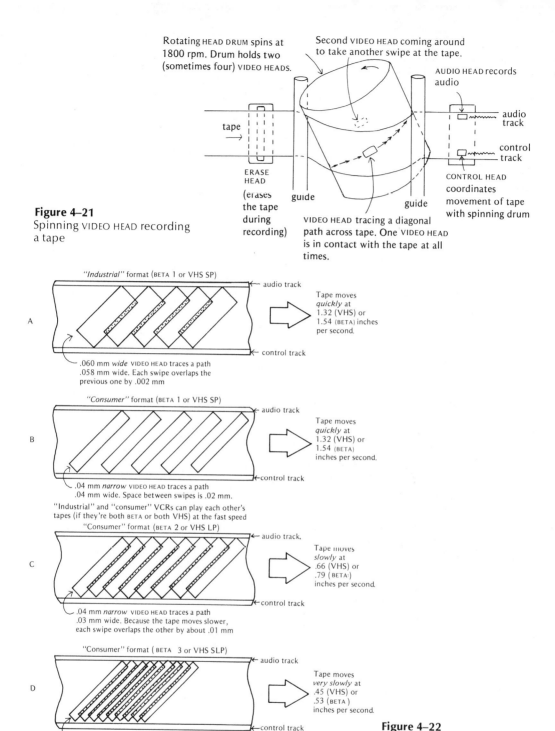

Figure 4–21
Spinning VIDEO HEAD recording a tape

Figure 4–22
Path of the VIDEO HEAD across the tape (nearly the same for both BETA and VHS formats)

118

squeeze more onto that tape. Industrial VCRs use a fast speed and fat VIDEO HEADS to make high-quality recordings (one case where fatheads are desirable). As we slow down the speed of the tape, as is done with VHS and BETA formats, the traces left by the wide VIDEO HEADS start to overlap (see Figure 4-22A). This degrades the quality of the picture. If we make the heads narrower (see Figure 4-22B) we regain some space between the tracks, but now we're using skinny VIDEO HEADS which can't make as good a picture. If we now slow down the tape some more, we use less tape but look at the overlap between the traces (see Figure 4-22C). Lousy narrow heads, lousy overlapping, lousy picture. Slow the tape down one more speed (Figure 4-22D) and the problems are further compounded.

DROPOUTS are another problem. No, I don't mean the high school flunkies your teenage daughter always seems attracted to. DROPOUTS are specks of snow on the TV screen caused by the VIDEO HEAD passing over a "bare spot" on the tape, where the magnetic powder has flaked off leaving no signal (and no picture) behind. DROPOUTS also occur when flecks of dirt or tape debris cover the magnetic powder in places, impairing contact with the spinning VIDEO HEADS. Figure 4-23 shows what a DROPOUT looks like on the TV screen. Better tape has fewer DROPOUTS, but all tape has some, especially at the beginning where the tape gets more abuse from threading. Slower, denser recordings are more sensitive to these scratches, defects, and debris, and therefore display more DROPOUTS.

Advances in equipment technology and tape manufacture have worked miracles at squeezing a watchable, stable picture out of these super-slow speeds.

The latest effort at packing more minutes into the cassette is being made by the tape manufacturers. No, they can't put a bigger reel in the cassette; it won't fit. But they can make the *tape* thinner, putting more feet on a reel. Good deal at last, eh!

Figure 4–23
DROPOUT as it appears on a TV screen

Sorry, Bunkey. This new tape is now *so* thin that it stretches and snags easier than the regular tape. So what's the moral of this story?

1. If you want the best picture and sound, use standard-thickness tape (T120 or L500 or lower numbered) and the fastest VCR speed. Same goes if you plan to copy, edit, or play your tape through a video projector.

2. If you plan to do much winding, rewinding, pausing, scanning, starting-and-stopping, editing, or frequent ejecting-and-inserting of your tape, use the standard thickness tape.

3. If you plan to record a long passage from beginning to end and play it back nonstop from beginning to end, feel comfortable in using the thinner tapes (numbers higher than T120 or L500).

Incidentally, newer VCRs seem to handle the thinner tape more gently with fewer hangups. Also, BETA VCRs seem to handle tape more gently than VHS decks. So, you may be less plagued with thin tape troubles on your BETA.

Figure 4-24 lists the various tape designations and their playing times. RCA happens to market its T60 cassette under the unusu-

Figure 4–24
Cassette running times

BETA	Length (feet)	Play time (hours: minutes) 1	2	3	Average retail price	Price per minute at middle speed ($)
L125*	125	0:15	0:30	0:45	12.00	.40
L250	250	0:30	1:00	1:30	13.80	.23
L370*	370	0:45	1:30	2:15	14.75	.16
L500	500	1:00	2:00	3:00	17.90	.15
L750†	750	1:30	3:00	4:30	21.20	.12
L830†	830	1:40	3:20	5:00	24.00	.12
VHS		SP	LP	SLP		
T15*	115	0:15	0:30	0:45		
T30	225	0:30	1:00	1:30	16.20	.27
T45*	335	0:45	1:30	2:15		
T60	420	1:00	2:00	3:00	18.12	.15
T90*	645	1:30	3:00	4:30	22.00	.12
T120	815	2:00	4:00	6:00	24.80	.10
T140†	1003	2:20	4:40	7:00		
T150†	1075	2:30	5:00	7:30		
T160†	1147	2:40	5:20	8:00		
T180†	1260	3:00	6:00	9:00		

Note: Most manufacturers include a little extra tape in the cassettes, yielding a little higher running time than listed here.
*Used by cassette-dubbing houses or available on special order.
†Extended length tape which uses a thinner base than standard length tapes, allowing a greater amount of tape to be spooled onto the cassette reels.

al VK-125 designation and its T120 cassettes under VK-250.

If you play a cassette all the way through, you're likely to get a pleasant surprise. The manufacturers generally put more tape in the cassette than advertised. A VHS T120 cassette generally has as much as 15 extra minutes in it at the SP mode. BETAS are known to have at least 5 extra minutes in their L500s.

When stocking cassettes, longer is not always better. Sometimes shorter cassette lengths help you organize certain subjects together. Also, if you have numerous segments recorded on the same tape, it won't take as long to wind from one to another if they aren't a mile apart on a long, long tape.

JUDGING AND
SELECTING VIDEOCASSETTES

The manufacturers There was a day when all the tape made came from one of three factories in Japan and all cassette shells were produced from molds approved by patent-holders Matsushita and Sony. JVC, RCA, Fuji, TDK, Sony, and Sears were all different names on the same product. Tapes sold through TV dealers for $27 each could be mail-ordered on discount for $17, yet still be the same.

This is changing. Millions of research dollars spent by BASF, TDK, Ampex, Maxell, Fuji, 3M, Sony, Memorex, and others are spawning new formulations, brands, and prices. One of the latest improvements is the so-called HG (for High-Grade) formulation introduced by Maxell, TDK, and Fuji. HG tapes have less "noise," evidenced by less graininess in the picture. An HG tape made at the SLP (slowest) speed would look as good as a normal tape at the LP (next faster) speed.

For the chemists among you, most BETA tapes use chromium dioxide for their magnetic powder, manufactured by Du Pont and BASF. Most VHS cassettes use cobalt-treated ferric-oxide powder, not unlike what's used in premium-priced audiocassettes.

How to judge tape and cassette quality The three most important things to examine when measuring the quality of a videocassette are: S/N (signal-to-noise) RATIO, DROPOUTS, and packaging. Let's take them one at a time.

1. S/N RATIO is a number, expressed in decibels (db), which tells how much desired signal you have compared to how much unwanted NOISE (graininess) you have. It is a fraction with "signal" on the top and NOISE on the bottom. Lots of good signal with very little NOISE makes the fraction into a bigger number, so higher db numbers represent "cleaner" sound and pictures. A 40 db S/N picture is annoyingly grainy. A 50 db S/N picture is about perfect. New, healthy BETA and VHS recorders have S/N RATIOS of about 43 db with good tape. HG tapes are about 3 db better than "normal" tapes.

Besides appearing as undue graininess on the TV screen, NOISE manifests itself very often as dark dashes of blue or black intruding on an otherwise smooth, pure, intense red scene.

In short, a high S/N RATIO listed in the tape's specifications or in a technical report, promises a higher quality, smoother, purer picture.

2. DROPOUTS are specks of snow that represent temporary loss of magnetic signal between the tape and the VIDEO HEAD. A DROPOUT can occur during recording, in which case it will remain in your program forever. A DROPOUT can appear during playback in places where the tape "sheds" or gets dirty, or a particular DROPOUT could appear just once during a playback and later get scrubbed off by the VIDEO HEADS, thus disappearing in subsequent playbacks. In any case, the fewer DROPOUTS the better. Technical reports sometimes count how many DROPOUTS per minute a tape has. Less than 20 small DROPOUTS per minute is hardly noticeable and is therefore a good number. DROPOUTS should not be counted during the first minute or two of a cassette because the threading process scratches the tape, causing frequent DROPOUTS there. This is also a good reason for storing your tapes wound or rewound; it keeps the threading-induced DROPOUTS before the beginning and after the end of your shows.

Some tapes increase in DROPOUTS as they are used and reused. Most, strangely, get better with time.

3. Packaging—if your cassette jams, it's useless. If it squeals or squeaks, something's rubbing in there and that's not good either. This will cause your picture to jitter (left to right *or* up and down) and often makes your sound waver or warble, like it was underwater.

Most brand-name cassettes are well constructed. Many off-brand, foreign, or "white label" cassettes are not, and will cause trouble.

Chapter 20 (paperback Volume II) goes into more detail about selecting and judging the quality of purchased and rented videocassettes. Home video magazines like *Consumer Reports*, *Video Review*, *Videoplay*, and *Video* can provide more up-to-date quality assessments than a book can. I will mention that the 1981 *Videophile* tape survey ranked Sony (HG), Fuji, and TDK tapes as best BETAS, with Ampex, Scotch, and DuPont as worst. For VHS cassettes, they ranked TDK (HG), Fuji (HG), and Maxell (HG) as tops and Memorex and Scotch as bottoms. *Video* magazine in 1982 ranked JVC (HG), Maxell (HG), and TDK (HG) as generally best.

Stay away from "white box" brands of videocassettes imported from Korea, Taiwan, and Hong Kong. These brands, being sold under names like "Forward" (with a label that looks like Panasonic's), or "Review" (with a lookalike DuPont logo), are cheap, but they're real dogs. Which brings us to the next topic.

Counterfeit blank tape It probably wasn't sold under the "El Cheapo" brand label. In fact it looked just like the name-brand tapes you're used to buying. The price was unbelievable. The internal construction is garbage, but you couldn't see that. You probably came across a bunch of them in a duty-free shop in Hong Kong, Taipei, or Singapore, or maybe in a junk shop or a flea market. Friend, you just bought a counterfeit tape, a fake with a familiar-looking label.

Take solace in the fact that there's a buyer like you born every minute. What some companies do is produce a low-quality (often unplayable) cassette and slap a familiar-looking label on it. If you look closely, you may notice a spelling change in the name, like Memex instead of Memorex, TKD instead of TDK, Maxwell instead of Maxell, Suny instead of Sony. Sometimes the cassettes will be labeled "B-type" instead of BETA or "V-type" instead of VHS (BETA and VHS are patented logos which can legally be used only on the real McCoys).

Your "bargain" tape will probably:

1. Jam in your VCR, perhaps requiring the machine to be dismantled before the tangled tape can be removed.

2. Muck up your machine with shedding powder.

3. Contain less tape than it says on the package.

When you think you see such a "bargain," examine the label *closely* for:

1. A misspelled brand name.

2. Smeared or fuzzy small print, or poorly registered or overlapping colors.

The genuine article is likely to have:

1. The manufacturer's name *and address* on the box, rather than simply "Made in U.S.A."

2. The actual BETA or VHS trademark, not "V" or "B-type."

Well-established, reputable dealers are unlikely to deal in counterfeit tape. If, perchance, you do get stuck with some:

1. *Take it back.* And complain *loudly*, if necessary.

2. Send a letter with the dealer's name, address, date of purchase, photocopy of the receipt, and cassette package to the *real* manufacturer. They will want to track down the counterfeiter and may appreciate your lead—and may find a way to thank you.

Cost The bitterness of poor quality lingers long after the pleasure of low price has faded away. Yet we all want the most play for the penny. Again, I must refer you to the aforementioned video journals (whose addresses are listed in Appendix 1) or to *Consumer Reports* or *Changing Times*, or *Consumer Union* for up-to-date value comparisons.

On the other hand, consider this alternate viewpoint: Mediocre video tape gives a picture somewhat similar to a distant, snowy TV station. If you get poor TV reception, *you're not going to notice* the imperfections in your tape. People with noisy mufflers don't notice squeaky springs. *The best-quality tape will make a difference only with a good-quality TV signal.*

Prices for the same tape may vary widely. Shop around. It is common to find prices ranging from 15% below list up to 100% above list. Figure 4-25 tabulates some common tape lengths and compares the retail and discount prices you're likely to encounter.

AVOIDING
ACCIDENTAL ERASURE

Picture yourself recording an "Eight Is Enough" rerun and looking down at the cassette box to discover that you're recording over priceless sequences of "Baby's First

Figure 4–25
Common videocassette prices

Tape	Average retail price	Average discount price
BETA		
L-125	$12.00	$ 8.00
L-250	13.80	8.50
L-370	14.75	8.50
L-500	17.90	9.00
L-750	21.20	11.00
L-830	24.00	18.00
VHS		
T-15		8.00
T-30	16.20	8.00
T-60	18.12	9.00
T-90	22.00	10.00
T-120	24.80	11.00

Steps" and "Grandpa's 91st Birthday Party." Or, you loan your "Neighborhood Theatre" masterpiece to your neighbor and get it back with his kid's recording of Saturday morning cartoons on it. Murphy's 108th Law of Recording states that junk recordings *never* get accidentally erased; prized ones *always* do.

Nothing short of copying all your tapes and keeping them in a vault will completely protect them, but two easy precautions will save you a lot of disappointments.

1. *Label everything as soon as it's recorded.* Don't guess at what's on your tapes. You're likely to record over your wrong guesses.

2. *Snap the* SAFETY TAB *off the videocassette* so the tape cannot be recorded upon. Figure 4-26 shows how. This procedure does not affect playback or any other feature of the VCR or tape. It just defeats the VCR's RECORD function. Covering the hole with a piece of scotch tape will restore the cassette's recordability.

CARE OF VIDEO TAPE

Longevity Magnetic tape is not an archival medium. Although the magnetism recorded on it will last easily 50 years, the plastic ribbon itself stretches and contracts with heat and cold and "relaxes" with age, causing the picture to bend at the top. This bending (shown in Figure 4-27) is called FLAGWAVING (the top of the picture flutters back and forth) or SKEW ERROR (the top of the picture skews to the side) or TAPE TENSION ERROR (the machine is misadjusted and is stretching the tape as it plays it).

Even under ideal environmental conditions your tape may begin to suffer "the

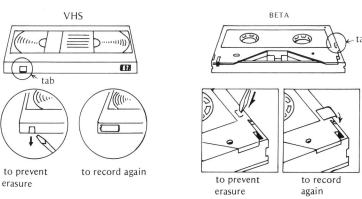

Figure 4–26
Preventing accidental erasure

To protect a recording from accidental erasure, break off the SAFETY TAB on the bottom of the videocassette (BETA) or the near edge of the videocassette (VHS) using a screwdriver. With the tab removed, the RECORD button won't go down and the tape is "safe" from erasure. If you change your mind and wish to record, cover the hole with a piece of cellophane tape.

Figure 4–27
FLAGWAVING. When tape shrinks or stretches, or is pulled too tight by the VCR playing it, the picture bends at the top. In extreme cases, the picture collapses altogether into a mass of diagonal lines like in Figure 1-7. Adjusting HORIZONTAL HOLD on your TV may help.

bends" after a year or so. Playing the tape once or twice a year will help, perhaps stretching this figure to 10 years. When the problem gets bad enough, the whole picture may twist into diagonal lines, like in Figure 1-7. Much early broadcast video tape was lost this way.

Conceivably a technician or a knowledgable VCR owner could open the machine and readjust the TAPE TENSION on the VCR, making a shrunken tape play again, and could then copy it. Once a playable copy existed, the VCR could be adjusted back again to its proper TAPE TENSION.

Super-thin tapes (numbers higher than T120 or L500) are more likely to stretch than normal tapes. Thin tapes are especially sensitive to stopping, starting, winding, and scanning, especially if done on older machines, and should be played straight through from beginning to end to promote long life.

A tape can be played between 100 and 200 times before the picture becomes noticeably degraded. PAUSING wears out the tape quickly.

Storage The lifetime of your recording is directly related to how carefully it's *stored* and how gently it's *used*. Here are the rules of storage:

1. *Store cassettes in a vertical position.* This prevents the tape's edges from getting "roughed up" against the inside of the cassette.

2. *Keep the cassettes in their boxes.* This keeps out dust and dirt, which causes DROPOUTS.

3. *Store tapes at average temperature and humidity,* about 70°F (20°C) and 50% or less humidity. Temperatures below −40°F (−40°C) and above 140°F (60°C) will permanently damage the tape. A car trunk or passenger or glove compartment on a hot day may easily exceed the 140°F maximum. Leaving a tape in the sun can also overheat it.

Mildew will grow on damp tapes.

4. *Keep tapes away from magnetic fields* such as hi-fi speakers, amplifiers, transformers, magnets, or big electric motors. The magnetism from these devices can partially erase your tape.

5. *Wind or rewind tapes before storage.* Don't store them partially rewound because you want the DROPOUT-causing threading process to occur at the beginning or end of your tape, not in the middle of your show.

Some experts advise playing cassettes all the way to the end for long-term storage. This provides very uniform tension on the tape. Also when you rewind the tape before playing it back, you "air it out," dehumidifying it, so it will slip easily through the machine when it plays.

6. *Don't leave recorded tapes threaded in the VCR for days or weeks.* This creases them.

7. *Remove the* SAFETY TAB *on any cassette you wish to make erase-proof.*

Handling

1. *Treat cassettes gently.* Dropping them rubs the delicate tape edges against the cassette housings, abrading them.

If you drop and damage a cassette don't try to play it, it may jam in your VCR. You can buy new cassette shells as kits and transfer the tape to the new housing.

2. *With VHS machines, use* PAUSE *rather than short* STOPS. It stretches the tape a little every time the VHS machine unthreads and rethreads the tape on STOP. On PAUSE, the tape stays threaded. On BETA machines, the tape stays threaded on STOP, so it's okay to STOP BETAS.

3. *Don't* PAUSE *a tape for more than a couple minutes.* After a while the spinning VIDEO HEAD will wear out the tape or crease it, causing DROPOUTS in that part of the program.

These lists could go on, including things like, "Do not smoke near VCRs or cassettes because smoke particles will make a film on the tape and machinery and clog the VIDEO HEADS," but the precautions, though accurate, start to get ridiculous. How many "rules" you follow will depend on how much of a perfectionist you are. A lot of owners knock their tapes around and store them half-wound on their sides atop a loudspeaker in their basement. Their tapes won't last years or look great, but many owners don't seem to notice. Two things *no* tape will tolerate, even if you're an indiscriminating viewer, are *heat* and *dirt*.

Radiators, hot cars, sunny windowsills, dusty workshops, and sandy, salty beaches will wreck a videocassette.

KEEPING TRACK OF
YOUR CASSETTE COLLECTION

When you first get your VCR, you'll probably record everything, *everything*! Almost every program will seem too valuable to be erased, so you'll take a deep swallow and purchase another batch of videocassettes to feed your insatiable monster. A year or two later you'll discover that you never watched 95% of those "priceless" recordings. In fact, you'll probably not remember recording half of them.

As you begin to discover what kinds of things you *do* enjoy viewing over (as opposed to storing and forgetting) you become ready to develop a cataloging system best for *you*. You may become a light video user, mostly using your VCR to catch a show or two while you're away. Your collection may consist of a few favorites plus some tapes which get constantly recycled with new shows. In this case, you don't need much more than a simple title printed on the spine of your videocassette jacket, and maybe a 3 × 5 card slipped into the box listing several shows and their COUNTER num-

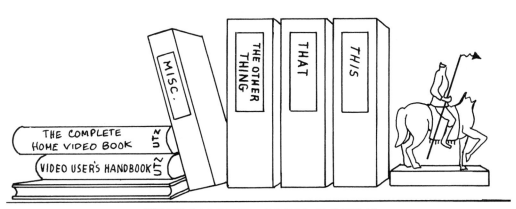

VIDEOCASSETTE FILING SYSTEMS

bers. Some folks use white china markers (available at art stores) temporarily to label the spines of their cassettes. The markings can be wiped off with a soft cloth when the tape is erased.

More involved videophiles may keep a collection of tapes, each numbered sequentially. A log book listing the cassette number could index, by COUNTER number, all programs on each tape.

True videofreaks may create a card file categorizing their programs by subject and date, and may actively pursue trading clubs (see Chapter 20 and Appendix 2 for more about trading). They may even record shows *they* don't particularly want, with the hopes they're getting something *someone* else will want—good for trading purposes. Video recording thus becomes an avocation, and inventory control becomes a necessity.

It is better to have a short pencil than a long memory. Here are a few systems for cataloging your recordings:

Logging and shelving by accession number Number your cassettes, on the jacket and on the cassette shell itself (in case they get separated), from one to whatever. Thereafter, each new tape gets the next higher number. Get a looseleaf or other type of notebook and number each page. List each tape's contents with COUNTER numbers on the corresponding page. You may wish to include length of show, date, time of broadcast, channel, tape speed, whether or not the program has the commercials edited out, original or rerun, the year the movie was made (usually seen at the movie's end), the cast, original recording or duplicate, recording quality, black-and-white or color, or other details on the page. You could easily slice out the *TV Guide* listing for the show and paste it on the page.

Logging and shelving by chronology number Photographers file their negatives using a numbering system which incorporates the production date into the number. The first two digits tell the month (01 for January, 12 for December), the next two digits, the year (82 for 1982), and the last two digits sequentially number the tapes recorded that month. Thus the number 068202 represents the second tape recorded in June of 1982.

Logging and shelving by subject This keeps all your Lon Chaney movies together, your family outings and your porn collection (assuming these categories aren't one and the same) shelved together. You may wish to number all your Sheriff Lobo classics 1001, 1002, etc., while your weddings, "family theatre," and other home brew take numbers like 2001, 2002, etc. This system has room for nine categories and 999 tapes per category, or 99 categories with no more than 99 tapes in each, take your pick.

Some collectors prefer to catalog by category, such as Drama, Comedy, Musical, etc. This works okay *if* you record *new* drama on the remaining *old*-drama tapes. If you record a comedy right after your sports show (something your programmable VCR will happily do for you), you'll have trouble categorizing and shelving the cassette.

Logging and shelving by title Here you organize everything alphabetically by title. This makes finding movies a snap. If you put two movies on a tape, the system gets complicated. Perhaps page 1 of your logbook could list the movies alphabetically and tell where the "second" movies are recorded. Another problem with organizing by title is that a lot of reshuffling may result as you record new shows and sandwich them between the others.

Storage boxes and labels The boxes your cassettes came in are fine except that when you shelve all the various brands side-by-side the mix of colors and trademarks starts to look a bit busy (oh yes, very gauche for the *true* videophile). Also, the space available for titling is a bit cramped. Wouldn't it

Cassette storage box. (Left, courtesy of Amaray International Corp.) (Right, courtesy of Reliance Plastics & Packaging Corp., Bloomfield, N.J.)

be nice if someone sold just cassette *boxes,* all looking the same, with big spaces for you to write your programs on?

It would be nice—so nice that several companies are doing it. Here are a few:

- Sturdy plastic boxes with clear plastic label holder.
  Price, about $2
  Amaray Sales Corp.
  2251 Grant Road
  Suite H
  Los Altos, CA 94022
- Cardboard or plastic sleeves, much like the original cassette packages, except they're blank.
  Price, about $.50
  Reliance Plastics & Packaging Corp.
  108-18 Queens Boulevard
  Forest Hills, NY 11375

Perhaps you could get by with just blank labels you could stick onto your present cassette boxes and cassettes to keep them looking more uniform. Also, when you erase a program, recording a new one, you could neatly affix a new label rather than erasing, scraping, or blowtorching off the old label.

Here are two (of many) companies that sell such labels:

- Comprehensive Video Supply
  148 Veteran's Drive
  Northvale, NJ 07647
- Meyer's Printing Co.
  500 South Third Street
  Minneapolis, MN 55415

These labels cost about $10 per 100. Shop around your stationery stores for a better deal on something which may serve just as well.

One final tip on organizing tapes: When numbering cassettes, put your number on *both* the box *and* the cassette. Place the cassette's number so it will be visible through the recorder's window while the tape is in use. This way, you'll not lose track of what tape you're playing *now.*

Common VCR Ailments and Cures

You followed the directions (you thought). Everything is connected and rarin' to go. Only it doesn't. Before you head to your dealer or to the repair shop with your beast, there's a lot you can try in order to diagnose and solve your VCR's problem yourself.

After a while you will learn the eccentricities of your machine and will intuitively head right for the trouble when your VCR goes kaput. Until that time, you're going to have to use logic and patience to coax the recalcitrant recorder to life. You will find that 80% of your problems are because you failed to throw a switch, turn a dial, or connect something up the way it should be. Another 10% of your problems will be minor defects, maybe in one of your connecting cables, perhaps a broken wire. Another 5% of the time it's a dirty or broken switch somewhere—still an easy thing to replace or fix. Then 5% of your time the machine fails because *it* has a problem and needs technical attention.

Let's look at some common VCR problems and how to solve them.

MACHINE CONTROL
PROBLEMS

In these examples, the machine's buttons don't seem to do what they're supposed to.

1. *VCR fails to operate;* POWER *light does not come on.*
  (a) Check to see if the VCR is plugged in.
  (b) Is the socket it's plugged into operated by a wall switch somewhere? If in doubt, check the outlet using a lamp.
  (c) TIMER may be switched to AUTO, or OFF, or ON and waiting for a scheduled time. Check which position manually activates the VCR, or try flipping it to another position and watching for signs of life.
  (d) Turn TIMER/SLEEP switch OFF.

2. POWER *light comes on but tape doesn't move when you switch to* PLAY.
  (a) Check PAUSE; it should be OFF.

  (b) Check the REMOTE socket; if anything is connected there, *it* may be PAUSED. (An earphone accidentally plugged into the REMOTE PAUSE socket will PAUSE the machine too).
  (c) Is the DEW light on? If so, you'll have to wait until the machine dries out before it will work. Try the hair dryer trick mentioned in the Getting More out of Your VCR section of this chapter.
   Is the machine cold (thus condensing humidity)? Move to Arizona where it's warm and dry.
  (d) Is the cassette properly loaded? Are you trying to use a BETA cassette in a VHS machine? That will never do.
  (e) On BETA VCRs, is the STANDBY light on? If so, wait till it goes off.
  (f) Maybe the tape is loose in the cassette; it should be taut. Often VCRs sense loose tape and switch themselves to STOP. EJECT the tape and give one of the tape hubs a turn with your fingers to tighten the tape.
   Or, press REWIND, STOP, WIND, STOP, then PLAY. The rewinding and winding may tighten the tape.
  (g) Could it be that the tape is at its end and needs to be rewound? Some cassette machines have an AUTO OFF feature which stops the tape at the end of play. Some also have an AUTO OFF light that lights up to appraise you of this condition.
  (h) Are you using a portable VCR in the cold? Perhaps your battery died, or got too weak. When a 12-volt battery drops down to about 10.4 volts, the VCR senses the weakness and shuts itself down. Perhaps it's so cold, the oil in your VCR became gooey, slowing down the VCR's wheels. The VCR may sense this and shut itself down to avoid damage. The VCR may also sense whether its transistors are too cold to function, and will shut itself down.

(i) Do you have a camera connected to the VCR? Perhaps its PAUSE trigger is pulled.

3. *Buttons won't depress.*
   (1) Switch the POWER ON.
   (b) Make sure you have a cassette in the VCR.
   (c) Check everything in example 2 above.

4. RECORD *button won't depress or activate.*
   (a) Check the cassette SAFETY TAB. If it's been removed, the tape cannot be recorded. See Avoiding Accidental Erasure earlier in this chapter.
   (b) Check everything in examples 2 and 3.

5. AUDIO DUB *button won't depress or activate.*
   (a) Same problem as example 4; most likely SAFETY TAB.

6. FAST FORWARD *does not function.*
   (a) Maybe you're already at the tape's end. Rewind.
   (b) Check everything in examples 2 and 3.

7. *VCR stops in* REWIND *before end of tape.*
   (a) Turn the MEMORY or SEARCH switch OFF.

8. *Cassette won't insert into VCR.*
   (a) Is it the right FORMAT (VHS or BETA) cassette?
   (b) Are you holding it right? The trap door part points *away* from you; the reel hubs face *down*.

9. *Cassette* EJECT *doesn't work.*
   (a) Switch POWER ON, press STOP, wait a moment, then try again.

10. *Cassette tape plays okay, but displays no picture in* PAUSE.
    (a) Some older VCRs don't give a STILL FRAME picture in PAUSE.

11. *The tape moves slowly or moves backwards when you press* RECORD *or* PLAY.
    (a) The belts in your VCR are tired, slippery or broken. Take two aspirin and call your repairman in the morning.

SIGNAL CONTROL PROBLEMS

In these examples, you can't get a picture or sound where you want it to go. Often the culprit is a misflipped switch. If you do much connecting and disconnecting of cables, it may be a cable plugged in the wrong socket.

If the cables are moved often, the connectors sometimes get loose or the wires fray inside the plug. After trying the obvious (switches, correct connections), try wiggling the cable at the plug end, or rotating the plug in its socket (except for multipin plugs) and watch your TV for flashing and listen for crackling. A frayed wire or dirty or bent plug may make contact intermittently when wiggled, cutting your signal on and off. When you doubt a cable, try another if you have one. A successful substitution will prove the first cable defective. They can usually be fixed easily (if you're the handy type) or replaced fairly cheaply (except for the multipin cables—they're worth fixing because they're expensive).

Sometimes switches get dirty inside. When you switch, the TV shows a lot of breakup and hash, and the audio may crackle. Flipping the switch back and forth a dozen times may clean it some, but a deteriorating switch will become more and more intermittent and unreliable. If you're handy, there are sprays you can use to clean a dirty switch. Otherwise, a technician may have to clean it for you.

Now let's see what else can go wrong, go wrong, go wrong . . .

1. *Can't get regular TV programs on the TV set.*
   (a) Switch the VCR's OUTPUT SELECT, or ANTENNA, or VTR/TV switch to TV.
   (b) Is the VCR connected to the antenna? Is the TV's ANTENNA input connected to the VCR? They should be.
   (c) If you're using the VCR's tuner to feed your TV, make sure *it* is FINE TUNED

and your TV is FINE TUNED to the VCR's channel (3 or 4), and your VCR's INPUT SELECT is on TV or TUNER and its OUTPUT SELECT is on VTR.

2. *Can't get VCR to play through the TV.*
(a) The VCR's OUTPUT SELECT must be in the VCR, VTR, or CASSETTE position.
(b) Check the connections between the VCR and the TV.
(c) Is your TV tuned to the same channel as your RF CONVERTER?
(d) If you're sending video and audio straight to a TV monitor/receiver, make sure the monitor/receiver is switched to LINE or VTR, not to AIR or TV.
(e) Check to see if the TV even works by switching the VCR's OUTPUT SELECT to TV or ANTENNA and tuning the TV to various broadcast channels. If nothing comes in, perhaps the VCR-to-TV connection is bad or the TV is bad. If stations *are* received, then the TV and its connections are okay. Switch the OUTPUT SELECT back to the VCR (or VTR or CASSETTE) mode and check channels 3 or 4. Check to see that the RF MODULATOR isn't missing.
(f) Wiggle the cables while playing a tape and watch the TV for flashing to check for loose connections.

3. *Unable to record TV programs.*
(a) Flip the VCR's input switch to the TV or TUNER position:
- VHS—switch INPUT SELECTOR to TUNER
- JVC—switch RECORD SELECTOR to TV
- BETA—flip CAMERA/TUNER switch to TUNER

(b) Make sure the cassette's SAFETY TAB is still there.
(c) Unplug the camera and microphone from the deck. They sometimes automatically override the TV inputs.

4. *No sound while recording a TV broadcast.*
(a) You left the microphone plugged into the VCR, overriding the TV.
(b) Check separate AUDIO IN connection for the same thing.
(c) If portable VCR uses a separate tuner with separate AUDIO and VIDEO connections, the AUDIO wire may be loose.
(d) Maybe you have sound but you can't hear it. Can you get sound while playing other tapes or from off-the-air broadcasts? If not, maybe your TV has an earphone plugged into it, cutting off its speaker.
(e) Are you using a TV monitor instead of a receiver? TV monitors don't have sound.
(f) Turn up the volume control on the TV. If the sound crackles on and off as you do, work the knob back and forth a couple of times or tap it. You may have a volume control that has dirt inside it. If you get the crackling volume control working for the time being, leave it. Later, have a technician look at it. It's generally easy to fix.
(g) If you're using a monitor/receiver and sending separated audio and video to it, check your audio wire.

THERE IS SOUND BUT NO PICTURE

5. *No picture while recording a TV broadcast; sound is okay.*
(a) The camera or a separate video input is still connected to the VCR.
(b) On portable VCRs, check that the

tuner's video cable is connected to the VCR's VIDEO IN.

6. *Camera or microphone signals won't record.*
  (a) Switch the VCR's input switch to CAMERA or VIDEO (away from TV or TUNER).
  (b) Are the microphone and camera plugged in? Is the camera turned on?

SPECIAL AUDIO PROBLEMS

Chapter 7 will dig deeper into this subject, but here are a couple common gremlins to chase away:

1. *A screech or howl comes out of the TV.*
  (a) Turn down the volume on your TV receiver. You've got what they call FEEDBACK. See Figure 4-28. No, FEEDBACK isn't the chef's position on a football squad. What happens is this: Sound goes in the microphone, goes to the VCR, goes to the TV receiver, and comes out the speaker loudly enough to go back into the microphone again. Around and around the sound goes and where it stops—well, it stops where you break the cycle by turning the volume down or moving the mike farther from the TV.

2. *Echoey, hard-to-hear sound.*
  (a) Move the mike closer to the subject.
  (b) Move to a room with more carpet, curtains, and furniture. You have too many echoes from hard bare walls.
  (c) Turn the volume down on your TV; some of the sound is getting back to your mike.

3. *Microphone sound is weak, tinny, has hum or hiss.*
  (a) The mike is not right for your VCR. Read Chapter 7, and try another mike.
  (b) The mike may be defective or of poor quality.
  (c) The connection may be bad.

4. *Using* AUDIO IN, *the sound distorts or sounds raspy.* Your audio signal is too strong. The sound may also have lots of hum and hiss.
  (a) Reduce the volume from the source.
  (b) Attenuate the signal using methods described under Audio Connections earlier in this chapter or in Chapter 7.

Figure 4–28
Audio FEEDBACK

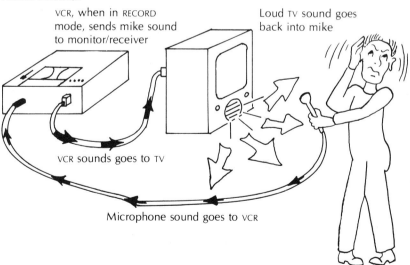

PICTURE QUALITY PROBLEMS

1. *Picture is distorted, usually with a band of snow or* NOISE *through it.* Figure 4-8 shows a mild case. Worse cases may look like Figure 1-22. The picture may jiggle a lot, even though you're *not* watching "Charlie's Angels." Adjust the TRACKING control. This is perhaps the most frequent adjustment you will need to make on a VCR. Usually the picture is best when the TRACKING control is set in the middle at FIX, but tapes made on another cassette are most likely to require that the TRACKING knob be turned to some position other than FIX. Turn the TRACKING control until the breakup moves off the screen at either the top or bottom and the picture remains stable.

Old, stretched, hot, cold, or damp tapes won't track well at all.

2. *You play a prerecorded videocassette and the picture jitters and rolls.*
  (a) Adjust TRACKING.
  (b) Adjust your TV's VERTICAL HOLD.
  (c) If you bought the tape commercially, it may be COPYGUARDED—its sync signals are messed up to keep people from duplicating it. These signals are messing up your VCR or TV. Send the tape back and ask for a non-COPYGUARD edition.
  (d) If a friend gave you this copy, which he made from a COPYGUARDED tape, naturally it shouldn't play. That was the tape producer's whole idea.
  (e) Cheap, COUNTERFEIT, PIRATED, or Nth GENERATION (copies of somebody elses copies which were previously copied from somebody else's copy) tapes are likely to be of low tape quality, low cassette-shell quality, or low recording quality, and they just won't play well.
  (f) Clean the HEADS with a CLEANING CASSETTE (process described in Chapter 13, paperback Volume II). Your CONTROL HEAD, which synchronizes the motors and tape movement, may be dirty. Also the CAPSTAN (the part that steadily pulls the tape through the machine) may be dirty. If you're handy, open the machine and clean these parts manually to do a better job (process also shown in Chapter 13, paperback Volume II).

3. *No color or poor color when playing a tape.*
  (a) Adjust the TV's FINE TUNING.
  (b) Perhaps the VCR's FINE TUNING wasn't properly adjusted when the recording was made. If so, the program will never play in color.
  (c) Are you sure this is a color program?
  (d) Turn up the COLOR control on the TV set.
  (e) Check to see the TV's ANTENNA input switch is set to 75Ω if you're using coax cable or 300Ω if you're using twin lead.
  (f) If you're using straight video into a TV monitor, make sure the set's properly terminated (as explained in Chapter 1).

4. *Using AFT worsens reception.* You've tuned your VCR manually with the AFT OFF and everything looks great. You switch the AFT ON and things get worse, or perhaps you lose color.
  (a) Your AFT circuits need adjustment by a technician.
  (b) In the meantime, leave it OFF and you'll get by.

5. *While playing a tape, the picture blanks out or temporarily distorts or shrinks.*
  (a) Is there a power-hungry motor, heater, microwave oven, or other appliance on your circuit sapping power? If this power interference occurred during the recording, the glitch is there to stay. If it happens only during playback, the tape's okay, go back and view again.
  (b) If there's an air conditioner or big motor nearby, it can interfere with your video signal.
  (c) If it looks like TV interference is being picked up by your VCR's tuner, check Chapter 2 for ways to cure TV interference.

6. *Tape plays at the wrong speed.* The picture may roll, bend, tear, or collapse into diagonal lines looking like Figure 1-7.

    (a) Is your SLOW MOTION feature engaged?

    (b) Perhaps the machine is damp inside and the DEW sensor failed to stop the machine. Let it warm up a bit and try again.

    (c) Your VCR's insides may be dirty. Run a CLEANING CASSETTE through (process described in Chapter 13, paperback Volume II) and see if it helps.

    (d) Is your portable VCR's battery dying? Doublecheck the AC power adapter connection. Does the power meter say it's getting enough juice?

    (e) Are you using a cheapie or COUNTERFEIT cassette? It could be rubbing inside.

    (f) Try another cassette tape. If the problem persists, perhaps the VCR's drive belts are old, tired, slippery, or in need of replacement.

    (g) You would get speed problems if your AC power fluctuated in frequency. Do you have weird power? (I have a grandmother with weird powers).

7. *The picture bends or flutters at the top during playback.* FLAGWAVING (or TAPE TENSION ERROR or SKEW ERROR) is shown in Figure 4-23. It is usually caused by the tape being too tight, too loose, too stretched, or too shrunken as it plays through the machine.

    (a) Adjust the HORIZONTAL HOLD on your TV set.

    (b) Adjust the SKEW or TAPE TENSION control on your VCR if it has it (most don't).

    (c) If the problem happens on all tapes, have your VCR adjusted inside. If your TV is very old or the tube-type, the picture will always look worse than it has to. A technician can modify the TV's AFC TIME CONSTANT to improve the picture. If you have another TV, try that one.

    (d) If the tape and/or the VCR are cold, let them stabilize to room temperature.

    (e) If the tape was recorded long ago, it probably has shrunk. Without adjusting the VCR, it will probably never play right. One thing you can try is winding the tape all the way to the end. Then, rewind it all the way to the beginning. Now try to play it. The winding may stretch or relax the tape a little, making a marginal tape playable.

    (f) Super-thin tapes stretch easily, especially in hot weather, causing FLAGWAVING.

    (g) The problem may be similar to #2 or #6 earlier. Try those maneuvers.

8. *The picture looks very grainy or even snowy on playback only.* It looked fine when it was recorded (eliminating the possibility that you recorded a snowy picture to start with) and the sound is okay. Perhaps it looks like Figure 1-22.

If the picture is very snowy, it could be due to a worn or dirty VIDEO HEAD. You can't easily see the dirt on a VIDEO HEAD, but a tiny speck of it can clog this very sensitive device and render it inoperable. Usually a clogged HEAD will not completely obliterate the picture—some faint image may show through the snow. If this is what you see, you can be fairly sure your problem *is* a clogged HEAD.

What do you do for a clogged HEAD (besides taking nasal decongestants and getting lots of bed rest)? You clean the HEAD with a special "cleaning cassette." Following the manufacturer's instructions, insert the cleaning cassette into the machine just as you would a cassette with video tape in it. Play the special cassette for about 30 seconds, then remove it without rewinding it. The HEADS should now be clean.

Chapter 13 (paperback Volume II) gives further details on this procedure.

9. *The TV used to work all right before you got the VCR, and the VCR worked all right in the store, but together they display interference or herringbone patterns on the TV screen or in the recordings.* The TV or the recorder is sending out a weak signal which is interfering with the other's tuner. Move

Common VCR Ailments and Cures   **133**

the two farther apart, maybe 3 feet. This may solve the problem. Also, avoid long antenna cables coiled up behind the equipment. Use good quality, well-shielded cables and avoid excessive lengths.

10. *You used to get sharp cable TV pictures before you bought the VCR and a few other goodies, like switchers. Now you get a ghost on some channels.* Nearby TV stations are broadcasting strong signals directly into your equipment while the cable company is sending you the same programs on the same channels slightly delayed (because their signals tranverse so much wire). The unwanted direct signal "leaks into" your wires and into poorly shielded equipment to cause ghosts. Seal these invading signals out by using COAX or SHIELDED TWIN LEAD everywhere, and well-shielded, high-quality switchers and other accessories.

PROGRAMMABLE
TIMER PROBLEMS

Here you've set your timer to catch Monty Hall making another of his unforgettable Deals, and when you come back to play your gem . . .

1. *You play your tape only to find a blank recording.* You forgot to switch the input selector to the TV or TUNER position.

2. *You play your tape only to find a picture with no sound.* You left a microphone or some other audio source plugged in.

3. *You play your tape only to find sound without picture.* You left the camera plugged into the VCR.

Future Developments

With VCR sales of 4 million and cassette tape sales of 35 million in 1982, you can expect the masterful manufacturers to be spending a few megabucks trying to improve what they've got. VCRs will get better and more loaded with features. Conversely, some VCR makers may give up the more expensive VARIACTOR TUNING and PROGRAMMABLE timers and go back to the old twistable channel tuners and one-event start/stop timers, while lowering the VCR prices to about $500.

The race for longer, thinner tapes is well underway. New strengtheners and recording powders may make these tapes tougher and more reliable than they are now. So-called METAL TAPES are under refinement by several manufacturers. These tapes, having recording surfaces of metal evaporated in a vacuum onto their plastic bases, offer higher recording densities, improved S/N RATIOS, and sharper pictures.

Smaller is better. Technicolor was lightest with its 7-pound VCR and tiny CVC (Compact Video Cassette) cassette. Now JVC's NR-C3U and Sharp's VC-220 untip the scales with 5½-pound VHS-C recorders costing $750 and $1000 respectively. Unfortunately, these minis only record 20 minutes per cassette. Panasonic's $1000 PV5110 weighs three pounds more but uses a normal VHS cassette lasting 2, 4, or 6 hours.

Perhaps the next step in VCR manufacture is the CAMCORDER, the combination VCR/camera in a single box. Such a camera can be made smaller by replacing the bulky vidicon tube (the "eye" of the camera) with a CCD (charge-coupled device), which can be made very small. The VCR part would also be miniature, using ¼-inch METAL TAPE moving very slowly through a miniature videocassette, using a totally new FORMAT recording 20 minutes per cassette. Sony's latest prototype, the one-piece Betamovie, uses the standard BETA FORMAT, and weighs only five pounds. To play back tapes, you have to use a separate BETA VCR, however. Hitachi, not to be outdone, is planning to introduce *its* version, the Mag Camera, which is supposed to sell for $1400 and hold

2 hours of recording on a ¼-inch videocassette and not require a separate player unit. Another 2-hour CAMCORDER which will need a separate adapter for playback is Matsushita's Micro Video System, scheduled to weigh 4.6 pounds when it's born in 1984 or thereabouts. And Sanyo is preparing its Handy Video CAMCORDER to weigh in at 6.4 pounds. JVC's HR–C3U VHS–C Compact Video Recorder can be attached directly to a lightweight camera, making a one-piece CAMCORDER. The VHS-C cassette is the size of a deck of cards, and records 20 minutes on a cassette. Using an adapter, the cassette can then be played on any VHS player. In August '83, Sony unveiled its new VCR with stereo hi fi sound. By invisibly coding the audio signal in with the video, they obtain a super 20Hz-20KHz frequency response (as opposed to the typical 70Hz-7KHz) and very low distortion *at all speeds*. The VCR and tape is compatible with standard BETAS. To keep up, I'll bet VHS makers follow suit by 1984.

VCRs in a Nutshell

I asked my wife (and typist) "How do you summarize a 50-page chapter?" "With as few words as possible," she answered with conviction. So with concerted brevity, here goes:

The VCR's controls function much like an audiocassette recorder's controls.

A VCR hooks up between your TV antenna and your TV. With its OUTPUT SELECT in the TV mode, your TV "sees" the antenna signal directly, like it would normally. With this switch in the VTR position, the TV (once turned to channel 3) "sees" what the VCR is recording or playing back. A VCR can be told to "see" different things too. Switching its INPUT SELECT switch to TV or TUNER, feeds it the antenna signal. With its INPUT SELECT in the VIDEO or LINE mode, the VCR "sees" its VIDEO input, which could be a camera or another VCR. On some models, the VCR "switches" itself when you plug something into the VIDEO or MIC inputs.

Direct video and audio inputs provide the best picture and sound quality. RF signals degrade the picture by about 10%.

For the best picture and sound quality, record at the fastest speed on your VCR. Use standard length tapes if you plan to stop/start, wind, or eject the tape frequently.

Weak audio signals (like from microphones) should be fed to the sensitive MIC input, whereas strong audio signals (like from a cassette tape player, hi-fi, or another VCR) should be fed to the AUDIO IN sockets.

The DEW INDICATOR on your VCR stops the machine when its mechanism is too damp to slip the tape over the heads easily and safely.

The COUNTER tallies how much tape you've used and can help you index where certain things are on a tape. You need a conversion chart (often homemade) to translate this arbitrary number into "time used" or "time remaining."

The TRACKING adjustment keeps the picture smooth and stable. When a band of snow creeps into the picture (usually while you're playing someone else's tape), adjust TRACKING.

AUDIO DUB allows you to add a new sound track (music or narration) to an already-recorded tape, while erasing the old sound.

PROGRAMMABLE TIMERS will allow you to leave home for a week and have the VCR record various shows from different channels while you're away. Most such devices are a bit complex to run and go awry if you're not careful programming them and checking your switches. When using a timer, check your INPUT SELECT switch, TIMER ON/OFF switch, channel selector, tape speed and length of tape left, cassette SAFETY TAB, and record time settings. Many timers run amok if the electric power fails.

To swap tapes with your friends, you must both have compatible VCRs. Primarily, both machines must be the same FORMAT (both VHS, or both BETA) and the tape must be played back at the same speed as when it was recorded.

To ensure good off-air recordings, monitor their beginnings and FINE TUNE the VCR's tuner for the best picture. When opportunities arise, play back a sample of your recording to see if it's coming out okay.

PAUSE is a handy way to cut out commercials, edit out the chaff in your own productions, or to butt sequences or programs together. Don't PAUSE for more than a couple of minutes; you'll wear out the tape in that spot.

And a few more precautions:

- Don't stick your fingers into the VCR's mouth (it chokes easily).

- Keep the VCR away from magnetic fields and intense heat.

- Don't block the ventilation slots on the top of your console VCR or operate it with the dust cover on. Don't operate it on a deep pile rug (blocking the bottom vents).

- Operate your console VCR only in the *horizontal* position.

- If liquid gets spilled into your VCR, unplug it immediately. Take it to the repair shop for a sponge bath.

- Do not disassemble your VCR unless your warrantee is up and you think you're pretty handy with electronics.

Portable VCRs, for weight reasons, have no tuner or have a separate tuner. Portables run on batteries or AC POWER ADAPTERS (which also charge the batteries) or on automobile batteries through cigarette lighter hookups. To save battery power, switch the VCR OFF when you expect more than a few minutes between shooting scenes. VCRs and cameras often consume power when on "standby" and not actually recording. When the delay between scenes is short, use the camera's trigger to PAUSE the tape.

When shooting outdoors, keep your batteries warm, but don't let your tape get hot. If you must bring a "cool" machine to a warm environment, seal it in a plastic bag to keep out condensation until the machine warms up. Seal up your VCR from sand, salt air, and water spray when recording at the beach. Always be aware of the sun's glare when recording outdoors. One look directly at the sun and your camera will be blinded. Use polaroid or neutral density filters and/or high f-numbers when recording in the bright sun. All these dangers aside, the outdoors is where the action—and your best scenes—are likely to be.

If you have several TVs and one antenna (or cable TV with an UP CONVERTER) there are several ways you can wire up your system so that any of the TVs can watch either broadcast TV or the VCR's program.

Shop around for the best price, but buy name-brand videocassettes. The standard-length VHS tape is a T120, which plays 2 hours at the SP (fastest) speed, 4 hours at the LP speed, and 6 hours at the SLP (slowest) speed. The standard BETA tape is the L500, which plays 1, 2, or 3 hours at the BETA-1, 2, or 3 speeds.

A tape's quality can be measured primarily by its S/N RATIO (how grainy the picture is), DROPOUTS (how many specks of snow invade the picture), and packaging (the tape's ability to shuttle smoothly without binding or jamming). Recent HG (High-Grade) tapes show improved S/N RATIOS. Counterfeit and off-brand videocassettes are likely to jam, perhaps choking your VCR on spilled tape.

To avoid accidental tape erasures, break off the SAFETY TAB on the cassette. To keep track of your tapes, label the cassettes *and* cassette boxes. Keep a logbook and an organized filing system if you possess many tapes.

To care for your video tapes:

- Store them in a cool, dry place.
- Avoid hot places like radiators, hot cars, or places that get direct sunlight.
- Avoid dust, dirt, and moisture.
- Never touch the tape (your fingers are greasy).
- Avoid magnetic fields (motors, speakers, transformers, magnets).
- Handle cassettes gently.
- Store tapes either wound or rewound, not half-wound.
- Store cassettes upright and in their boxes.
- Don't try to play a cold tape in a warm room or warm machine.

Tape shrinkage or stretching, or maladjustment of the VCR's internal TAPE TENSION controls, or an old TV's too-short TIME CONSTANT circuit will give you FLAGWAVING in your picture as it plays back. Adjust the TV's HORIZONTAL HOLD to straighten the problem out. Otherwise, check if *all* your tapes FLAGWAVE, and if so, have a technician adjust your VCR or your TV.

Most video ailments are caused by user error. Read your instructions carefully (you knew I'd work that in somewhere, didn't you?). Most of the time one of the many switches on your VCR is in the wrong position. Much of the time, some cable is plugged into the wrong socket. Often, a bad connection or broken wire is the culprit. Occasionally, the VCR is plain busted. Before you haul it off to the repair shop, trace your signals, from their sources, through their wires, to their plugs, into the right sockets, through the appropriate switches, out the output sockets, and through more cables to the monitoring devices. Be patient and logical. These machines don't give up their secrets easily. If recording doesn't work, try playing a tape to see if *that* part of your system is functional. Similarly, narrow down any problem area first ascertaining what *does* work. Find ways to test single components or individual cables. Wiggle cables at their plugs. And when all else fails, quit for a while; let your frustration subside. Attack again later with renewed spirit and patience. And if all this fails, then get an ax and bash the living transistors out of it, letting the IC chips fall where they may.

VCRs in a Nutshell

Chapter 5
TV Cameras and Lenses

HIS MASTER'S PRIVACY

How do you turn a vivacious, chattering party into a pack of zombies? Yep, bring out the camera. Another question: How do you double the cost of your video hobby? Buy a color TV camera to go along with your VCR. Once more: How do you build giant muscles while on vacation? Bring your portable VCR and camera along. And you *still* want a camera to produce your own shows? Very well then, Cecil B., here goes.

How TV Cameras Work

And now for the mini science lesson you've learned to abhor: Remember how a TV set made its picture by zigzagging an electron machine gun across the screen, turning "shoot-don't shoot" video signals into light and dark places on the screen? Of course you do. Well, the camera works the same way in reverse.

The camera lens focuses the image of a scene onto a light-sensitive VIDICON tube in the TV camera. The inside surface of the tube is electronically sprayed with electrons which stick to the place where the lens image is projected. Wherever light strikes the tube, the electrons get "bumped off." Thus the light-and-dark parts of a scene create "bare-and-full" places on the VIDICON'S surface. If you could see electrons, you would see a tiny picture painted with electrons, sitting on the face of the tube. Inside the camera tube, an electron gun zigzags in the same pattern as the TV's electron gun did. Only this gun knocks off the remaining electrons as it scans back and forth. As they are shot off, the electrons are collected, counted, and turned into a video signal.

So following the whole process from camera to TV: Light-and-dark parts of a scene get projected onto the camera tube. The light areas dislodge electrons from the surface, creating an image made of electrons. The electron gun in the camera tube sweeps across the surface, knocking off the remaining electrons, which get measured and turned into a video signal. This signal goes to your TV and tells its electron gun to "shoot-don't shoot." Thus light-and-dark parts of the original scene become "shoot-don't shoot" commands for the TV's electron gun, which creates a light-and-dark image on the TV screen. SYNC circuits in the camera make a signal (which becomes part of the video signal) that starts the camera tube's electron gun at the top of the picture, as well as starting the TV tube's electron gun at the top of its picture at the same moment. Both guns coincide with their zigging and zagging to create a TV picture that looks just like the original. Figure 5–1 diagrams the process.

Color pictures are a bit complicated. Remember your color TV whose picture tube was made of narrow, colored phosphor stripes, and how the TV's electron gun zapped certain phosphors to make certain colors? The home-type TV camera has a VIDICON tube with tiny stripes, each sensitive to a different color. The tube senses the image's color (CHROMANANCE) along with the black-and-white (LUMINANCE) image and sends the combined signals to your VCR or TV.

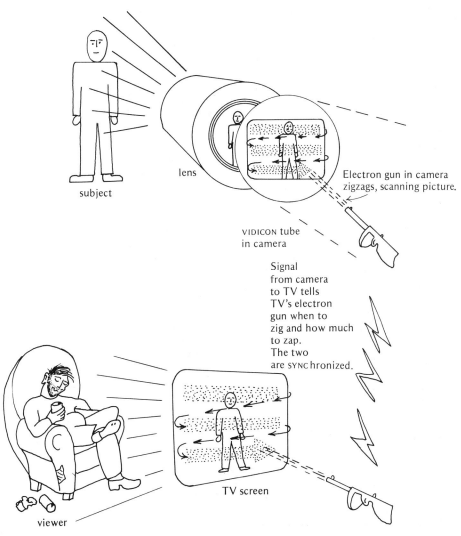

Figure 5–1
How the camera and the TV make the picture

Words to Know

VIDICON
Light-sensitive electronic tube, used in many home-type TV cameras, that converts light images into video signals.

VIEWFINDER
The part of the TV camera that shows you what you're aiming at. The three kinds of VIEWFINDERS are listed below.

OPTICAL-SIGHT VIEWFINDER
Simplest VIEWINDER, much like what you had on your Brownie box camera years ago, this little scope has several tiny lenses of its own (separate from the TV camera's lens) aimed in the same direction as the TV camera's lens. Looking through it would show you approximately the same view that the camera is seeing.

OPTICAL THROUGH-THE-LENS (or TTL) VIEWFINDER
Much like that found on your 35 mm single lens reflex camera, this scope uses a prism to split off some of the camera lens' image for you to see.

ELECTRONIC VIEWFINDER
This is a little TV screen attached to your camera which samples some of the camera's picture and displays it.

LENS IRIS
Controls how much light gets through the lens to the camera. For bright scenes, you reduce the light allowed into the camera by rotating the IRIS ring to a *high* f-STOP or "f-number" (the numbers are etched on the lens). For dimly lit scenes, you "open" the lens to the *low* f-numbers to admit more light.

ZOOM LENS
A lens that can "blow up" an image, making it look closer or farther away.

FIXED FOCAL LENGTH, STANDARD LENS
A lens that doesn't ZOOM, and gives a "normal" picture, not a close-up, not a wide view.

ELECTRIC FOCUS OR POWER FOCUS LENS
A lens which you focus by pushing a button that activates an electric motor which focuses the lens.

AUTO-FOCUS LENS
A lens connected to a tiny computer which senses the distance between the camera and the subject and focuses the lens for you.

Operating a TV Camera—
The Basics

There's not much to using a simple TV camera other than connecting it up, plugging it in, turning it on, letting it warm up, uncovering the lens, aiming the camera, adjusting the lens IRIS, then focusing the lens while looking through the VIEWFINDER.

CONNECTING IT UP

Figure 4-4 showed the most common camera/VCR connections. Color and black-and-white cameras, with or without VIEWFINDERS, all hook up the same way. Unless you're using an industrial or indoor camera that needs separate power, video, and audio wires, your camera's only connection may be its umbilical cord with its multipin connector at the end. If you're using a portable camera with a console (nonportable) VCR, you may need a separate AC POWER ADAPTER or CAMERA CONTROL UNIT, which will power the camera, accept its multipin plug, and connect up to your VCR's VIDEO IN, AUDIO IN, and REMOTE sockets.

TURNING IT ON

If yours is an indoor camera that plugs into the wall outlet, it will have its own POWER ON/OFF switch. If the camera is portable and connects to a console VCR via an AC POWER ADAPTER, then the ADAPTER'S POWER switch turns the camera on. If your camera hooks directly to the VCR, then the VCR turns it on, either when you switch the VCR's power ON, or when you press PLAY or RECORD.

AIMING THE CAMERA

Once you uncap the lens, the camera can "see." It becomes immediately vulnerable to damage by bright lights. Remain ever cautious about where you aim it.

Get the kind of "shot" you want, then pull the camera's trigger (if it's a portable) to start the VCR recording. When you lose your good shot or you're done shooting, pull the trigger again to PAUSE the recording. Try to keep your picture centered, level, in-focus, and steady. The next chapter will detail how to get professional-looking shots and camera angles.

ADJUSTING THE LENS IRIS

This little ring on the lens controls how much light the lens lets through to the camera's VIDICON tube. In dim light, turn the ring to the *low* f-numbers indicated on the ring. In bright light, rotate it to the *high* f-numbers. Much of the time you can shoot at f2.8. If your VIEWFINDER has an indicator warning you of insufficient light, or your ELECTRONIC VIEWFINDER displays a faded picture, "open" your lens to a *lower* f-number. If the indicator reads "too much" light, or your ELECTRONIC VIEWFINDER displays a too-contrasty picture, turn to higher f-numbers. More on this later.

FOCUSING THE LENS

There are scillions of lenses, each with its own personality. These differences will be covered in detail later, but here are a few basics to get you started. There are FIXED FOCAL LENGTH or STANDARD lenses, ZOOM lenses, ELECTRIC FOCUSING lenses and AUTO-FOCUSING lenses.

With the STANDARD lens in Figure 5-2, you focus by reaching around in front of the camera and turning the focus ring on the lens until your subject looks sharp in your VIEWFINDER. If you haven't a VIEWFINDER, use the distance numbers etched on the lens to guide your focusing.

With the ZOOM lens in Figure 5-3, you can focus simply by turning the focus ring, but if you focus this way your picture may go out-of-focus as you zoom in or out.

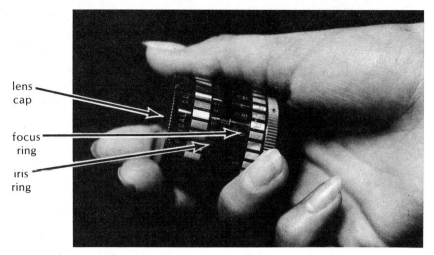

Figure 5–2
STANDARD or FIXED FOCAL LENGTH lens

Figure 5–3
ZOOM lens

To make your lens stay in focus as you zoom, do this (perhaps while the VCR is PAUSED):

1. Zoom in all the way to the closest shot possible of your subject.
2. Now focus the lens.
3. Zoom back out to the shot you want.

From here on, unless the subject moves closer or farther from you, your picture will stay sharp, even while you zoom in and out.

With ELECTRIC or POWER FOCUS lenses the technique is exactly the same, except instead of reaching around to the front of the camera to turn the focus ring, you press a button. You still will have to zoom in (usually by pressing a button) all the way before

Akai VC-XI Automatic focus portable color TV camera (Courtesy of Akai America)

focusing if you want your picture to stay sharp throughout your zoom range.

With AUTO-FOCUS lenses the camera focuses for you. Simply press the FOCUS button any time while shooting and the lens does its thing.

Features and Controls on the TV Camera

Naturally, you'd expect different *kinds* of cameras to sport different kinds of features. So before we dive headlong into a sea of gadgets, let's skim the surface of this topic a bit.

KINDS OF TV CAMERAS

In broadcast and industrial TV, there are hundreds of cameras, each designed for a specific need. There are high-quality broadcast studio cameras which cost over $40,000 each. There are LOW LIGHT LEVEL cameras designed to work in starlight or moonlight. There are medical TV cameras with tiny lenses built onto long snouts, designed for probing the tubes and cavities of the human body. There are cameras which can make blacks white and whites black. Years ago, these cameras permitted someone to shoot a roll of movie film and project the *negatives* in front of the camera and get a *positive*, proper image (saving a step in processing the film). There are, of course, color and black-and-white cameras. There are self-contained, automatically controlled cameras which require nothing but aiming, and there are industrial TV cameras which allow you to manipulate and adjust a dozen different aspects of the TV picture, such as left-to-right reverse, upside-down, negative, brightness, contrast, sharpness, and synchronization with other cameras.

The home video user is likely to come across just a few main variations on the TV camera (but many different models). There are black-and-white versus color cameras. There's the industrial or indoor-use camera versus the portable. And there are different kinds of VIEWFINDERS that come with TV cameras. Let's start with the most common home-type TV camera, the portable color camera.

PORTABLE COLOR CAMERAS

Color portable TV cameras range in price from about $600 to $1500 (black-and-whites range from about $150 to $800). They connect directly to the portable VCR through a

RCA CC007 color camera with pivotable ELECTRONIC VIEWFINDER and AUTO FADE feature. (Courtesy of RCA Consumer Electronics Div.)

Sony HVC-2200 color camera with pivotable ELECTRONIC VIEWFINDER and POWER ZOOM lens. (Courtesy of Sony Corp. of America)

multipin umbilical cable. This cable provides power for the camera, and carries video, audio, and other signals between the camera and the VCR. Most have built-in microphones and come with some kind of lens.

Automatic gain control (AGC) Just as you can adjust the brightness and contrast of your TV's picture, a camera can adjust the brightness and contrast of its picture. This feature, sometimes abbreviated AGC or called AUTO TARGET is common to all home-type color portable cameras and many other kinds too. In bright scenes, it will darken the image; in dim scenes it will add contrast to the image. There is a limited range over which the camera's AGC can make adjustments for scene brightness. Beyond these limits one must "open" or "close" the lens IRIS to increase or reduce the light allowed into the camera.

Automatic iris The lens IRIS controls how much light the camera "sees." On some cameras you manually adjust this lens control to adapt to bright or dim lighting conditions. Cameras with AUTOMATIC IRIS will perform this adjustment for you, sensing the amount of light admitted through the lens and opening or closing the lens accordingly.

Besides the convenience of not having to adjust the IRIS as you move from scene to scene, there are other benefits to AUTOMATIC IRIS. This feature protects the camera on occasions when you aim the camera at something too bright. Instead of "burning" streaks into your precious VIDICON tube, the AUTO IRIS will cut down the light level, protecting the tube.

There are times when AUTO IRIS can get fooled. For such cases most cameras will allow you manually to override this control. A typical situation is when a dark subject is standing against a light background. The AUTO IRIS may adjust the lens to give an excellent rendition of the background while the subject comes out dark and murky. By manually "opening" the lens further, you can get your subject to look good (the important part) while overexposing and sacrificing the unimportant background.

Color temperature COLOR TEMPERATURE describes the warmth (redness) or chill (blue-ness) of a scene. For example, have you ever noticed how cold and sterile offices lit with fluorescent lights sometimes look? Or have you looked into a darkened room illuminated only by the light from a black-and-white TV set and noticed how stark and bluish everything seems? On the other hand, have you noticed the warmth in a home lit by incandescent lamps, or the warmth of a supper lit by candlelight, or the richness of the whole outdoors during an August sunset? These differences are caused by COLOR TEMPERATURE in the light.

Under different lighting conditions, the color of things changes drastically, even though you may not be aware of it with your naked eye. The eye of the camera, however, sees these differences and makes them even more pronounced. A face that looked red and rosy when lit by a sunset will look deathly pale when photographed on a foggy day. Somehow, the color camera must be adjusted to compensate for these differences in lighting so that colors will look familiar and proper. This is called COLOR BALANCE.

Some color cameras have a built-in set of colored glass lenses that counteract the "coldness" of the light and bring it into proper balance. The COLOR TEMPERATURE control on most cameras is a four-position thumb-wheel. Position 1 is for shooting scenes under studio lamps or outside during a sunrise or sunset—all "warm" light conditions. Position 2 on the wheel is for fluorescent lamp lighting. Position 3 is for bright or hazy sunshine. Position 4, assuming you are shooting outdoors, is for the "cold" light of cloudy or rainy days. Adjusting the wheel balances the color for the lighting conditions you face.

Some less expensive cameras will not have a wheel with built-in lenses, but will have separate colored filters which snap onto your lens.

White balance Every time you use your camera or change lighting conditions (like moving from indoors to outdoors, or even from scene to scene sometimes) you have to "teach" your camera what color *white* is.

You remember how a color picture is the composite of three pictures, one green-and-black, one red-and-black, and one blue-and-black. A certain mix of these three primary colors is needed to make pure white. If the mix is off, you get tinted white. Sometimes things that are supposed to be white (a white piece of paper on a desk) turn out not to be white at all (the desk is in an orange room, casting orange light on the paper). Still, you'd like the paper to look white on camera, so by adjusting WHITE BALANCE you adjust that mix of primary colors to *make* it white.

WHITE BALANCE or WHITE LEVEL SET or WHITE SET is often adjusted thusly:

1. In the area you plan to shoot (and in its light), first place a white card in front of the color camera, close enough to fill the camera's VIEWFINDER screen.

2. Adjust IRIS and COLOR TEMPERATURE controls to their proper settings.

3. Find the WHITE BALANCE meter, either on the camera or in the VIEWFINDER.

4. With the card still filling the screen, adjust the WHITE BALANCE control (or controls—there may be two, one perhaps labeled RED, the other BLUE) to move the WHITE BALANCE meter needle to its proper position. Some manufacturers want the meter needle centered, others as low as possible, others high—check your instructions.

Sometimes the COLOR TEMPERATURE and WHITE BALANCE controls are coupled into one control called by either name or something else like COLOR BALANCE, COLOR TONE, or TINT.

If you used your camera instructions to wrap your last trout catch, and don't know which way to set your needle, then try this:

1. Hook up a color TV to your camera/VCR system.

2. Adjust your TV's color so it looks nice while viewing a strong TV station with lots of flesh tones.

3. Switch the VCR to feed the camera signal to the TV and now observe the color of the white card on the TV. If it's not white, adjust your WHITE BALANCE controls until it *is* white.

Automatic white balance AUTO WHITE BALANCE adjusts the controls for you at the push of a button. Still, you have to hold the white card in front of the camera while pushing the button. Some cameras come with a milky white lens cap which serves normally as a lens cap, and during WHITE BALANCE adjustments substitutes for the white card. Simply aim the camera at the area you plan to shoot and, with the white cap on the lens, push the AUTO WHITE BALANCE button and hold it a few seconds. Done.

Fade If you turn your lens IRIS all the way to the high f-numbers, you'll decrease the light allowed into the camera, making the picture FADE to gray or black. On some lenses the IRIS can *close* (position marked C), admitting no light at all, essentially FADING OUT your picture entirely.

Some cameras have this feature built into them so that, by pushing a button, you can FADE OUT from your picture at the end of a scene and FADE IN to a new picture at the beginning of the next. By PAUSING your VCR after a FADE OUT and setting up your next scene, then UNPAUSING and FADING IN on the next scene, you create a smooth, professional-looking transition. It's less abrupt than the instantaneous CUT from scene to scene that you get when just PAUSING.

VIEWFINDERS

Not all TV cameras have VIEWFINDERS. Many security and industrial cameras do not. Most portables do. Some color portables even have detachable VIEWFINDERS, handy for tight spaces. Some systems allow you to buy your camera with a cheapie OPTICAL VIEWFINDER and later trade up to an ELECTRONIC VIEWFINDER for your camera. Detachable ELECTRONIC VIEWFINDERS are handy when you need to conserve battery power while shooting—simply disconnect the ELECTRONIC and substitute the OPTICAL to save-a-watt.

VIEWFINDER eyepieces, like binoculars, often have to be adjusted to your eye (otherwise, you may think your camera's picture is blurry when it's really your VIEW-

Magnavox B/W Video Sound Camera, Model 8211 with OPTICAL-SIGHT VIEWFINDER (Courtesy of N.A.P. Consumer Electronics Corp.)

Panasonic TV camera model PK-530 with THROUGH-THE-LENS OPTICAL VIEWFINDER (Courtesy of Panasonic)

picture is fuzzy or off-center. You can tell exactly what's in a picture as you zoom.

TTL VIEWFINDERS often have a few lights or meters built into them to indicate whether you have enough light, whether your battery is weak, and whether your VCR is paused or recording. Some TTL VIEWFINDERS have lights to tell you whether your WHITE BALANCE is properly adjusted.

TTL VIEWFINDERS have some disadvantages too. Since the light coming through the lens is divided between the camera and the finder, the minimum amount of light needed to make a picture rises about 10–20%. This is rarely a big problem, but if you do a lot of indoor shooting where light tends to be dim, you may have to fuss with extra lights more often with TTL-finder cameras.

Another disadvantage to the TTL is the fact that with rare exceptions (like the JVC BX-68U) the TTL VIEWFINDER is *built into* the camera. You can't later upgrade to an ELECTRONIC VIEWFINDER like you usually can with the external OPTICAL SIGHT VIEWFINDERS. So you're stuck with your TTL-finder camera once you've bought it.

For a film camera, a through-the-lens system is the ultimate, but for video it's only "next best." Since cameras "see" things a little differently than we do, the OPTICAL finder doesn't show us how our picture *really* looks. This can only be done by viewing the picture on a TV screen. A tiny TV monitor that attaches to your camera is an ELECTRONIC VIEWFINDER.

Electronic viewfinder These 1½" TV monitors display the picture exactly the way it is being sent to the tape. Focus, framing, IRIS, and zoom all manifest themselves in the picture you see.

Color is the one thing you don't see on your ELECTRONIC VIEWFINDER. All ELECTRONIC VIEWFINDERS are black-and-white, even on color TV cameras. That's not a mistake. Black-and-white VIEWFINDERS give a sharper picture than color (of prime importance when focusing). They're smaller, lighter, much cheaper, and adequately perform their basic mission—to display what you're shooting.

Like many TTL VIEWFINDERS, ELECTRONIC VIEWFINDERS have indicators for light level, pause, battery, and other functions of the camera and VCR.

One great advantage of the ELECTRONIC VIEWFINDER is that it can display the image *played back* from your VCR. So after recording a sequence, you can rewind the tape and play it through your VIEWFINDER to see how you're doing.

Some manufacturers build the VIEWFINDER into the camera, which streamlines the package and makes it look much like a super-8 movie camera. Others, usually the more expensive ones, have separate VIEWFINDERS which detach for transporting and power-saving situations, hitch up to either side of the camera (for left-eyed or right-eyed people), and can be adjusted in many directions, freeing the camera operator to hold the camera above or below him. Some finders, such as the Sony 2200, can be removed from the camera and attached to an extension cable for remote viewing—handy if your camera's up in a tree or in the

FINDER's picture). Aim the camera at a distant object, focus the lens at infinity (∞) and zoom out all the way. That will make a picture that *should* be sharp. Now adjust your eyepiece to make the VIEWFINDER picture sharp (and comfortable) for your eye. Done.

Optical-sight viewfinder This inexpensive little range finder clips onto your camera and, like a weak telescope, shows you where your camera's aimed. It's advantage is its price. It's disadvantages are:

1. If not perfectly aligned (most aren't) you don't see exactly the scene the camera sees; you may be a little off.
2. You don't automatically know if you're focused because an unfocused camera lens won't display a fuzzy image on this VIEWFINDER. This may not be a problem if you have a camera with an inexpensive FIXED-FOCUS (unadjustable focus) lens.
3. You don't know if your IRIS is properly adjusted. The image may look good to you, but not to the camera. Again, this doesn't concern you if you have an inexpensive lens lacking an IRIS control.
4. You can't tell where you're zoomed to. Usually, OPTICAL SIGHT VIEWFINDERS have lines etched on them to show the zoomed-in and zoomed-out angle of view, but you end up guessing a lot. Again, cheapie lenses usually don't zoom anyway, thus posing no problem with these VIEWFINDERS.
5. Since the OPTICAL SIGHT VIEWFINDER is a ways off from the camera lens, it's not looking "straight at" subjects which are close to the lens. This sighting difficulty is called a PARALLAX error and is diagrammed in Figure 5–4. Again, camera systems with cheapie OPTICAL SIGHT VIEWFINDERS are likely also to have cheapie lenses that don't focus too close anyway, so this problem may never occur. (Notice the recurrent theme that the cheapie VIEWFINDER very nicely matches with the cheapie lens?)

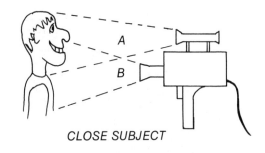

CLOSE SUBJECT

A OPTICAL SIGHT VIEWFINDER'S VIEW

B CAMERA'S VIEW

Figure 5–4
PARALLAX

6. If you forget to uncap your lens before shooting, your OPTICAL SIGHT VIEWFINDER won't remind you. It's image is independent of the camera lens.
7. You can't check your recordings by playing them back through an OPTICAL VIEWFINDER.

Through-the-lens viewfinder This VIEWFINDER, much like your 35mm SLR (Single Lens Reflex) slide camera displays the image which actually comes *through the lens* (ergo its catchy name). You see what the VIDICON tube sees. You know immediately if your

Sharp model XC-35U color TV camera with adjustable ELECTRONIC VIEWFINDER (Courtesy of Sharp Electronics Co.)

middle of a cattle drive or some other hazardous location.

Usually the heavier, higher-quality cameras come with ELECTRONIC VIEWFINDERS. These cameras, too heavy to be steadily held by the trigger handles, sit on the shoulder much like the professional cameras do. The VIEWFINDER is attached near the front of the camera, improving balance and steadiness even further.

MICROPHONES

You'll read more about mikes in Chapter 7, but here are a few basics to get you started.

Nearly all portable TV cameras have a built-in ELECTRET microphone to pick up conversation and other sounds. Their sound quality is not bad, but they do have one fault. Since your camera is likely to be 6 feet or more away from your subject, that means the built-in mike is also 6 feet away. This results in echoey speech and a distracting amount of background noise, such as doors slamming, dogs barking, traffic, wind, even the camera operator breathing.

You could solve all these problems by recording in a soundproofed tomb using a corpse for a camera operator (no breathing sounds). Or, you could use a separate microphone, such as a lavalier, and hang it around your talent's neck. The plug goes into the MIC input of the VCR and supersedes the camera's built-in mike. This way, the subject speaks directly into the mike (a foot away) and the sound is clearer. If the talent moves around too much for an attached mike, you could use a SHOTGUN MIKE.

A SHOTGUN MIKE, named so because of its long, shotgun-like barrel, "listens" in one direction only. A small one, attached to your camera, aims wherever the camera aims, and thus picks up the sound of whatever the camera sees (unless you're aiming at the moon). Unlike the little ELECTRET mike which "listens" in all directions, the SHOTGUN mike picks up only what it's aimed at, greatly reducing the loudness of distracting background noises. Unfortunately, SHOTGUN mikes are fairly expensive.

Features and Controls on the TV Camera   149

TV camera with mike boom (Courtesy of Curtis Mathes)

One other problem with the built-in ELECTRET mike is that it picks up the clicking and whirring of the camera operator pulling the trigger, focusing (especially with motorized ELECTRIC FOCUSING cameras), and zooming. Some systems get around this problem by putting the mike out on a little telescoping boom, getting it farther away from these noises.

INDUSTRIAL AND INDOOR
BLACK-AND-WHITE CAMERAS

If you're buying a TV camera for surveillance (viewing your front stoop or the baby's room) or for a little casual video play around the house, you may not need an elaborate, expensive portable color camera with VIEWFINDER, built-in mike, and other bells and whistles. A decent portable color camera is likely to cost $1000. A simple black-and-white industrial camera costs as little as $300 and gives *twice* as sharp a picture as its color counterpart. *Twice* as sharp! *And* it will work in much dimmer light. *And* it's generally smaller, lighter, and more rugged than a color camera.

The disadvantages of a black-and-white industrial camera are:

1. No color. (Ugh!)
2. They're small but, because they use house current, are not as portable (imagine running a seven-mile extension cord down to the beach).
3. Today's research dollars are going into color cameras. Black-and-white technology is stagnating.

There are a lot of used black-and-white industrial cameras around in schools, businesses, and garage sales, often in excellent condition. Figure 5-5 shows a popular model used in schools. They're comparatively easy to fix, so spending $50 on an old

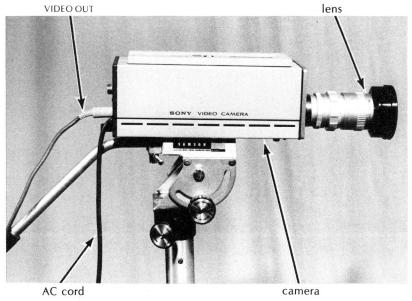

Figure 5–5
Industrial black-and-white TV camera

one and investing $45 in a new VIDICON tube for it may still be a good deal. Industrial cameras have standard C-MOUNT couplings, permitting you to screw on an assortment of lenses.

Although most industrial black-and-white cameras have the same automatic controls as the portables and colors, the older cameras and the studio models are usually bristling with manual controls for just about everything. Anyone who remembers driving a Model T and having to set manual spark advance, choke, throttle, and shift gears, will find these old cameras easy to learn and operate. So here's a list of some of the controls you're likely to see on an industrial camera and what they do, just in case you end up with one without directions (often the case with used gear).

An interesting note (before you skip over this section completely): Nearly all cameras have the following controls built into them somewhere, perhaps adjusted internally, perhaps totally automatic. If your portable or color camera's preset or automatic controls get out of whack, you may recognize some of the symptoms here. If you're handy at fixing TVs and your warranty is up, you may want to try adjusting your camera before taking it to the shop. Even if you take the camera directly to the repairman, you may be better able to describe the problem by being familiar with these camera controls.

Inputs and outputs Figure 5-6 shows the rear of the very common Sony AVC-3200 camera. It is similar to many black-and-white cameras. One switch is labeled VIDEO/RF OUT. Normally this switch is set to VIDEO and the video comes out the socket labeled VIDEO/RF. In special cases, however, you may wish to send the signal from the camera into a simple TV receiver through the receiver's antenna terminals. To do this, merely flip the switch to RF, and now RF (not video) will come out the VIDEO/RF socket. Connect a 75-OHM coax cable from this socket directly to the TV set's 75-OHM antenna input, or through a matching transformer (as described in Chapter 2) to the set's regular antenna terminals.

Another switch is labeled SYNC and has positions marked IN (for INternal) and EXT (for EXTernal). In the IN position, the camera

Figure 5–6
Industrial black-and-white camera with VIEWFINDER attached

generates its own sync (INSIDE itself), mixes it with the video, and sends both to the VIDEO OUT and through a coax cable to the TV monitor or VCR. When the SYNC switch is set at EXT, the camera does not generate its own sync signal—something else, usually an external SYNC GENERATOR, must make the signal for it. Since it is inconvenient to run a separate set of wires from a SYNC GENERATOR, most cameras are equipped with a multipin socket labeled EXT SYNC, for the cable that carries the sync *to* camera and the video *from* the camera. Unless you're running several cameras at once in a studio, multi-pin sockets and SYNC GENERATORS won't concern you. Switch the SYNC switch to IN and forget it. Run a coax cable from the camera's VIDEO OUT to your VCR's VIDEO IN.

On/Standby/Off On the Sony AVC-3200 camera, you'll notice a switch that says ON/STANDBY/OFF. This switch turns on the power. When the camera is getting power, the pilot light over the switch lights up. In the STANDBY mode, the camera's circuits are warmed up and are ready to give you a picture the instant the switch is flipped to ON. When the camera is in STANDBY, a shutter automatically protects the VIDICON tube from light (and from accidental burn-in).

On the side of the VIEWFINDER, you'll notice the usual TV monitor controls. They affect only the image on the VIEWFINDER, not the image being recorded.

Let's take a look at the back of a semipro camera and twiddle some more knobs. Figure 5-7 shows what you're likely to find on one of these old dinosaurs.

Mechanical focus The MECHANICAL FOCUS knob moves the VIDICON tube closer or farther from the lens and results in a change of focusing for the camera. Once set, this mechanical control is seldom readjusted unless the camera has been shaken, jarred, transported, or taken apart.

When we were discussing the lens earlier, you were told that the proper way to focus was to: (1) zoom in, (2) focus the lens, and (3) zoom out to the shot you want. Now that you have a camera with MECHANICAL FOCUS, there is another step to the focusing procedure. This step can be dispensed with during recording if the camera is not bumped or jostled, but it's a good habit to include it in your preshooting preparations.

To focus properly during setup:

1. Zoom in all the way.
2. Focus the lens.

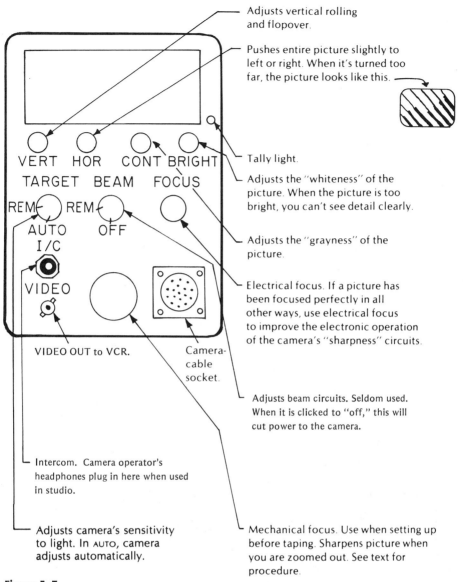

Figure 5–7
Rear of a semi-professional black-and-white camera, showing controls

3. Zoom out all the way.
4. Adjust MECHANICAL FOCUS for the sharpest picture.
5. Repeat steps 1, 2, 3, and 4 because they all interact.
6. Finally, zoom to the shot you want.

When used correctly, MECHANICAL FOCUS insures a super-sharp picture.

Target TARGET adjusts the camera's sensitivity to light. Industrial cameras with VIDICON tubes usually have an external TARGET control. Cameras nowadays require infrequent TARGET adjustment, and that is done internally by a technician.

In all but rare cases, the TARGET would be left in an automatic mode. This allows the camera to make its own adjustments internally. If there is no automatic mode available, the TARGET is adjusted to give a picture with good contrast and smooth, white whites. (TARGET adjustment affects only the white values in the picture.) Adjust the TARGET to the lowest position that will give the desired results—running it too high wears out the VIDICON tube.

Beam BEAM works in conjunction with TARGET to adjust the camera's sensitivity to light. This, too, is operated at the lowest level practical to yield a good picture. Once set, BEAM seldom needs adjustment unless TARGET is changed.

If you are adjusting a camera for the first time, or if the BEAM and TARGET have been messed up, do the following:

1. Light a scene for the camera to "see." With the power off, uncap and manually focus the lens to where you guess it should be focused. Then zoom the lens all the way out.

2. Turn BEAM and TARGET down all the way.

3. Turn on the camera and warm it up.

4. Turn TARGET up about halfway.

5. Slowly turn up BEAM until a picture appears on the VIEWFINDER, monitor, or whatever.

6. Adjust BEAM so that the whites are nice and white throughout, but don't stray *too far* beyond where the whites start to look good. The object is to get a nice picture with the least BEAM possible.

7. If the picture is too bright or contrasty, turn the TARGET down. If the picture is too dull and gray, turn TARGET up somewhat. Once you touch TARGET, BEAM gets messed up, so go back and redo step 6. Then redo step 7 as well.

Adjusting BEAM and TARGET is tricky. Their maladjustment could damage the VIDICON tube. Therefore, these corrections are frequently left to technicians. Figure 5–8 shows some BEAM and TARGET misadjustments. Even minor BEAM/TARGET adjustments make subtle but important changes in the quality of the TV picture. Examine carefully the bottom cases in Figure 5–8.

Good BEAM and TARGET

TARGET too low (faded)

Figure 5–8
How a camera's BEAM and TARGET adjustments look on a TV monitor

Figure 5-8 (*Cont.*)

TARGET too high. May cause whites to fill in, but problem not always noticeable. Try to keep as low as possible without getting a faded picture

BEAM too low (whites fill in: increased lag)

BEAM too high. May cause defocusing, but problem not always noticeable
Try to keep as low as practical

Minor BEAM/TARGET adjustments
BEAM slightly low for this particular TARGET setting. Note loss of detail and gray scale in the whites

TARGET and BEAM adjusted perfectly

"Sticky" picture when panning. Problem may also be caused by insufficient light while in the AUTO TARGET mode

Flare on white parts of picture as it moves

Figure 5–9
High TARGET or low BEAM

High TARGET and low BEAM cause an effect known as "lag" or "sticking" or "comet-tailing" (see Figure 5-9). When shooting a moving object, a faint ghost will remain for a second where the object *was*. Pictures seem to "stick" on the screen. In bad cases, a white FLARE will follow bright moving objects across the screen, looking much like a trailing white scarf blowing in the wind. This problem may be alleviated by increasing BEAM a little. If lag happens to be your problem and your camera is in the AUTO TARGET mode, the solution is either to increase BEAM a little, or to open the IRIS on the lens, or to put more light on the subject. There is no one solution—it's trial and error. Better quality TV cameras are less prone to exhibit lag or sticking.

Electrical focus This is not the same as ELECTRIC FOCUS or POWER FOCUS, the pushbutton, motorized lens focusing system found on the fancier cameras mentioned earlier. The ELECTRICAL FOCUS control adjusts the camera's "sharpness" circuits. It seldom has to be touched except when BEAM or TARGET have been readjusted. If you know you have focused the lens properly, and if you have also done the MECHANICAL FOCUS routine but your picture is still fuzzy, it's time for an ELECTRICAL FOCUS. Just turn the knob, using trial and error, until you get the sharpest picture possible. Figure 5-10 shows what happens when ELECTRICAL FOCUS is badly maladjusted.

Other controls If you find a knob called PEDESTAL on your antique camera, it's for adjusting the lightness of your camera's picture, like adjusting BRIGHTNESS on your TV set.

You might see a knob labeled GAIN or VIDEO GAIN. This controls the contrast of your

Figure 5–10
Badly maladjusted ELECTRICAL FOCUS

camera's picture, much like the CONTRAST control on your TV set.

A VIDEO REVERSE switch will turn all your black-and-white pictures into white-and-black (negative) pictures. A SWEEP REVERSE switch will make the picture swap its left and right sides, giving a mirror image. A VERTICAL SWEEP REVERSE will flip your picture upside-down.

If you find a TV camera with a knob marked FRAMUS on it, get some bed rest, lay off the funny cigarettes for a while, and seek professional psychiatric help.

HIS MASTER'S LENS

More About TV Camera Lenses

The lens is the camera's window to the world. This window could be a $50 STANDARD or FIXED-FOCAL LENGTH (it doesn't zoom) lens or a $2000 ZOOM lens. The price depends on the complexity and quality of the lens. The principles for using and caring for both kinds of lenses will be the same.

For both fixed-focus and zoom lenses, light enters the lens, becomes concentrated or magnified by the optical elements in the lens, passes through an IRIS which attenuates the light, and is then focused onto the VIDICON tube. All of these factors are quantifiable and are used in describing the lens and its applications.

THE f-STOPS

The f-STOPS measure the ability of a lens to pass light through it. The IRIS ring, a rotatable collar somewhere on the lens (see Figures 5-2, 5-3, and 5-13) has numbers etched on it—typically these are 1.4, 2.0, 2.8, 4, 5.6, 8, 11, 16, and 22—which are called f-STOPS. The f-STOPS represent the SPEED of the lens. SPEED is a photographic term describing how much light is allowed through the lens. The lower the f-number, the "faster" the lens is; that is, the more light it allows through for a bright, contrasty picture. In general, you run a lens "wide open" (set at the lowest f-number) if you're shooting inside with minimal light. Outdoors in bright sunshine you "stop down" to a small lens opening (a high f-number) to allow limited light.

You may have noticed that it runs contrary to sensible logic for the *high* f-STOPS to let in the *least* light while the *low* f-STOPS let in the most light. Others have noticed this anomaly and tried to correct it by establishing t-numbers ("t" standing for "transmission"), which gets bigger as the lens lets in more light. Unfortunately, traditions die slowly, leaving the well-entrenched f-STOPS the likely standard for the foreseeable future.

Why aren't the numbers simply 1, 2, 3 instead of weird decimals? Because the f-STOP numbers are the result of a mathematical formula:

$$\frac{\text{Focal length}}{\text{Diameter of lens}} = \text{f-STOP}$$

Figure 5-11 diagrams this relationship. Common lens sizes make the f-STOP numbers come out weird, plus there's a lot of tradition mixed in here.

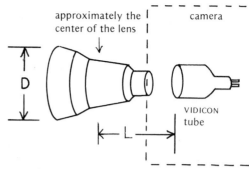

lens focused on
far-away object

D = diameter of lens
L = focal length of lens

$$\text{f-STOP} = \frac{L}{D}$$

If the lens focal length is 25 mm (millimeters) and the lens diameter is 12.5 mm, then the f-STOP would be 25 ÷ 12.5 = 2. "Closing" the IRIS of the lens blocks off part of the lens opening, effectively making D smaller. Reducing D to 6.25 mm gives us an f-STOP number of 25 ÷ 6.25 = 4.

Figure 5–11
f-STOP measurement

An f-STOP of 1.4 allows in twice as much light as f2. In fact, as you "click" off each f-STOP on the IRIS—1.4, 2, 2.8, 4—you double the amount of light to the camera. So f4 admits eight times as much light as f1.4.

In summary: The IRIS is adjusted to the f-number that allows in *enough light* for your camera to "see" to make a good picture but *not too much light,* which will result in too contrasty a picture or poor DEPTH-OF-FIELD. So what's DEPTH-OF-FIELD? Read on.

DEPTH-OF-FIELD

DEPTH OF FIELD is the range of distance over which a picture will remain in focus. Good DEPTH-OF-FIELD occurs when something near you and something far from you can both be sharply in focus at the same time. Poor DEPTH-OF-FIELD is the opposite. Things go badly out of focus when their distance from you is changed.

Generally, one wants to maintain good DEPTH-OF-FIELD so that all aspects of the picture are sharp. There are times, however, when, for artistic reasons, one would prefer to have the foreground of the picture (the part of the picture that is up close) sharp while the background is blurry. Such a condition focuses the viewer's attention on the central attraction in the foreground, while making the fuzzy background unobtrusive. In such cases, poor DEPTH-OF-FIELD is an advantage.

The mechanism for adjusting the DEPTH-OF-FIELD is our old friend, the IRIS. Low f-numbers give poor DEPTH-OF-FIELD, while high f-numbers give excellent DEPTH-OF-FIELD.

Notice how you never get something for nothing. As you improve your DEPTH-OF-FIELD by increasing your f-number, you simultaneously reduce the amount of light permitted through the lens. While f2.8 lets in plenty of light for a brilliant, contrasty picture with poor DEPTH-OF-FIELD, f22 lets in very little light for a gray, dull picture with excellent DEPTH-OF-FIELD. What can one do to get the best of both worlds?

1. Try to get as much light on the subject as possible.

2. Decide where to compromise. Usually, people go for the bright-enough picture

Figure 5–12
IRIS settings with lens focused at 5 feet

f2.8—contrast to spare but poor depth-of-field

f4—a good compromise

f8

f16—good depth-of-field but the picture is too faded

and sacrifice DEPTH-OF-FIELD. See Figure 5–12 for examples of various IRIS settings.

3. One can make the camera and VCR work harder to compensate for the insufficient light associated with the high f-number. On industrial TV cameras you can increase the MANUAL TARGET (and correspondingly readjust the BEAM). On old industrial VTRs you can increase VIDEO LEVEL on the VTR. Cameras and VCRs with automatic controls will make these adjustments for you. All of these adjustments will give a brighter, more contrasty picture; but the picture will become more grainy with the increase in these levels.

A good compromise in the outdoor setting is to provide plenty of light, setting the lens at f4 for moderate DEPTH-OF-FIELD and moderate contrast. Indoor shooting may generally require f1.4 to f2.0. Bright daytime shooting outdoors may permit f8 to be used.

FOCAL LENGTH AND ZOOM LENSES

The FOCAL LENGTH of any lens is measured in inches or millimeters (mm) and describes two attributes of the lens:

1. How far from the outside end of the lens the image gets focused (length L in Figure 5–11). Actually, nobody gives a hoot about this measurement.

2. How wide an angle the lens will cover.

A WIDE-ANGLE FIXED-FOCUS LENS is likely to have a FOCAL LENGTH of 12mm or less. It will display a wide field of view like the "long shots" shown in Figures 6-1 and 6-2. It is useful for surveillance of large areas, or for shooting in cramped quarters where you can't easily back up far enough to "get it all in."

FIXED FOCAL LENGTH GENERAL PURPOSE lenses for TV cameras typically have a FOCAL LENGTH of about 25mm and will display a medium field of view like the "medium shot" shown in Figure 6-1.

A TELEPHOTO lens will have a FOCAL LENGTH of 50mm or more and will display a narrow field of view like the "close-up" shown in Figure 6-1. The TELEPHOTO lens gives close-ups of objects far from the camera. *The greater the FOCAL LENGTH, the greater the magnification of the picture, the narrower the field of view, and the flatter the scene looks.* You are probably familiar with telephoto shots of baseball games where the players in the outfield look like they are standing on top of the pitcher, who himself looks inches away from the batter.

A zoom lens has a variable FOCAL LENGTH (as opposed to a FIXED FOCAL LENGTH) and may range from wide angle to telephoto. One example of how a zoom lens's range of FOCAL LENGTHS can be expressed is: 12–72mm, (the lens can give a wide angle shot like a 12 mm lens and can be zoomed in to give a shot like a 72 mm lens). This lens has a ZOOM RATIO of 6:1 (six to one), (you can ZOOM it in to six times its lowest focal length and make something look six times closer). This ratio can also be expressed as 6X.

Such lenses cost around $300-plus and are typically found on the better equipment. A lens with a zoom ratio of 3:1 can be had for $200-plus and is common on inexpensive portable TV cameras. Generally, the greater the range of the zoom, the higher the cost will zoom.

FOCUS

You have learned that focusing is done by rotating part of the lens. This process always involved looking through the VIEWFINDER (or a TV monitor), or a TTL VIEWFINDER to see your results. If you have no VIEWFINDER or are forced to guess when things are properly focused, do this:

Observe the focus ring (Figures 5-2, 5-3, and 5-13). On it are etched numbers representing the distance in feet or in meters at which an object will be in-focus. Estimate the distance to your subject and turn this ring to the appropriate number and you will be roughly focused.

You have already learned the relationship between DEPTH-OF-FIELD and IRIS settings, and you appreciate the fact that it is much harder to get something into focus if the DEPTH-OF-FIELD is very narrow. But there are times when this narrow DEPTH-OF-FIELD can *help* you focus.

Take the example in which you have good DEPTH-OF-FIELD and you are trying to focus. You turn the focus ring and the blurry picture becomes: less blurry, less blurry, pretty good, fairly sharp, sharp, maybe sharper, maybe not sharper, still fairly sharp, a little blurry—so you start turning back in the other direction. There is a range where the picture doesn't change much while you turn the focus ring. Which position is right?

Now take the example of poor DEPTH-OF-FIELD. You turn the focus ring and the blurry picture becomes: blurry, blurry, less blurry, good, perfect, good, blurry again—so you start tuning back. Here the picture zaps into focus and out again. There is no guesswork as to where the right focus position is, there's just one narrow range where the picture is good.

The moral of the story is: *For accurate focusing, open the lens to the lowest f-number, focus, and then return to the higher f-number.* Team this method up with the focusing methods described for zoom lenses and you have a *100% super-accurate, punctilious, micro-precise, but takes-a-while-to-do regimen for focusing* which goes like this:

1. Zoom in all the way on the subject.

2. Open the IRIS all the way.
3. Focus the lens.
4. Zoom out to the shot you want.
5. Reset the IRIS to the appropriate f-number.

Obviously, the above list of steps is only useful when you have plenty of time and a subject which is either dead or tied down.

What do you do when you have a subject who moves around a lot? How do you stay in-focus? Here are some possibilities:

1. If you are a good focuser, just stay alert and adjust for every movement. Most of us aren't good focusers, however.

2. Flood the subject with light so you can use high f-numbers for wide DEPTH-OF-FIELD.

3. Stay zoomed out. Focusing inaccuracies are most noticeable in close-ups. When you are zoomed out, nearly everything appears sharp.

4. Try to get the subject to move *laterally* to you, not toward or away from you. Since the subject stays roughly the same distance from the camera, you won't have to refocus, just aim.

5. Try to use big subjects so you can zoom out or stay farther away from them. Why are big, zoomed-out subjects easier to focus? In order to fill a TV screen, little objects must be magnified. You do this by zooming in or by moving the camera closer to them. A zoomed-in lens blows up all the little focusing inaccuracies, especially if an object is itself deep or is moving toward or away from the camera. A camera close to an object also exaggerates the focusing problems. When an object 3 feet away from a camera moves 1 foot, you get a very noticeable 33% focusing error. When an object 30 feet away moves 1 foot, the error is a minor 3%. Combining both the zooming and the nearness concepts, we find that zooming in on a postage stamp held in somebody's hand 3 feet away will display formidable focusing problems as the hand moves. But zooming out on a giant poster of a postage stamp held 30 feet away poses no focusing problems, even if the poster moves a few inches.

6. Last and least, try to confine your subjects by seating them, tranquilizing them, encumbering them with microphone cables, or marking a spot on the floor where they must stand.

CLOSE-UP SHOOTING

Normally, if you try to shoot something closer than 4 or 5 feet from your camera,

CLOSE-UP SHOOTING

the picture will be blurry. If you must get closer to your work, the following options are available:

Close-up or MACRO lenses These are lenses made especially for close work. Some are able to focus on things only an inch away. Figure 5-13 shows one.

In their "normal" (non-MACRO) mode, most of these lenses work like any other lens, focusing down to 4 feet or so. To change them to MACRO, you throw a "safety catch" (which prevents the lens from going into this mode accidentally) and then refocus on a very near object. Many MACRO lenses can be refocused easily from the super-close-up back out to normal distances, opening the possibility of "arty" or creative transitional effects (see next chapter).

Unfortunately, MACROS may cost about $100 more than non-MACROS. Also, in the MACRO mode they will not stay in focus during a zoom (normal lenses, if properly focused, will stay focused throughout their zoom range).

Close-up lens attachments Your regular zoom or FIXED FOCAL LENGTH lens can be made to focus on closer objects by the mere addition of a close-up lens attachment as shown in Figure 5-14. Attach these by unscrewing the lens shade and screwing on the close-up lens attachment in its place. The curved surface of the lens should face *away* from the camera. *Do not* screw close-up attachments down tight. They easily seize up and become hard to remove. The lens shade may now be screwed onto the close-up attachment. Leave this slightly loose, too. One close-up lens attachment can be screwed onto another, thus making their combined power greater. *With a close-up lens attachment on your zoom lens, you can zoom the full range without going out of focus,* assuming that you are the right distance from your subject and that you followed proper focusing procedure to start with.

There are a variety of close-up lens attachments on the market. The buyer is usually interested in two things: (1) the magnification of the lens, and (2) its compatibility with one's present lens.

1. Magnification. Close-up lens attachments don't appreciably magnify an image per se. They merely permit you to bring your camera closer to your subject without going out of focus. They essentially make your camera nearsighted.

The power of a close-up lens is measured in DIOPTERS. The bigger the diopter number, the stronger the lens, and the closer your camera can "see" with it on. The weaker ones are +1 and +2; the stronger ones, +5 and higher. Where normally we could shoot from infinity up to 4 feet, a +1 diopter close-up lens lets us shoot ranges from about 3 feet to 1½ feet; +2 gets us from 1½

Figure 5-13
MACRO zoom lens (Courtesy of Vivitar Corp.)

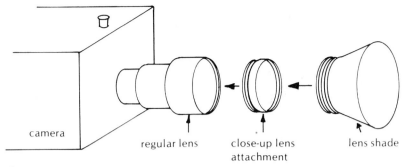

Figure 5–14
Close-up lens attachments

feet to 1 foot, and +3 from 1 foot to ¾ of a foot (9 inches). Up to +3 diopters, your image will stay sharp throughout the zoom range. Above +3, it becomes increasingly necessary to refocus as you zoom in order to keep a sharp picture.

Close-up lens attachments are generally threaded so that after one is screwed onto your regular lens, another close-up lens attachment can be screwed onto the first attachment—sort of a piggyback arrangement. In such cases, the diopters are additive: A +2 lens added to a +3 lens is equivalent to +5 diopters in power. When stacking lenses, the highest diopter lens should be nearest the camera.

Adding lenses tends to decrease sharpness and brightness, especially in the corners of your picture. To combat the blurriness caused by using several lenses at once, flood the subject with light and stop your IRIS down to f16 or more.

2. Compatibility. Most zoom lenses are threaded so that the lens shade can be screwed off and a close-up lens attachment screwed on. Photo-equipment stores sell such attachments for about $25.

Wider-diameter lenses cost more. Little Cosmicar FIXED FOCAL LENGTH lenses (such as the one shown in Figure 5-2) and the like can be fitted with close-up attachments for about $10 per lens.

Photographers have been using close-up attachments for years. If you take your TV camera lens to a local photo-equipment store, you can probably be suited with a close-up lens in minutes for between $10 and $80, depending upon the size and quality of the lens needed.

TELEPHOTO AND WIDE ANGLE CONVERTERS

Another goodie you can attach to the outside of your camera lens is the TELEPHOTO or WIDE ANGLE CONVERTER. Whatever focal length your STANDARD or zoom lens is, these attachments can make it greater or smaller, thus yielding a wider view or a more magnified view of your subject.

If you're a cheapskate inventor like me, scrounge up one of those peephole viewers made for the front door. Tape it to the outside of your camera lens and mask out stray light with cardboard and tape and voilà, you've made a super WIDE ANGLE (FISHEYE) lens.

LENS FILTERS

A FILTER blocks something unwanted and lets through something wanted. Lens FILTERS attach to the outside of your lens and block out unwanted colors, brightness, or glare.

More About TV Camera Lenses 163

"2X TelXtender" TELEPHOTO CONVERTER

Normal view

View with "2X TelXtender"

"Curvatar" WIDE-ANGLE CONVERTER

Normal view

With "Curvatar"

TELEPHOTO and WIDE ANGLE CONVERTERS (Courtesy of Spiratone, Inc.)

Polarizing filters Like your Polaroid sunglasses, this lens attachment reduces glare and reflections from water, snow, highways, windows, and sky. It very slightly darkens the whole scene, doesn't affect color rendition, yet specifically knocks out shine and reflected light. Once attached to your lens, it *must be rotated to the right position* to work properly. You can tell what's right by looking through your VIEWFINDER and watching the highlights disappear and come back as you rotate. POLARIZING FILTERS do remarkable jobs at making lakes and streams appear more transparent (less reflective) and making the sky look richer and less hazy.

Colored filters Amber FILTERS help sharpen aerial and long-distance outdoor photography (as do POLARIZING FILTERS, too).

A red FILTER will make a blue sky very dark. On a black-and-white camera a red FILTER can make day seem night-like.

Neutral density filters These lens attachments are like dark glasses for your camera. They reduce the amount of light through the lens, much like the IRIS does. They don't change color rendition and are very handy if you do a lot of beach or ski slope shooting where the light may be too bright for your camera even with the F-STOP at its highest number.

Filters

COLOR CAMERA FILTERS:

Skylight (1A), or Haze (UV)—Nearly clear lens absorbs ultraviolet light. Removes excessive blues from open shade. Excellent for protecting camera lens from blowing sand, water, fingerprints, etc., and may be left on all the time.

Polarizing—Reduces outdoor reflections in glass, chrome, water. Deepens blue sky and whitens clouds. Reduces bright light. Take off when not needed.

#85—Orange FILTER makes camera think it's seeing indoor incandescent light even though it's outdoors. Use on cameras without indoor/outdoor COLOR TEMPERATURE adjustments.

Graduated—Half the lens is tinted, then gradually changes to clear glass for the other half. A blue tint at the top half of the lens, for instance, would create a blue sky while the lower foreground looked normal.

BLACK-AND-WHITE CAMERA FILTERS:

Yellow (K2)—Yellow. Lightens yellow, darkens blue, increases contrast slightly. Reduces haze, especially on distance shots.

Orange (G)—Goes one step further than above FILTER, darkening skies, sunsets, seascapes.

Red (25A)—Goes the whole route, darkening sky to black, increasing contrast. Good for pseudo-night-scenes.

> *Green (X1)*—Darkens reds, lightens greens, emphasizes flesh tones.
>
> *Neutral Density (ND–X1, X2, X3 . . .)*—Simply reduces the light in a bright scene. The higher the "X" number, the darker the FILTER.
>
> FILTERS must be bought in sizes which fit the diameter and threads of your present lens. If this is a difficulty, or you already have FILTERS from your photo equipment, buy a STEPPING RING which adapts your lens to your FILTER.

ATTACHING
AND CLEANING LENSES

The lens is generally attached to the camera by screwing the base of the lens into a threaded hole in the face of the camera as in Figure 5–15. About a third of the home TV cameras and lenses use what is called a C-MOUNT, which indicates standardization of the hole size, lens-base size, and thread size. Consequently, any C-MOUNT lens should fit any C-MOUNT camera.

C-MOUNTS are usually found on cameras which have the standard 1-inch or ⅔-inch wide VIDICON tubes. Consequently, C-MOUNT lenses have a rear opening 1 inch or ⅔-inch wide to match the VIDICON. Cheaper cameras may use a ½-inch VIDICON, and thus a ½-inch based lens. These are called D-MOUNTS.

Some cameras, usually the ones with power zooms, etc., or often the cheapie models, have the lenses built into the camera. These are not removable, exchangeable, or upgradable. You're stuck with them.

When changing lenses, be very careful not to touch or soil the VIDICON tube just inside the hole in the camera's face. Dirt on the tube will show up in your picture. If the VIDICON tube is dusty (and the tiniest flakes of dust make a big difference), gently brush it off with a soft CAMEL'S-HAIR BRUSH like the

Figure 5–15
Attaching lens to camera

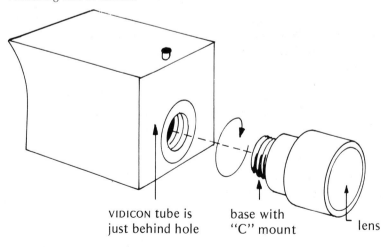

VIDICON tube is just behind hole

base with "C" mount

lens

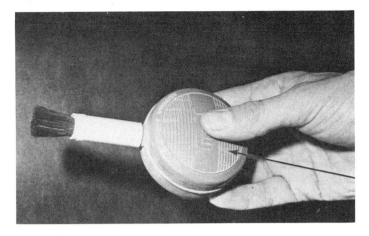

Handle is bellows that blows air when squeezed.

Figure 5–16
Lens-care brush

one shown in Figure 5–16. A stream of dry air may also blow off lint and dust particles.

The same care is appropriate for TV camera lenses. The outside lens, however, always seems to have a way of attracting fingerprints. Since oily prints do not brush off easily, one must resort to gentle wiping with lens tissue. If that doesn't work, try a clean rag dampened with soapy water as a last resort. Take care to rub gently lest you scratch the thin blue or amber coating on the lens. The coating on the lens helps the light go through the glass in the lens.

Lenses are pretty delicate and like to be pampered. Like most TV equipment, they abhor bumps, water, sand, oil, smoke, and food. The lens cap protects them from some of these hazards when the camera is not in use. Wise camera operators, through judicious handling, can provide this protection while the camera is in use. Lenses like to travel in their protective cases (which are usually sold with the lenses) rather than be attached to their cameras where they stick out, knocking into corners and catching in doorways.

How can you tell whether that fleck of dust you see on the monitor screen is from the camera's lens or from its VIDICON tube? Rotate the lens, unscrewing it slightly from the camera mount. If the dirt rotates also, it is on the lens. If the spot of dust remains stationary, it is on the VIDICON tube. If the dirt is on the lens, is it on the outside glass or on the glass at the base? After snugging the lens back into its mount, try rotating the focus ring of the lens. If the dirt remains stationary, it is probably on the base glass. If it rotates, it is on the outside glass.

Now let's go back for a closer look at the C-MOUNT and lens interchangeability. Any C-MOUNT lens will fit any C-MOUNT camera, including movie cameras. There are even adapters which allow C-MOUNT lenses to be interchanged with the "P"-type, "T"-type, or "bayonet" lens mounts on 35mm slide cameras. Just because the lenses fit, however, doesn't mean they all work equally well with all cameras. There's a little complication called "format," which is of interest to you if you swap lenses around between dissimilar cameras. Format refers to the size of the image the lens projects onto the sensitive face of the camera's VIDICON tube. This little image must be the right size to fit the face of the tube. The two common formats are 1 inch and $2/3$ inch. Generally speaking, portable home-type cameras usually use $2/3$-inch VIDICON tubes. They need $2/3$-inch format lenses. Cheapo home TV cameras use $1/2$-inch wide VIDICON tubes using a $1/2$-inch format lens. Then there are the various formats for home 16mm cameras and super-8mm cameras. If one wanted to, one could successfully use the bigger lens on the smaller camera, if the lens mount fit. There would be image to spare. If you tried a ½-inch lens on a ⅔-inch camera, however, the image would be too small and the camera would "see" the dark edges of the lens in the corners of its pictures, a phenomenon called VIGNETTING (see Figure 5-17).

Another interesting thing happens when you use a slide camera lens or a film lens on your TV camera. A "normal" lens is one which fills the screen with a 3-foot wide object 7 feet away. A "normal" lens for a 35mm camera has a FOCAL LENGTH of 50mm. A "normal" lens for a TV camera is 16mm. So if you put a slide camera lens on your TV camera, the lens acts like a moderate TELE-PHOTO lens. "Normal" for a still camera is generally a FOCAL LENGTH 3.6 times greater than "normal" for a TV camera, so by multiplying back and forth you can calculate that if you put your TV camera's 12.5mm-75mm six power zoom on your slide camera, it would act like a 45mm-270mm ultra TELE-PHOTO. These calculations apply for the TV camera with the common ⅔-inch VIDICON tube. Other sizes give other figures. Super-8 and 16mm movie camera lenses also give different fields of view.

Tripods and Dollies

Really smooth camera movements require a tripod. A tripod also alleviates arm fatigue during long shooting sessions like weddings or sports. And during "trick" shots it holds the camera still enough for edits to be unobtrusive (more on this in the editing chapter).

Figure 5-18 shows a tripod with a HEAD and DOLLY. If you decide to build a mini-studio in your basement, this will be a necessity. The HEAD part at the top (not to be confused with the heads of a VCR, which are something else altogether) connects to your camera and allows you to aim it using a long handle. The tripod part usually has an adjustable PEDESTAL raised and lowered by a crank to allow for high or low shots. The tripod legs often telescope so that the camera can be raised an additional 3 feet overhead. On the bottom is the DOLLY, a set

Figure 5–17
VIGNETTING

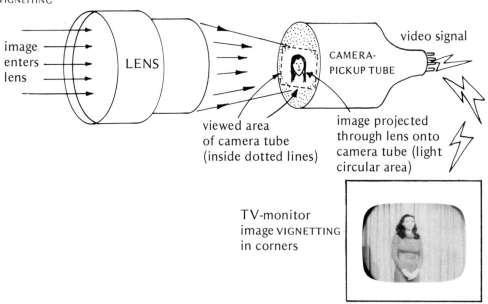

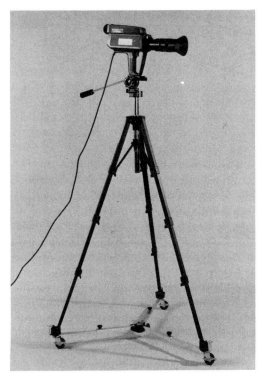

Figure 5–18
Tripod with HEAD and DOLLY (Courtesy of Comprehensive Video Supply Corp.)

ing its movements smooth and precise. A DOLLY probably isn't much use to you unless you do a lot of moving across a smooth floor.

Usually the moving parts on a tripod system can be locked firmly into place. That means the DOLLY wheels can be locked so the camera doesn't roll, the PEDESTAL crank can be locked so gravity doesn't wind it down with a *zwoop*, and the HEAD has controls to stop up-and-down and side-to-side camera movements.

If you leave the camera, *tighten the* HEAD *controls.* The weight of the camera makes it want to tilt. If it squirms loose while unattended, it could tilt down abruptly and smash its lens against the tripod. Also, if someone kicks the camera cable, that could swing the unattended camera around and pan it into a bright light, causing a burn-in on your precious VIDICON tube.

When you *do* decide to move the camera, be sure to *loosen the controls* first. Otherwise the flimsy parts of the HEAD and PEDESTAL will strip and wear out quickly.

Attaching the camera to the HEAD is sometimes tricky. There is a threaded hole of wheels which allows the tripod to glide smoothly over the floor.

There are less expensive, and less flexible units available. The typical portable tripod has three-part telescoping legs which extend to 4 feet or shrink down to about 18 inches when closed. Smaller, flimsier models may lack the PEDESTAL and crank and the HEAD may be wobbly. They're really meant for still photography and don't smooth out camera moves a heck of a lot, but they are better than nothing and are very light and compact. Incidentally, unless you're the type that would mount a diamond in a plastic setting, you may not want to trust your $1200 color camera to a $29 "precariouspod." Professional film and videomakers will spend several hundred dollars on a sturdy tripod and a FLUID HEAD. The HEAD has oil-damped components inside it, mak-

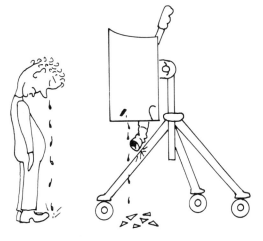

LOCK THE CONTROLS WHEN
LEAVING THE CAMERA
UNATTENDED

Tripods and Dollies

in the base of the camera or the base of the trigger handle. In the HEAD is a captive bolt that shouldn't fall out. Somehow that bolt has to screw into the hole in the camera's base. It's not easy to get the bolt started straight. Starting it crooked will strip the threads, so be patient.

Lighter, more portable, and cheaper than the tripod is the MONOPOD, a one-legged tripod. Although it can't stand up by itself, it takes a lot of weight off your arms and steadies your shots quite well as you hold it upright. It moves with you quickly with a minimum of setup (for when you're shooting wild scenes of that tornado ripping the town apart a block behind you).

And then there's the CENTIPOD, a furry little worm with 100 legs. You'll only see them used with cameras that have a FRAMUS.

MONOPOD (Courtesy of Comprehensive Video Supply Corp.)

Camera Cable Compatibility

Your portable TV camera has a cable and a multipin plug. Your portable VCR has a multipin camera socket. Simply plug the camera in and you're ready to go, right?

Only if the VCR and camera are made for each other by the same manufacturer. Most manufacturers use 10-pin DIN connectors so that the plug from most cameras will fit into most VCRs, regardless of make. But *just because the plug fits doesn't mean the VCR's video wire is connected to the camera's video wire, etc. This wire mixup could actually damage your equipment.* Sony uses an unusual 14-pin K-type connector, totally incompatible with the other manufacturers. This incompatibility is a bad deal for everybody. Buyers of video equipment feel obligated to buy the same brand camera as their VCR so the two will mate, even though they would prefer a different camera or different VCR. Also, if the camera or VCR becomes broken or obsolete, they can't replace just that unit with the latest model or the best brand.

There are several solutions to this dilemma.

1. Wait until the manufacturers standardize their camera/VCR connections. Don't hold your breath. Akai has at least come close with its model VC-X2 color autofocus camera with a four-position "universal switch" which makes the camera compatible with 11 VCR brands.

2. Get a knowledgeable technician or repair shop to rewire your camera's plug so that the right signals go to the right places. This may involve lopping off the plug and wiring on a new one of the proper type. This method is chancy (what if they make a mistake with those 10 or 14 wires?) and most likely voids your camera warranty.

3. Buy a BREAK-OUT BOX or an AC POWER ADAPTER which accepts the camera plug in one end and accepts standard audio, video, etc., cables at the other end. But these are cumbersome and expensive.

4. I saved the best for last. Buy a CAMERA-TO-VCR ADAPTER like that shown in Figure 5-19. Comprehensive Video Supply sells an assortment of them for about $80 each. Their address is:

- Comprehensive Video Supply Corp.
  148 Veterans Drive
  Northvale, NJ 07647

Below is a list of which Comprehensive adapter works with which cameras and VCRs:

for $110 which conforms their cameras to several other manufacturer's VCRs.

Technicolor makes several $60 adapter cables for their model 212 VCR. Their model 1912 adapts their VCR to Sony 14-pin K-type cameras, their 2012 adapts to Hitachi, their 2112 to JVC, and their 2212 adapts to Panasonic and several Akai cameras. Their model 2412 switchable camera adapter cable can link the Technicolor to 29 models of VCRs from 7 makers.

Three other places to shop for adapter cables are: The Cable Works, RMS Electronics, and Marshall Electronics (addresses in Appendix 2). Marshall makes its Camera Mate CM1014, a $129 microelectronic

VCR	CAMERAS						
	Panasonic Magnavox Quasar RCA	JVC G71US GX77U	JVC GX33U GX68U	Hitachi GP-4B,4D,5A VK-750,770	Hitachi GP-5 NEC NCI-2100	Akai	Sony Sanyo Toshiba
Panasonic Magnavox Quasar RCA	—	V-431	V-436	V-441	V-446	V-451	V-461
JVC	V-421	—	—	V-442	V-447	V-452	V-462
Hitachi	V-422	V-432	V-437	—	—	V-453	V-463
Akai	V-423	V-433	V-438	V-443	V-448	—	V-464
Sony Sanyo Toshiba	V-424	V-434	V-439	V-444	V-449	V-454	—

The above adapters will carry audio, video, and remote pause signals from the camera to the VCR. They carry power, video (for playback on ELECTRONIC VIEWFINDER), and sometimes audio from the VCR to the camera.

Before I start to sound like an advertisement for Comprehensive (although their large catalog exemplifies their name), there are others who make CAMERA-TO-VCR ADAPTERS:

Sony makes a CMA-1010A adapter cable

10-to-14-pin camera/VCR interface which cures a lot of little incompatabilities.

Common TV Camera Ailments and Cures

As before, step one is to home in on the problem. If your camera has an ELECTRONIC VIEWFINDER and you can send your camera's signal to the VCR and monitor the results on

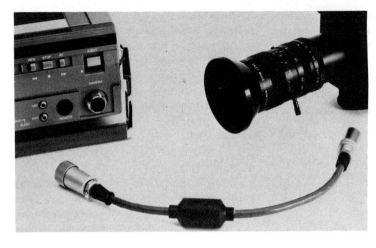

Figure 5–19
CAMERA to VCR ADAPTER
(Courtesy of Comprehensive Video Supply Corp.)

your TV, you're well equipped to track down the elusive camera gremlin. For instance, if your ELECTRONIC VIEWFINDER shows a fine picture and your TV set doesn't, the problem probably isn't in your camera, it's in your VCR or TV or the cables connecting them.

PICTURE PROBLEMS

No picture on ELECTRONIC VIEWFINDER or TV set

1. Is the camera turned on?

2. Is the VCR turned on?

3. If using an AC POWER ADAPTER with your camera, is *it* turned on?

4. Is the lens uncapped? This isn't as silly an oversight as you may think.

5. Is the lens IRIS open enough? Turn it to low f-numbers.

6. Is this camera designed to work with this VCR?

7. If using an industrial camera, are the BEAM and TARGET turned up enough?

Good picture in VIEWFINDER; no picture on the TV set

1. Wiggle each end of the camera cable near the plug. If the picture flashes on and off the TV screen, you have a broken wire—very common on simple camera setups. Substitute another camera cable.

2. Are the VCR's switches set to the right positions?

3. Is this camera designed to work with this VCR?

4. If using an AC POWER ADAPTER, are all its wires connected properly to the VCR?

5. If using an industrial camera, is the GAIN control turned up?

Good picture in the VIEWFINDER; the picture on the TV rolls

Check the VERTICAL HOLD and the HORIZONTAL HOLD on the TV. The VERTICAL HOLD controls roll, but if the HORIZONTAL HOLD is also messed up, you may need to adjust them both for a stable picture.

Pictures roll on both the TV and the VIEWFINDER

1. Check VERTICAL HOLD on both your ELECTRONIC VIEWFINDER and TV set. If that doesn't help, check battery power—is it low?

2. Does this VCR go with this camera?

3. You may have a SYNC problem. If the camera is the industrial type, check for a switch marked SYNC. It should be in the INT position.

4. If using a gas-powered generator, or if using an INVERTER (a device which changes DC, like from your car's battery, to AC) to power your equipment, make sure these sources:
  (a) Can provide at least 110% of the wattage your total video equipment needs.
  (b) Control the frequency of their alternating current to 60 cycles per second plus-or-minus one-half cycle.

Picture on VIEWFINDER or monitor screen "draws in" on all sides, yielding a shrunken picture

Your voltage is low.

1. Check your battery power.

2. If using AC power and long extension cables:
  (a) Shorten the extension cords.
  (b) Turn off any unnecessary power-using equipment on the extension cord.
  (c) Turn off any high-wattage lights plugged into the same circuit.

Picture is faded on both the TV and the VIEWFINDER

1. Perhaps there is insufficient light on the subject.

2. Is the lens IRIS closed too much? Turn to the lower f-numbers.

3. Is the camera's FADE button depressed?

4. Is the lens covered with water or salt spray, or did you leave a dark FILTER attached to your lens?

5. On industrial models, is the TARGET turned too low (if the camera is in MANUAL TARGET mode)? Or is the GAIN turned too low?

Picture is good in the VIEWFINDER, but faded on the TV

1. On industrial cameras, the GAIN may be low.

2. The TV's BRIGHTNESS or CONTRAST may be too low.

Camera's picture is fuzzy on both the VIEWFINDER and the TV

1. Focus the lens.

2. Are you using a MACRO lens or a close-up attachment while viewing a distant subject?

3. Is your lens screwed into the camera tightly?

4. On industrial cameras:
  (a) Adjust the MECHANICAL FOCUS.
  (b) Adjust the ELECTRICAL FOCUS.
  (c) The BEAM and TARGET may be set too high. Lower them, adjust them to create a satisfactory picture, and then readjust the ELECTRICAL FOCUS.

Picture lags or "sticks" on both the VIEWFINDER and the TV as shown in Figure 5-9

1. Perhaps there is insufficient light on the subject. Increase the light or open the lens IRIS.

2. On industrial cameras, the MANUAL TARGET may be set too high or the BEAM set too low.

3. Perhaps parts of the picture are *too* bright or shiny.

4. Lag problems are quite common with inexpensive color TV cameras. There may be nothing you can do.

Ghost of same image stays on the TV and VIEWFINDER screens no matter where you point the camera (see Figure 5-20)

You have a BURN-IN from having the camera "see" the same contrasty picture for too long. Mild BURN-INS can be removed by:

1. Turning the camera off and waiting a day or so before you use it again.

2. Aiming the camera at a smooth white (not shiny) object, for example, an out-of-focus close-up on a well-illuminated sheet

Common TV Camera Ailments and Cures

Figure 5-20
BURN-INS

BURN-IN from aiming at the "THE END" sign too long. The image remains, even though the camera is now aiming at a new subject.

Figure 5-21
White parts of picture streak and sometimes turn black or gray

BURN-IN from aiming across a shiny object.

of dull white paper. The white image *must fill* the screen. The burn may go away in about an hour or so.

Marks or streaks on the VIEWFINDER and TV screens which don't move when the camera moves (see again Figure 5-20)

1. This could be a BURN-IN from panning across a very shiny or bright object, such as the sun, a light, or the reflection of these on chrome or glass. The cure is the same as described in #2 above—aim the camera at something smooth and white for an hour or so.

2. Check your lens to see if there is a hair on it. If you cover the lens and the streak re-

mains, it's a BURN-IN. If the streak disappears, your lens is probably dirty.

TV and VIEWFINDER show a picture with black (or sometimes white) streaks flaring from shiny or white parts of the picture (see Figure 5-21)

1. If using an industrial camera, is the GAIN, or TARGET, up too high?

2. Perhaps the object is too shiny for the camera. Soften the lighting or dull the shiny spot. Chapter 8, Lighting, tells you how to do this.

Picture won't stay in focus throughout zoom

1. Have you focused correctly (zoom in, focus, then zoom out)?

2. Do you have a close-up lens attached or is your lens in the MACRO mode? Return the lens to normal.

3. Are you trying to focus on something too close for your particular lens? Read the focus etchings on the lens; the lowest distance marked is the closest distance to which the lens will normally focus.

4. Has the camera been bumped or knocked around? If so, the VIDICON tube may have

174 TV Cameras and Lenses

jarred out of alignment. A technician can readjust this alignment, or, if you're handy and your warranty is up, you can open the camera and see the frame which holds the VIDICON tube. Industrial cameras often have this adjustment accessible on the outside; the knob is called MECHANICAL FOCUS. To adjust it, zoom the lens out all the way, aim at something distant, focus the lens at infinity (∞ on the lens), and then adjust MECHANICAL FOCUS.

OTHER PROBLEMS

When you pull the camera trigger, the VCR fails to start

1. Have you pressed RECORD/PLAY as you should?
2. Check to see that the VCR *itself* isn't PAUSED. Its PAUSE button must be OFF for the camera to work.
3. Is something (an earphone perhaps) plugged into the REMOTE input on the VCR? Unplug anything from the REMOTE jack.
4. Does this VCR go with this camera?

The camera trigger and its little PAUSE light seem to work opposite from how they're supposed to; everything else works okay

Your camera apparently goes with this VCR except the PAUSE part is wired up backwards. Some cameras have a switch on them which reverses the wiring and corrects this incompatibility. So do some VCR/camera adapter cables.

Weak or no audio is being recorded from the camera; picture is okay

To monitor your audio, connect up a TV to the VCR while the camera and VCR are in RECORD. If you don't hear sound from the TV's speaker, double-check the sound by trying an earphone in the EARPHONE or HEADPHONE OUT socket of the VCR. Once you're sure the sound's deficient, then:

1. Check the MIC IN or AUDIO IN. Nothing should be connected there.
2. Does this camera go with this VCR? Sony cameras will provide adequate video but not enough audio to different brand VCRs unless an adapter is used. The Marshall CM1014 Camera Mate may fix this, as will other VCR/camera adapter cables.
3. Is your VCR's INPUT SELECT in the VCR or TV mode? Switch it to VCR.

You can play a picture into your camera's ELECTRONIC VIEWFINDER, but the sound is missing

1. Make sure the tape has sound recorded on it to start with. Try playing the VCR into a TV.
2. Is the VCR the type that has the earphone in *it* while the camera has no speaker or earphone socket?
3. With earphone-equipped cameras and different-brand VCRs connected via adapter cables, sometimes the adapter cables aren't wired to send sound from the VCR to the camera. Plug your earphone into the VCR directly.

TV Cameras in a Nutshell

Portable TV cameras connect directly to portable VCRs via their multipin umbilical

cables. When the camera is a different brand from the VCR a CAMERA-TO-VCR ADAPTER CABLE may be necessary. When the camera is used with a console VCR, an AC POWER ADAPTER may be necessary. This unit will accept the camera's cable and will have standard audio and video connections for the VCR. Industrial TV cameras generally have standard video outputs which connect directly to the console VCR's VIDEO IN.

The IRIS control on a camera lens adjusts the amount of light admitted to the camera and ultimately the contrast of your TV picture. Use high f-numbers when it's bright and low f-numbers when shooting in dim light. High f-numbers yield a picture with better DEPTH-OF-FIELD, making careful focusing less necessary. Low f-numbers magnify focusing errors.

There is only one right way to focus a zoom lens:

1. Zoom in all the way on your subject.
2. Focus.
3. Zoom back out to the shot you want.

This regimen will ensure that your picture stays in focus throughout the zoom (unless the subject moves toward or away from you).

Portable cameras have mostly automatic controls. Some even have automatic IRIS and AUTO FOCUS.

COLOR TEMPERATURE must be manually adjusted to account for different lighting conditions. WHITE BALANCE adjusts the camera's color circuits so white things come out white.

Good lighting is the key to good pictures. Dim light will produce grainy, indistinct pictures with lots of "lag" or "stickiness." Colors will be bluish and muddy. Open the lens to a slow f-STOP and use extra lights indoors. Standard electric lights will make faces look yellow or jaundiced, but that can be counteracted with a COLOR TEMPERATURE adjustment on the camera.

Black-and-white cameras need less light to make a good picture than color cameras. Their pictures are twice as sharp and the cameras cost half as much.

The least expensive but least reliable VIEWFINDER is the OPTICAL SIGHT VIEWFINDER, much like the rangefinder on an old Brownie film camera. It aims you, but it doesn't tell you exactly what you're getting. Next best is the THROUGH-THE-LENS (TTL) VIEWFINDER, much like your 35mm SLR slide camera. What you see through the finder is what the camera sees. TTL finders usually include lights or meters in them appraising you of battery condition, pause, and light levels. If you want to be really sure of what's being recorded, you need an ELECTRONIC VIEWFINDER. This displays the camera's electrical signal during RECORD and displays your taped picture during PLAY.

Portable cameras have built-in microphones. These mikes are sensitive and work well but tend to pick up spurious sounds around the room. For improved sound, connect a separate mike to the VCR and hang it on or mount it near the person speaking. Another option is to use a SHOTGUN microphone, a mike which "hears" in mainly one direction and rejects sound from elsewhere.

Many industrial or indoor black-and-white cameras are available new and used at reasonable prices. Some have manual controls for picture brightness and contrast.

Cameras which don't have built-in lenses often have C-MOUNT lenses which can be removed and swapped with other C-MOUNT lenses.

Camera troubleshooting is made much easier if you can observe your camera's signal in *both* its ELECTRONIC VIEWFINDER and in a TV set connected to the VCR. If the set shows a picture and the VIEWFINDER doesn't, it's the VIEWFINDER's fault. If the VIEWFINDER displays a good picture but the TV doesn't, the camera's okay; maybe the camera cable, VCR, TV cable, or TV is at fault.

Camera problems which aren't due to a

wrong connection or a switching error are often the fault of the camera cable. Wiggle it and look for flashing on the TV set.

VCRs which fail to UNPAUSE when the trigger is pulled while recording have something connected in their REMOTE jack, or have their VCR PAUSE button pressed.

To ensure a long healthy life for your camera, follow these precautions:

1. *Never aim a TV camera into the sun or at any bright light.* Strong light may permanently damage the VIDICON tube. Similarly, beware of shiny objects that could reflect a bright light into your camera.

2. Never leave the camera pointed at a very contrasty subject for a long period of time. The bright parts of the image will BURN-IN the VIDICON tube and will remain in your picture later when you use the camera to shoot something else. For example, Figure 5-19 shows the BURN-IN caused by leaving the camera focused on a sign reading THE END in black letters on a light background. A faint image of THE END now appears on every picture that camera takes.

3. Don't knock the camera around. It is fragile and easily misaligned.

4. Avoid extremes in temperature. The heat in the trunk of a parked car on a sunny day could damage the circuits. In super-frigid weather, the zoom lens may get "sticky" and fail to rotate smoothly.

5. If traveling by air, don't check your camera with your baggage. Remember the American Tourister Luggage ad with the ape tossing around the suitcases?

Chapter 6
Advanced Camera Techniques and Effects

LIKE THE PROFESSIONALS

So far, we've studied only how to operate the equipment. Now that we can work our machinery, let's focus on how to use it to make professional-looking pictures.

Anyone can pick up a camera and make a boring, fuzzy video tape that plays like the typical home movie. Study the professional cinema and TV techniques and practice them—and I mean *study*, and I mean *practice*—and *your* recordings will become so slick and smooth, *your friends will actually enjoy watching them*. Get a reputation for producing nice stuff and people may actually *want* to take part in your little epics. Sounds unbelievable? Read on.

Our stairway to expertise will have three steps: (1) mechanics of skillful camera handling, (2) aesthetics of picture composition, and (3) camera "tricks" which add visual interest or create a mood. It is the nature of art that sometimes you will balance on several steps at once, using a camera "trick" to compose an appealing shot or using raw mechanical skill to create a desired illusion.

TV is all illusion anyway. Making your audience see something that really isn't there is part of the magic of television. A *bigger part of the magic, however, is making something look like it really is*. It doesn't just happen.

You'll be shown a lot of "rules" on the aesthetics of picture composition and camera movement. These "rules" are meant to be broken at times—but *after they've been learned*. You'll be master of your tools if you can "do it right" whenever you need to, then "do it your way" whenever you want to.

Once you become aware of these "rules," "tricks," and techniques, a funny thing will happen to you. You'll never be able to watch TV again without becoming conscious of the camera angles and shot composition. You'll see the "rules" being followed, and sometimes broken. You'll have become a gourmet of visual craftsmanship.

Handling the Camera

First, let's learn the lingo. Figure 6-1 shows the fundamental moves the camera can make. They apply whether you're using a tripod or holding the camera by hand.

Figure 6-2 shows some basic camera angles and the moods they portray.

178

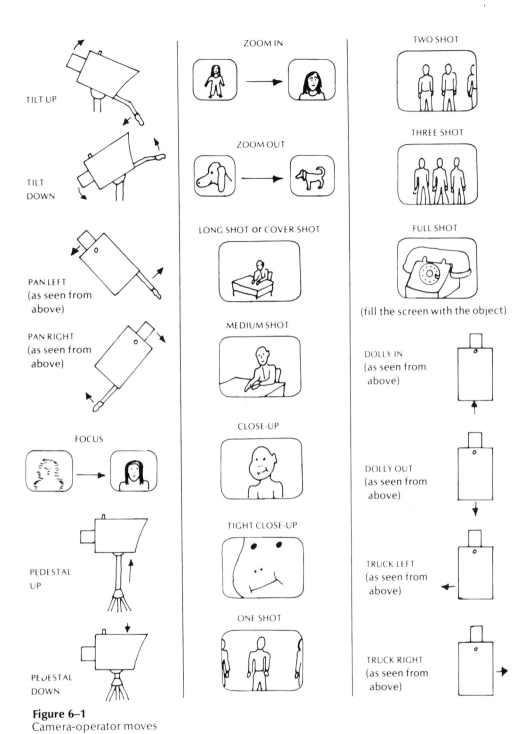

Figure 6–1
Camera-operator moves

Figure 6–2
Camera angles

Camera low

Camera high

Camera at eye level

People walking

Speak into

Tilted shots

Educational studies on perceptual motor skills favor performer's-eye view over observer's view

Zoomed out

Zoomed in

Zoomed way out. Subject looks insignificant and dominated by surroundings

> **Words to Know**
>
> TILT
> To aim the camera up and down on a vertical axis, like nodding your head "yes." TILT UP means to shoot higher, toward the ceiling. TILT DOWN means to aim lower, toward the floor.
>
> PAN
> To aim the camera back and forth on a horizontal axis, like shaking your head "no." PAN LEFT means to turn the camera to your left. PAN RIGHT, of course, means to turn it to the right.
>
> DOLLY
> To travel forward or backward across the floor with the camera. DOLLY IN means to move the camera forward, tripod and all, closer to the subject. DOLLY OUT means to pull back.
>
> TRUCK or CRAB
> To travel from side to side across the floor with the camera, tripod and all. TRUCK RIGHT means to travel to your right, and TRUCK LEFT tells you to go in the other direction.
>
> PEDESTAL
> To adjust the elevation of the camera above floor level. PEDESTAL UP means to make the elevation greater. PEDESTAL DOWN means to decrease the height of the camera.
>
> ZOOM
> The act of adjusting the camera lens to make the picture look closer or farther away without moving the camera itself. ZOOM IN means to make the subject look closer, more magnified. ZOOM OUT means to make it look farther away and smaller.

TIPS ON CAMERA MOVES

Every move you make during taping will be seen, and perhaps unconsciously become part of your message. This section will cover the mechanics of how to get the camera to move the way you want it to.

The steady camera A picture that bobs around, even a little, betrays amateurism *or* implies you're looking through somebody else's eyes. The camera, for instance, would follow someone down a flight of stairs, becoming the pursuer. Jumpy hand-held camerawork can also imply peril, reality (as in newsreels), and frenzy. Unless you intend to portray these moods, you'll want to keep the camera rock-solid while taping.

One common method is with a TRIPOD or MONOPOD. You aim your camera, tighten the tripod controls, then shoot with the camera motionless. All motion is in the scene.

If tripods are too cumbersome for you there are several methods for steadying a hand-held camera. Figure 6–3 shows the Tai Chi stance (from Oriental martial arts) which minimizes natural body sway while

Figure 6–3
Tai Chi position for holding camera

putting you in excellent position for smooth moves in most any direction. Stand with your legs 18 inches apart, slightly pigeon-toed with knees slightly bent. Keep your neck and the camera in close to your body. From the Tai Chi position you can PAN by turning at the waist. You can TILT using your whole torso. You can PEDESTAL DOWN by further bending your knees. And you can DOLLY in or out by proceeding to walk, bent over in the Groucho Marx style, sliding your feet along, letting your knees absorb all the ups-and-downs while your torso glides smoothly through the air. Perhaps it sounds more like Oriental Torture than good camera posture, and indeed it gets uncomfortable over time. But with a little practice this silly-looking stance will deliver nice-looking pictures.

And while we're on the subject of walking with the camera, here's a good habit to get into, one practiced by professional microscope and telescope users—stick one eye to the eyepiece, but *keep both eyes open.* It may be awkward at first, but a half hour of practice will teach your "unused" eye not to see. Thereafter you don't have to tire your face squinting one eye closed *and* you reap an unexpected benefit when you try to walk. Your "unused" eye starts keeping track of curbsides, low limbs, trip cables, and horse droppings as you march hither, thither, and yon.

A camera that rests on your shoulder will give a more stable picture than one with the common pistol grip. With the shoulder cameras, the camera gains stability by being pressed against you at several points: the eyecup (or forehead), the side of the face, the shoulder, the trigger-grip, and the lens. The pistol-grip camera, on the other hand, is held at the end of your bobbing arm. If using a pistol-grip camera, stabilize your arms by pressing your elbows against your chest. Press the eyecup against your brow and hold onto the lens. Again, uncomfortable, but stable.

It will be more restful and more stable if you can brace yourself or your camera against something while shooting. Sit down, lean against a wall, brace the camera against a pole, hold it against a car hood or a rock for rock-solid shots. A pillow or beanbag will do wonders at cradling a camera for steady shots.

For really low shots, you may need to lie on the floor, propping the camera on a cushion. Cameras with movable electronic viewfinders make it easy to look straight down into the finder while cradling the camera beneath you, perhaps at knee height. If you can connect up to a TV set to view your shots, you may not need the viewfinder at all, freeing you to hold the camera anywhere. Low shots are easily done by hugging the camera to your waist, and aiming it while viewing the TV screen.

TILT and PAN Think ahead. If you expect to aim the camera somewhere while recording, figure out where you want to go first. This avoids zigzagging and "searching" kinds of shots. If you can, PAUSE first and try out the move to practice it, and also decide on the picture composition. Then UNPAUSE and carry out your practiced move.

If using a tripod, think ahead to loosen the controls so that your camera will move effortlessly. If you expect frequent fast moves, such as with sports, loosen the PAN and TILT locks all the way. If you're doing slow, gentle moves, leave a little drag on the controls to damp out some of the jiggles.

If working hand-held, use the Tai Chi position and move your whole torso when you TILT or PAN.

DOLLY and TRUCK If using a tripod on a DOLLY over a smooth floor, the process of traveling is easy. Do think ahead to unlock your wheels and to sweep cables and obstructions out of the way so you don't drive bumpity-bump over them.

If traveling hand-held, it's best to first zoom out to a WIDE ANGLE. A TELEPHOTO shot greatly magnifies camera wiggles, while WIDE-ANGLE shots hide them. Use the Tai Chi posture to start from, then glide like Groucho.

Wheelchairs, shopping carts, and any number of wheeled or sliding vehicles can serve nicely as a temporary DOLLY. Be creative. I remember one camera operator needing a shot of a bowling alley and pins—from the bowling ball's point of view. He removed the trigger handle and set his camera on a carpet remnant, which he then pushed down the alley with a broom handle. Smooth! And a feast for the eyes.

Despite all the advertisements you may have seen showing smiling 115-pound models shooting a scene while also carrying the VCR on a shoulder strap, it just ain't so. Not only does that little VCR seem to double in weight every 15 minutes you carry it, but it unbalances you and hampers your moves. Either set it down when you shoot, strap it to your back, or have a friend carry it. That friend can also serve as a "guide dog" for you while you're walking (eye glued to the finder), as well as keeping your camera cord from dragging or tangling.

Avoid swinging the VCR around too much as you shoot. It won't hurt the machine (unless you bumped it hard) but it's likely to affect the quality of your picture. Inside the VCR, the spinning heads, etc., act a little like a gyroscope and resist changes in direction. Forced reorientation of the machine changes recording speed slightly, causing a momentary bend at the top of the recorded TV picture. When possible, set the VCR down, use a long cable, and move only the camera.

As long as it's not bumped, the *camera* doesn't care what position it's shooting in or whether it's moving. So when that curvaceous young gymnast invites you to record her on the trampoline, feel free to hop right up there with her (leaving your VCR behind), and shoot. The visual effect of her seeming to stand relatively still with everything else moving up and down will be too stunning to ever erase.

Focus and zoom

A. Remember, if time permits, that you focus on a subject by: (1) zooming in all the way first, (2) focusing for a sharp picture, and (3) zooming out to the shot you want.

B. If you're recording or if time doesn't permit proper focusing, you omit steps 1 and 3 above and just focus as best you can without extra zooming. Be aware that as you zoom in, you'll probably go out-of-focus.

C. Objects closer than 5 feet (or the closest distance etched on the lens barrel) probably cannot be focused clearly, so keep your distance.

D. When you're trying to focus in step A above, use this method: (1) Turn the focus ring until the picture changes from blurry to sharp to a *little* blurry again. (2) Since you've gone too far, turn the knob back again slightly. Once you're used to it, this technique becomes fast and accurate.

E. Practice, practice, practice. Step A above should take 3 seconds, and you will have a picture that is crystal perfect.

Know your lens controls. Know by "feel" which part of the lens does what. Know instinctively which way you turn the IRIS ring to open it. Know which way to turn the lens to zoom in.

A typical amateurish shot is the "false-zoom," a slight in-then-out move made because the camera operator didn't immediately know which way to twist the zoom lens to zoom out. *Know* which way to turn the lens; don't experiment while recording. The same goes for focus; if something comes toward you, *know* whether to turn the lens clockwise or counterclockwise to FOLLOW FOCUS.

What's FOLLOW FOCUS? That's a technique used by professional camera operators as they shoot moving objects. With a little practice, you can do it too. Zoom in on someone and have them walk toward you from 30 feet away. Try to keep their image sharp as they move. It isn't easy. With practice you can develop the skill of following a moving target, keeping it centered, keeping it focused, and even keeping it the right size on the screen (by zooming) as it moves around.

But if you can't get the hang of it, stay zoomed out on fast moving objects. That way your focusing errors won't be as noticeable and you'll have less trouble keeping your moving target on the screen.

Assuming you've mastered keeping your zoomed-in shots sharp and centered, *use them. Television is a close-up medium.*

TELEVISION IS A CLOSE-UP MEDIUM.

GET IN CLOSE.

LONG SHOTS, though easy to shoot, turn into monotonous mush on the TV screen. CLOSE-UPS capture the expressions, the detail, the excitement of a scene. Look back at Figure 6-2 and notice how the CLOSE-UPS are more interesting than the LONG SHOTS.

And now a word about that zoom lens of yours. Everyone who gets one loves playing with it. In-and-out, in-and-out, your audience's eyeballs feel like Duncan yo-yos. Zoom to your heart's content while practicing and get this urge out of your system. Then go out and force yourself to shoot without zooming. You want a CLOSE-UP of something? Then PAUSE, ZOOM-IN to a CLOSE-UP, then UNPAUSE your recording. That's better. You want to create a sense of travel, motion, or exploration? Then move the whole camera. That will create a real sense of travel, not the overworn zoom. Although a zoom and a DOLLY both can make things look closer or farther away, only the DOLLY changes perspective as it happens. You'll see an example comparing zooms to DOLLIES in Figure 6-7, coming up later.

One handy use for the zoom lens is to fill the TV screen with action when you *can't* move the camera. Picture a youngster up to bat. Pitcher and batter are both on the screen. The windup. The throw. It's a hit. You follow the ball into the outfield *zooming in* as you follow it. When the outfielder fumbles it, you'll be zoomed in close to *see* the fumble, rather than seeing a dot surrounded by the whole outfield. Here the zoom lens helps you fill the screen with action. In order to get that nice shot, however, you had to zoom *and* aim *and* refocus simultaneously, no easy task for the unpracticed.

Think ahead

1. Anticipate. Be ready to tilt up if someone is about to stand up. Be ready to zoom out if someone is about to move from one place to another. Being zoomed out makes it easier to follow unpredictable or quick movements. If someone is about to move to the left, start moving a little before your subject does. This will create a "space" for the subject to move into. Such a camera move also creates an unconscious anticipation in your viewers. They will expect the performer to move when the camera moved.

2. Have your controls unlocked if you expect to PAN, TILT, or PEDESTAL.

3. If you know you are going to have to DOLLY or TRUCK your camera, get the cables out of your path for easy travel.

4. If shooting a majestic landscape, plan your shots before you shoot. Decide on a place to start (perhaps a close-up on a flower), a place to go (a simultaneous zoom out and PAN to the right) and a place to end (the crowd alongside you hanging over the railing and snapping photos). Avoid aimless zigzagging across the same scene, even though your natural inclinations tell you to "show-whatever-you're-not-showing-now-because-you're-missing-something."

CAMERA PLACEMENT AND BACKGROUNDS

Two questions should come first to mind as you set up a camera shot: Where's my light, and What's in the background?

Lighting When you are driving into the sunset, it is hard to see the road. The sun glares into your eyes and makes you squint, and it creates reflections on your windshield. For cameras, the same is true, only worse. Bright lights near or behind your subject force your camera's automatic circuits to "squint," creating a very dark picture (see Figure 6-4). Light also reflects off the lens elements, creating white dots and geometric shapes (see Figure 6-5).

In general, try to keep all the light behind the camera, none behind the subject (with the exception of carefully controlled

Window shade closed

Window shade open

Figure 6–4
Excessive lighting from behind the subject

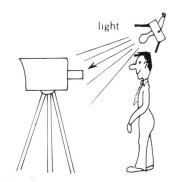

Figure 6–5
Lights shining into the lens

backlighting, which will be discussed in Chapter 8). In situations where you *must* shoot towards (not at) the light, the following steps may minimize the problem:

1. Use a bigger lens shade. The lens shade (shown in Figures 5-3 and 5-14) is the funnel-like scoop on the outside of the lens that shields the lens from ambient light. Or you could make a shade with some paper and adhesive tape.

2. Zoom in some in order to avoid as much of the extraneous light as possible. A tight CLOSE-UP of the face in Figure 6-4, for instance, would eliminate much of the glare from the window (but it may be easier to lower the windowshade than it is to maintain a good CLOSE-UP of a moving face).

3. Using extra lights, throw more light on the foreground (the face in Figure 6–4) to offset the background light.

Handling the Camera 187

4. The bright lights are fooling the automatic IRIS controls into "squinting," so turn the controls to MANUAL and adjust the camera to make the subject look good, even though the window behind the subject may appear overexposed or look washed out.

Note that if this light is too bright, a damaging BURN-IN can occur. *The above steps are applicable if the light is bright enough to affect the picture but too moderate to damage the* VIDICON *tube.* Things that are too bright for a camera *ever* to look at are: (1) the sun, (2) movie or TV-light, (3) any bright, bare bulb, (4) any chromed object reflecting light from any of the first three, and (5) mirrors or glass objects that are reflecting bright lights. Later on, in the Creative Camera Angles, Techniques, and Tricks section, you'll learn a couple of safe routes around these restrictions.

Things that a camera can stand to look at (but not for long periods of time) are: (1) an open window that looks outdoors (but not into the sun), (2) fluorescent lamps, (3) table lamps with translucent shades, (4) a flashlight or other weak light, (5) a *dimmed*

Figure 6–6
Lamp in background looks as if it were growing out of the subject's head

More distracting backgrounds. (Photos courtesy of Imero Fiorentino Associates, Inc.)

house light, if dimmed or diffused enough, (6) shiny automobiles on a hazy day; (7) white clothing, and (8) a TV-screen image.

Background The good camera operator should also take into account what is behind the subject. With just the wrong camera placement, the bush in the background could appear to grow out of the subject's ear, or a desk lamp could become your subject's antenna (see Figure 6–6). Avoid busy or distracting backgrounds unless they serve a purpose in your program.

Camera Angles and Picture Composition

DON'TS AND DO'S OF CAMERA ANGLES

Now we get deeper into the aesthetics of picture composition. Before, we merely concerned ourselves with keeping the subject sharp, close, centered, and level, without wiggling. If you can do that, you score 75%, but still Francis Ford Coppola you're not. To score 100% your TV screen must portray precisely the message you want it to. The image must embrace the subconscious nuances you desire, yet eliminate unwanted distractions. In Figures 6-2 and 6-6 we tasted a few "rules" of picture formulation. Let's look at some more "rules" of TV picture composition in Figure 6-7. Can you see why the "don'ts" should be avoided?

Figure 6–7
Don'ts and Do's of Camera Angles

DON'T

DO

Cut off a person at any of these natural divisions

Cut off feet

Include feet

Cut off neck

Leave in part of the body leading to the next part. The mind will complete the rest of the body

Figure 6–7 *(Cont.)*

DON'T	DO

Cut a person off where she contacts her surroundings. Here the talent looks like she's leaning on the side of the TV screen	Provide enough surroundings for shot to explain itself: the talent is in a chair.

Use the so-called RULE OF THIRDS placing important picture elements one-third down or one-third up the screen	Place most important picture element in the dead center of the screen
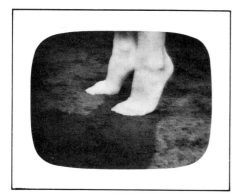	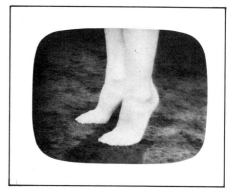
centered	⅓ up from bottom

Figure 6–7 *(Cont.)*

DON'T

DO

Place eyes in the middle of the screen.

Place eyes one-third down the screen.

Place the face in the middle of the screen. Too much headroom. Note also how shot cuts body at the bust, also a no-no

Place face one-third down the screen

Figure 6–7 *(Cont.)*

DON'T	DO

	The mouth and eyes are important. Here they follow the RULE OF THIRDS, where the eyes are one-third down the screen and the mouth one-third up. Here the missing chin and forehead will be mentally filled in by a psychological process called CLOSURE.

Provide insufficient headroom.	Leave just a little space so talent doesn't "bump her head"

Use many LONG SHOTS	Television is a CLOSE-UP medium.

Figure 6–7 (*Cont.*)

DON'T	DO

Try to imply movement by zooming	Dolly to give a real sense of movement. The dolly is more three-dimensional and more stimulating

Figure 6–7 (Cont.)

DON'T

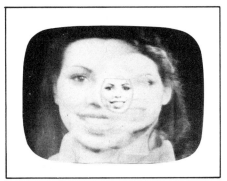

Get zoom happy using your fancy zoom lens to make yo-yos out of your audience's eyeballs

DO

If possible, cut from shot to shot when necessary

MORE DO'S OF CAMERA ANGLES

Seat people close together One of the oddities of television is how it distorts space. A 5-foot-wide "set" in your living room becomes the entire universe for your TV show. Your audience never sees the mess of lights, cables, and clutter just a foot "off-screen." To them, what they see is all that exists. Visit a TV newsroom which you've grown accustomed to viewing only on TV and you'll be shocked at how little there is to it.

One aspect of this distortion of space involves people in conversation. Normal Americans converse at about 3 or 4 feet from each other and sit even farther from each other. Not on TV. Three or four feet seems like across the room on TV. (See Figure 6-8.) Squeeze your people in tight in order to look "normal." Note that it may take some practice before your performers feel comfortable conversing from less than two feet apart.

Seat host to one side of guests This way you avoid giving the host "tennis neck" as he ping-pongs his attention first to the guest on his left, then to the guest on his right. Camera and viewers alike go bouncing back and forth to follow the discussion.

Placing the host between the guests has another disadvantage. If you later edit together the TWO-SHOTS which include the host and one guest (as in Figure 6-9), then the host and the other guest, the host snaps from the left side of the frame to the right

Camera Angles and Picture Composition 195

Figure 6–8
Seat the talent close together

Talent may look fine in the studio but . . .

. . . are too far apart for a good medium shot

Figure 6–9
Panel discussion with host between guests

First the host is on the left . . .

. . . then the host is suddenly on the right

side. First he's looking to the right; suddenly he's aimed to the left. Figure 6–10 is more comfortable.

Angle the guests This maxim applies to any shot with two or more persons in it. People facing each other nose-to-nose suggest an adversary relationship. It's a great way to portray an argument, but in a panel discussion the shots imply disagreement or debate. Conversely, people lined up facing the camera look like a team of contestants. They appear ready to perform individually for the camera, but not ready to react with each other. The most comfortable-looking seating arrangement has people angled toward each other. See Figure 6–11 for an example.

Avoid slightly off-camera looks Have your performers play directly *to* the camera, or react to each other or their props, disregarding the camera completely. Direct eye contact is very engaging, making the viewer feel as if the performer is talking to *him* alone. Profile and off-angle shots make the viewer feel like an observer to the action. Both kinds of shots have their advantages, but avoid splitting the difference. The slightly off-camera shot in Figure 6–12 makes the performer look insincere and shifty-eyed, or it makes the performer look like he's reading something.

Use a familiar object to create a sense of scale Extreme CLOSE-UPS and majestic landscape shots suffer one problem in com-

Figure 6–10
Panel discussion with host to one side of guests

Figure 6–11
Angle the guests

Nose-to-nose implies adversary relationship

Line-up gives no relationship between guests.

Angled is comfortable

Direct eye-contact engages viewer

No eye-contact. Viewer observes the actions of others

Slightly off-camera look loses impact, implies insincerity

Figure 6–12
Avoid slightly off-camera looks

the foreground to add dimension and meaning to the landscape in the background. Often a bush, a tree, a person, or an old wagon wheel serves nicely to create scale, balance, and visual variety in a scenic view.

CREATING MOODS AND IMPRESSIONS WITH THE CAMERA

How you show something tells as much of your story as *what* you show. Look again at Figure 6–2. Just by changing camera height, a subject changes mood from dominance to servility. Tilted shots create suspense. CLOSE-UPS involve the viewer with the action. Here are some more mood-creating shots.

Progress versus frustration A jackrabbit is racing to the right. The camera PANS along with it, but slowly falls behind so that the bunny moves forward in the frame. That's progress. A mountain lion pursues to the right. Here the camera PANS slightly faster than the lion can run, leaving more space in front of it. That's frustration. Score: jackrabbit one, lion nothing.

A runner in a telephoto shot approaches, and approaches, and approaches, in a vain attempt to reach the camera. Frustration. Shot at wide angle, the mild-mannered accountant strolls by the camera and appears to loom forward as she nears the camera lens. The look is one of decisive action. Progress.

Suspense Our hero creeps backwards and the audience cringes waiting for him to back into something awful. He looks up slowly. Something terrible is bound to drop on him. He pokes his head into an air duct. What's going to grab his sweet face? He draws open the curtains. What monster will leap from behind them? As he sleeps, a large shadow slips across the bedroom wall. What is it? He washes his hands. He looks up, and through the sink window, inches from his face, there's a . . . . He sits at the

mon: The viewer can't appreciate the smallness or largeness of the subject without a "visual yardstick" to gauge it by. Figure 1–8 used a hand to define the size of a plug. Figure 6–13 uses a familiar object in

Figure 6–13
Foreground figure to give scale to landscape shot

dressing table and looks up to the mirror. What unholy creature will he see behind him? He casually shaves while we DOLLY IN from behind him, stopping just over his shoulder. What unwelcome surprise awaits him in seconds?

In each case, your attention is drawn to what you don't see, out of the camera's view, behind the curtain, in the mirror, or behind the camera as it moves closer. For a crash course in suspense shots, see the movie *Alien*. It uses every trick in the handbook.

Authority Look again at Figure 6–2. When the camera peers down on someone (or something), the camera is dominant. The subject looks small, weak, lonely, helpless, inferior, or inconsequential. Such shots are great for portraying pathetic, helpless kids lost in a crowd, throngs of humanity trapped by a monster, or a single fugitive chased and outnumbered.

When the camera views the subject at eye level, the shot is neutral and egalitarian. It is used in news, talk shows, interviews, and on-camera narrations where no special status is given to the performer.

A bit more complimentary and polite is the slightly-below-eye-level shot. It adds a subtle air of stature to the performer.

When the camera looks up at someone or something, the viewer feels dominated. The subject takes on an air of power or authority. Shoot a woodchopper, an attacker, a hero (not the sandwich—the person), a monster, a bulldozer, or even a statue from this angle to add that bigger-than-life look.

When shooting children (with a camera, not a cannon), unless you are trying to portray them as weak or helpless, move down to their level. Even though it's normal for us to "look down" at little people, the camera at this angle assumes a hostile posture of dominance. Squatting down to the kids' level makes things look "normal" again. If shooting MEDIUM SHOTS of adults conversing with children, either have the youngsters stand on something or have the adults bend down to them so that both their faces share roughly the same level on the TV screen.

Anger, secrets Pose your performers almost nose-to-nose and have them shout.

Voilà, anger. As they move about, keep their heads very close together.

This closeness *without* the shouting implies intimacy or secretiveness.

Speed Keep the camera low to the ground (so you can see it rushing toward you) and use a wide-angle lens to portray speed. A camera on a skateboard moving at 1 mile per hour looks like it's moving at 60. A camera shooting from the roof of a tall truck at 60 miles an hour looks like it's only traveling at 20.

I once shot a traffic safety tape by mounting a TV camera on the fender of a car and driving it around a congested campus. The scenes made it look like I performed speeding maneuvers and dare-devil near-misses at every turn. Potholes loomed up like canyons. Casual pedestrians looked like kamikaze jaywalkers. Yet all these shots were taken at a calm 15 miles per hour or less.

Night Naturally, if you shoot in the dark you don't get a picture. The object is to make the scene *look like night*, but with plenty of light for the camera.

If shooting outdoors in daylight, pick a cloudless day with blue sky and harsh shadows. Place a red filter in front of the lens to make the blue sky very dark. Either shoot with a black-and-white camera or misadjust the camera's color controls to undo the redness you've just added to the scene. Use a neutral density filter or stop-down your lens to darken the scene more. Keep performers in the shadows or in "puddles" of light as they move from place to place. Shoot with the sun halfway or three-quarters behind the performer, creating frequent silhouettes. The sun will assumedly "moonlight" the scene. Be very careful not to shoot directly into the sun unless you have a sufficient neutral density filter in place and your lens is at a high f-number. This whole process is very difficult to do well.

If shooting a "night scene" at night outdoors, use one or two harsh lights, creating distinct shadows. Try to BACKLIGHT the performers (a process explained in Chapter 8) to form a white ridge outlining them. Again, watch out for burn-ins.

If shooting a night scene indoors, create harsh shadows and puddles of light by using one or two bare light bulbs hidden from the camera by props. BACKLIGHT the performers to create silhouettes. If you give your actors lanterns or candles, you have an excuse to beam some light on their faces to catch expressions, etc. The face light doesn't *really* have to come from their lanterns, you could carefully beam some light on their faces from off-camera, making it look like the light came from the lantern.

Creative Camera Angles, Techniques, and Tricks

POPULAR ALTERNATIVES TO THE SIMPLE SHOT

There are lots of ways to show the same thing. The real skill comes in knowing when fancy shots will add spice to a scene and when they will distract the viewer and ruin the scene.

TO FOCUS SHIFT (or to PULL FOCUS)
Something (or someone) is in the foreground. Something else (or someone else) is in the background. Usually, one of these two has to be out of focus for the other to be sharp. Okay, use this to your advantage. As two people talk, focus back and forth on whichever person is speaking. It's better than PANNING back and forth, and better than a LONG SHOT, trying to get both performers into the picture at once. Perhaps the foreground person's expression is paramount (tears welling in the eyes) while the background person is secondary, even distracting. So focus on the foreground, making the background blurry. Then when the background person's action becomes important (staggering to the cupboard and pouring

another drink) *that* is pulled into focus, diverting attention from the face. Yet both characters can be seen at once as you shoot past the tearful face 4 feet from the camera, and in the other half of the screen get a LONG SHOT of the drinker 20 feet away. Attention can be focused on one or the other performer, yet the viewer can *keep track of* both.

FOCUS SHIFT is also a popular way to display a long line of something—soldiers, flowers, gravestones, fenceposts, etc. Position yourself next to one member of the lineup and shoot toward the farthest member. Focus on the closest (or farthest) at first, and as you change focus, different members will, in turn, become sharp, then blurry as other members become sharp.

Low f-numbers and longer focal-length lenses make this focus-defocus process easier.

If you have a macro lens, you're equipped to carry out a super FOCUS SHIFT from 3 inches out to 20 feet. Picture how this could work with a beginning title. You letter your title onto a foot square piece of clear glass, plastic, or acetate transparency (like those used in schools for overhead projectors). You macro focus on that title, inches away from your lens (make sure you make the title the right size for this). In the background, so out-of-focus you can't see it, is your first scene. You refocus from the title to the distant scene as the action begins. The title will go so far out-of-focus that it almost disappears, making a nice cheapie dissolve effect.

The same technique may be used for 35mm slides taped to the front of the macro lens, or for FOCUS SHIFTS from miniature soldiers or plane models out to your live actors.

Mirror shots This is another way to slip twice as much into your picture without resorting to dull LONG SHOTS. Imagine this—a CLOSE-UP of a woman putting on lipstick. The camera slowly zooms (or DOLLIES) out, revealing that her face is a reflection in a mirror. You see the side of her head in the foreground. As the camera pulls back further you see someone else reflected in the mirror. It's a man getting dressed across the room. As he speaks, she turns, now facing toward the camera. He steps closer, tying his tie as he walks. Now your camera has two people talking *to* each other, *facing* each other, yet *you see both faces at once.* Nice trick. How else can you get two full face shots at once?

Mirrors are useful any time you want to show both sides of something, like an engine, a statue, or some complex device. They're useful for dramatic effects (she looks up and in the mirror sees someone coming through the window). Occasionally they serve as just another way of formulating a shot.

A mirror hung at an angle over the kitchen stove allows you to get straight-down-into-the-pan shots without the risk of having your camera fall into the tomato sauce. Conversely, a mirror slipped under the car allows you to document those rusty rocker panels on your late-model chariot.

Mirrors are not the only things that reflect images. Imagine the CLOSE-UP of a man's face slowly turning toward the camera. He's wearing those mirror-like sunglasses. You know what he's looking at when the twin likenesses of a sexy sunbather appear in the lenses.

Windows and water reflect too. Windows have an added advantage of being able to mix two pictures together for you, the figure behind the glass and the one reflected by it. In fact, one tricky way to dissolve from one picture to another without an expensive video fader and two cameras is to: First illuminate the figure behind the glass while keeping the reflected subject in the dark. Then dim the behind-the-glass scene while simultaneously brightening the reflected scene. The old scene will melt away, being replaced by the new. Makes a nice trick shot for a sci-fi drama, too.

Mirrors don't even have to be flat to work. Nice effects can be had by shooting a CLOSE-UP of someone's face reflected in a shiny Christmas tree decoration, teapot or a chrome bumper. Remember all those serving dishes you got as wedding presents, but never had occasion to use in the 31 years you've been married? Hallelujah, the time has come! Use them now for twisting the world into fun-house contortions.

Again, be careful not to reflect any bright lights into your camera when using mirrors. They may cause burn-ins.

Parallel movement Joggers, bicyclists, water-skiers, snowmobilers, and horseback riders all pose the same problem for the camera. They're fun to watch, but they're gone an instant later. To really catch the action, move with them. Video tape that jogger from your car (preferably with someone else driving). Shoot that water-skier from the back of the tow boat. By moving along next to the action, it stays with you while the background slips by.

Adding movement to still objects Photos and models don't move. That doesn't mean the camera can't make them come alive. Imagine a CLOSE-UP on a *photo* of a parade, PANNING across the picture to the sound of a marching band. The people will *seem* to be marching.

Imagine a slow *zoom-in* on a painting, starting first with the whole town but ending with a "stroll" down the main street.

Picture the museum's dinosaur display filling the screen. Cut to a CLOSE-UP of one of the models. Take the shot from slightly below the creature so it looks domineering and dangerous. TRUCK the camera around the model, making it seem to turn its head.

ZOOM-IN on a still photo of a horse race with the sound of crowds cheering in the background. Shake the camera as you zero-in on the lead horse. You can almost see the mud flying from the hooves.

To the beat of the rock music, quickly zoom in-and-out, in-and-out on the singer's face from the record jacket. Dance him around at the microphone by tilting the camera back and forth to the music.

Wall shadow In the movie *Goodbye Girl*, the young lady is forced to accept a male roommate in her apartment. She's had it with his guitar playing and barges into his bedroom to complain. Seems he strums in the nude. How would you shoot such a scene without going X-rated?

You could use the tried-and-true guitar-in-front-of-the-genitals shot. That's okay, but what do you do for your next shot? In the movie, they aimed the camera at the flustered lady standing by the door with this shadow of a man projected upon the wall. It all seemed very natural and the scene was creative and novel.

You can use this technique to show sinister shadows, lantern-cast shadows, or just for a stylistic view of something before you actually see it (like a shot of the bicycle shadow on the pavement followed by a TILT UP to the bicycle). Shadows make nice surprise revelations. Consider the lady's shadow cast upon a doorway. Then into view comes the shadow's owner, a fellow dressed ridiculously in women's clothing in order to sneak past a detective.

ARTY TRICKS

Here you're limited only by your creativity. Try experimenting with these ideas and soon you will be developing ones of your own.

Flashy opening Ever notice how when you first power-up your camera, all kinds of exotic colors and ghost-like images fill the screen for a moment, then melt away to reveal your picture? Use this. PAUSE your recording, set up your camera for a scene, then turn the power off. After a 10-second cooldown, start the VCR recording and turn on the camera. The dazzling images will unfold into your opening scene as your camera "warms up."

On industrial cameras, you can get the same effect by turning the beam control down. Everything will get chalky-looking, start to look like a cartoon, then melt into a blob. Figure 5-8 shows a couple examples.

LAG and COMET TAILING The dancer swings her arms and a streak follows them like a flowing sleeve. She slips across the floor leaving a faint trail behind her which evaporates in a second. This is COMET TAILING, caused by (1) bright objects, (2) dark backgrounds, and (3) very low light level. What you're doing is exacerbating the camera's natural LAG caused by insufficient light on contrasty scenes. Industrial cameras can be adjusted to display a lot of LAG in their pictures. Turn down their beam controls slightly and use a contrasty scene, like a white dancer with a black background. Figure 5-9 shows a couple of examples.

LENS FLARE Figure 6-5 showed what happens when you aim too close to a light. The geometric patterns and spots that you get are called LENS FLARE and are usually reduced by the lens shade. But what if you *want* LENS FLARE for artistic effects? Well, remove the lens shade. Play around with your zoom and IRIS further to enhance the effect. But watch out you don't shoot directly into a light and burn-in your VIDICON.

If you'd like to get one of those dreamy sails-across-the-sunset shots or sun-eclipsed-by-the-glider views, it's still possible—if you're careful.

Place a dark neutral density filter over your lens and stop-down your lens to f16. Take a quick zip past the sun with your camera and see if it's leaving a streak. If there's no evidence of temporary burn-in and your picture is altogether too dark to use, open your lens one-stop-at-a-time and repeat the experiment until you notice the sun just starting to leave a mild *temporary* streak. That's as far as you can go safely. Such a shot should yield nice halos, rings, rainbows, and geometric patterns, yet probably a dim silhouetted shot of that sailboat or glider. This technique works well with sun reflections on water, sunsets, tall-buildings-with-the-sun-just-behind them, the sun peeking through the trees as you look up from a stroll on a wooded path, and almost-into-the-sun shots.

If you *must* shoot directly into the sun, I'd suggest borrowing or using a film camera for the shots. The film camera won't be harmed and the film can later be transferred to video tape (see Chapter 9). This way you take no chances.

Gunbarrels, drainpipes and kaleidoscopes
Remember the title scenes of James Bond 007 where you're looking *through* a gun barrel at James? You could even see the spiral rifling of the barrel's interior. You can do this or something similar by placing tinfoil or cardboard over your lens and placing a pinhole in its center. Then aim your camera through the gunbarrel or whatever, and *flood* your subject with light. Because the pinhole "stopped-down" your lens to perhaps f300, you'll have nearly infinite depth-of-field but will need lots of light.

Chrome-plated sink drain pipes are interesting to shoot through. Kaleidoscopes, too. You can pull the end off your kid's scope (or wait till three days after Christmas when he's done this himself) and get kaleidoscopic views of the livingroom and outdoors.

Silhouette Hang a large, white seamless sheet from a wall, extending it in a smooth curve to the floor. Perhaps place another on the floor, overlapping with the first. Darken the room and flood the sheets with light. Have your performer stand in front of the sheets, *but not in the light*. The strong backlighting will create a silhouette effect.

"Painting" with the camera Most of the warm red sunrises and sunsets you've seen on TV weren't so warm as they looked. The camera shot the scenes with its COLOR BALANCE controls misadjusted to "paint" the scene redder than it really was. You can "cool" a scene by "painting" it blue with the opposite adjustment. Science fiction scenes can be done with this effect, making all the colors unusual and foreign. Faces become green, water becomes magenta. You can imitate the sepia tone of old-time movies by adjusting your camera's color controls for a perfect WHITE BALANCE while *aiming it at something light blue*. This will trick the camera into making whites look sepia.

Lost horizon The viewer who can't see the horizon can be easily fooled, much like the way you were fooled at the fun-house in that tilted room where *everything* was tilted—except you—and you soon became tilted, too.

Shoot straight down on someone climbing on his belly (panting and groaning) over a gently sloped cliff of rocks. The viewer will think the cliff goes straight down. Add the appropriate glances "downward" with an occasional "slip" of the foot and the face-pressed-against-the-rock look to add credibility.

Walls that look like floors and vice versa are splendid sets for such camera tricks. And imagine someone hanging onto a window ledge, ready to jump 30 floors—or is it really 30 inches? The camera will never tell.

Picture a rope in the gym, backed by a nondescript cinderblock wall. Have an athlete friend climb up a ways, get turned head-down, then slide slowly down toward the floor. Only shoot this all with the camera upside-down. Imagine your viewers seeing someone slither to the "top" of a rope and teeter at its free end like in an Indian Rope Trick.

Nail up a sheet of wall paneling—at a 30° angle. Prop a table or desk and chair at the same angle, stick in your performer holding himself at a 30° tilt, and place your camera at 30° from level. Now have your actor pour a glass of milk. Guess where the milk will go.

To make a car or runner seem to speed powerfully up a hill, tilt the camera so the road is at an uphill angle. Shoot with a neutral background, avoiding telltale trees, signs, and houses. This shot is very common in car and truck commercials. The runaway wagon, on the other hand, will look like it's careening down a much steeper slope than it really is with the help of a camera tilt in the downhill direction.

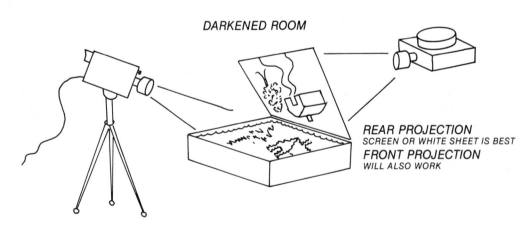

DREAMY EFFECT FROM REFLECTION OFF WATER

DARKENED ROOM

REAR PROJECTION
SCREEN OR WHITE SHEET IS BEST
FRONT PROJECTION
WILL ALSO WORK

A tilted camera can make an airplane "dive" to earth. A back-and-forth tilt can make a boat endure great sea swells.

Reflections and refractions Ever think of shooting through wavy glass for a dreamy or underwater effect? Edmund Scientific, an optics company, sells all kinds of crazy lenses. So do many photo shops. Kroma Studios in Santa Monica sells Kroma glass which creates colorful rainbow effects as you shoot through it.

Try shooting through a glass of water or a fish tank or just above a hot radiator or parallel to hot pavement for some nice effects.

How about shooting at a steep angle into a puddle of water, a pan of water, or a lake, to catch ripply reflections before you tilt up to the actual subject?

For a dreamy effect, PROJECT a scene onto a screen or even a white sheet in a darkened room. Place a pan of water in front of the screen image and shoot the reflected image (with ripples in the water). To make the reflected scene right-side-up, invert the slide in the projector.

CENTER FOCUS and STAR PATTERNS

CENTER FOCUS creates a dream-like image where the outside edges of the picture are

Camera lens effects (Courtesy of Spiratone, Inc.)

"Center Sharp" lens attachment

Resulting CENTER FOCUS image

"Multimage 5C" prism attachment

Resulting multiple image

"Crostar 1SQ" lens attachment

Resulting STAR PATTERN image

Creative Camera Angles, Techniques, and Tricks

blurry but the center is sharp. It's all done with a lens attachment.

STAR PATTERNS, also done with lens attachments, create straight lines going through each sharp point of light in the scene. Stylistic views of rippling water in the sun, or stage lights above a rock group are the most common applications. The pattern can emphasize a twinkle in an eye or a glint of a tiger's tooth.

Either effect can be used to portray the point of view of a drugged, crazy, crying, frenzied, or dreaming character, showing their distorted view of the world.

A cheapie version of these effects can be made at home. Slip a layer of nylon stocking over the lens; all the focus will go soft, like in a portrait or a semi-dream. Tape a piece of glass to the lens or screw on a lens attachment, smear a fine film of vaseline over it, and you'll get a similar effect. Leave the center ungreased and you have a CENTER FOCUS effect. Wipe the grease up-and-down and you'll get sideways shafts of light coming from bright spots in your scene. Half the lens wiped one direction and half the other will create crosses of light emanating from lamps, reflections, and twinkly places. Try a circular pattern. Experiment.

VIDEO FEEDBACK What do you suppose would happen if you aimed your TV camera at your TV set which was displaying your TV camera's picture? What your camera saw, your set would show and your camera would then see and your set would then show and around and 'round the signals would go in a FEEDBACK LOOP.

The visual effects are so limitless that one could sit all night tilting the camera a little, zooming a little, twiddling the TV's color and brightness controls as well as the camera's white balance controls, and then sticking one's hand in front of the TV screen to see what happens. You're unlikely to see exactly the same effect twice, and twice as unlikely to be able to reproduce a given effect at will.

Video feedback

Tilted cameras produce pinwheels and kaleidoscope patterns which spin, freeze, reverse direction, and break up into separate pinwheels. Iris changes create shrinking and growing blobs of light. Color these blobs with your TV and camera's color controls.

And when you've taped a couple hours

of this fantasy, go back and dub in an appropriate sound track. Hard rock or heavy classical scores set to the same visual piece can create completely different moods.

The effect of watching such a recording is much like gazing into the fireplace while listening to the stereo. Your eyes are riveted but your mind is free to wander. The name given to these kinds of abstract TV productions is VISUAL WALLPAPER.

It's great with Fritos and wallpaper-paste chip-dip.

Surreptitious Recording

If something is important enough to risk your life or (more important) your equipment for, that's your business. The camera operator is usually the one attacked when people want to hide what they're doing. One way to catch people off guard is to focus quickly for medium distance, zoom out all the way, and hold the camera by your side, perhaps under your shoulder, perhaps at arm's length, held by the camera handle. The VCR is still running, but no one knows it. The camera is still taking pictures (but not of the sun or other bright objects, we hope). Since the camera is not up to your eye, people think it's turned off. Try to divert attention away from the equipment by turning yourself away from the camera. You might set the unit down (running) on a table. You could pretend to switch it off and then button up the carrying case. Just make sure that your "dirty tricks" are worth the possible consequences if you're found out.

If you happen to be the target of such a sneaky camera operator, take a peek at the lens. Is it covered? Camera operators almost instinctively cap their lenses when they have finished shooting. Nevertheless, the camera could be recording audio with the lens capped. Spy into the camera's viewfinder if you can. If it's lit, the VCR is still on. Can you see the tape moving? Is the RECORD button pressed, or is a RECORD light on somewhere?

There are special "peephole" lenses which are very small and permit cameras to hide behind a hole in a wall or behind clutter with just the tiny lens snout sticking out. I remember once investigating storeroom thefts and wondering how to hide the camera and its regular lens. I hit upon the idea of stuffing the camera into a cardboard box of plastic drinking cups and sticking one cup over the lens. The image through the base of the clear plastic cup was excellent and no one noticed the half-opened box of cups peering out from the shelf.

It's best to use black-and-white cameras with wide-angle standard lenses for detective work like this. The black-and-white cameras work well in low light and give sharp pictures, the wide lenses pick up all the action, and if your "spy" gets stolen, you're not out a $1200 color camera and lens.

And after all this, you play back your tape the next day and see the thief in action—wearing a mask.

Shooting Sports Events

The owner of a small Santa Barbara restaurant spends his days video taping local sand-lot football games, city league games, and other sporting events. At night he plays his tapes over a video projector to a packed house. The name of his flourishing restaurant is the Instant Replay.

Besides entertainment, video can be a useful training tool as the players study their moves and formations. CLOSE-UPS can show one player's body position, angle, and form. Team shots reveal the dynamics of the group. Here are some shooting techniques unique to sports.

CAMERA ANGLES

For field sports, get as high as you can (without using drugs). Mount your camera atop a school bus or on some kind of tower. Face away from the sun. If the team is practicing moves for later review, the camera's view is of paramount importance. Have the practice moved to where the camera can see best, like on a lawn near a building. Here you can shoot from a window or the roof.

Feel free to zoom in or out on the action while the players are moving. Don't bother zooming in to a CLOSE-UP of a player standing still.

For football, place the ball carrier to the rear of the frame with the blockers in the front. As he passes the line of scrimmage, gradually center him so you get the action both ahead of and behind him. If the runner makes it through the secondary defense men, gradually let him outrun your PAN, thus positioning him in the front edge of the screen so you can reveal as much of the rear action as possible. On passing plays, cover the passer till the ball is thrown. Then PAN to the intended receiver while zooming out some. As the ball is caught, continue zooming out enough to get the other players in his zone. On punts, cover the kicker. When the ball is kicked, stay with the kicker a moment to watch for roughing, then PAN to the receiver (you'll probably have time for this as the ball generally stays in the air a while).

The general technique in sports videography is to catch the *main* event while recording the secondary action as well. No one cares what's happening *behind* the kicker during an extra point attempt, so you frame him off-center with space in front. Meanwhile, you're including the charging defense at the other edge of the viewfinder.

SHOOTING TECHNIQUES

If you're shooting games for entertainment and coaching purposes, don't forget to include a quick shot of the scoreboard after a touchdown, or the down indicator or referee signals, just to clarify what's going on and provide a visual respite.

If your VCR makes good PAUSE edits, try to cut out the chaff, like the huddle breaks. Start the tape rolling when the call of the play begins.

When you get near the end of a cassette, eject it during a break in the action and start a fresh cassette. You don't want to run out of tape in the middle of the game's best play (and Murphy's 108th Law says the play you missed was the best play of the game).

Perhaps this is a large enough dose of camera angles and techniques for one sitting. We'll be coming back to more camera and shooting strategies later, in Chapter 11 (paperback Volume II), when we cover editing techniques and how to get one shot to flow smoothly into another.

Camera Techniques in a Nutshell

Learn all the controls on your camera so you can use it like an extension of your body.

Make your camera moves deliberate and smooth. Keep your camera steady, level, and in-focus.

TV is a close-up medium. Stay in tight to your subject for the most visual impact.

Don't overuse your zoom feature. Make the camera move for more stimulating motion effects.

Follow the RULE-OF-THIRDS. Horizons, eyes, heads, and other important features belong *not* in the center, but a third of the way down (or sometimes up) the screen.

Leave room on the screen for people to walk into or talk into.

When choosing a camera angle consider (1) where your light is coming from, and (2) what's in the background. The light should generally be behind you and the background should not be distracting.

People and things look farther apart on TV than they do in real life, so squeeze them close together when shooting.

To capture the grandeur of a landscape and to provide visual variety, include a familiar object in the foreground of such shots.

The higher the subject is in relation to the camera, the more powerful and domineering it looks. The lower it is, the more helpless it looks.

There are a lot of alternatives to the simple camera shot. Creativity balanced with singleness of purpose is the formula for successful teleproduction.

MIRROR SHOTS and FOCUS SHIFTS are two ways to display far-apart characters on the screen at the same time.

Special-effect "trick" shots can be made by "breaking" some of the rules of camera operation like: using a camera's picture before the camera is "warmed up," shooting contrasty scenes in dim light, shooting toward the light without a lens shade, and shooting through layers of glass, gauze, or grease.

Shoot sporting events from up high. Try to get both the main action and the secondary action into the shot at once.

Chapter 7
Recording Audio

HIS MASTER'S VOICE

Unless you're producing Old-Time Silent-Movie Classics, you're going to need sound. Although this book is about video, audio is half the show and deserves a chapter of its own.

The Basic Basics

In Figures 4-21 and 4-22 you saw where the VCR's AUDIO RECORD HEAD put the audio signal on the tape. How did the audio signal get into the VCR to start with? If you recorded a show off-air, then your VCR's tuner took care of everything, separating the antenna's RF signal into video and audio and sending the audio to the AUDIO RECORD HEAD automatically. If you used a portable TV camera to record a scene, most likely the built-in mike on the camera took care of everything, sending the right amount of audio signal to the right place automatically. It's when we start playing around with extra microphones, record players, stereos, and other goodies that we need to know more about audio and how it works.

The microphone picks up sound vibrations and turns them into a tiny electrical signal which travels down a wire into your VCR. Automatic controls in the VCR adjust the volume of the sound being recorded.

What kind of mike should you use? The mike on your camera is pretty good. Any mike that has a plug that fits the VCR's MIC IN is also likely to work. Try it. You can't hurt anything by trying it. Even if the plug doesn't fit, you can buy an adapter plug which will mate your mike to your machine. *Must* you always use a particular mike for certain occasions? Not necessarily. If the mike is designed for hanging around the neck and the performer wants to hold it or put it on a stand, it will still work. The mike will still pick up sound. If the mike is designed for stand use and the performer wants to hang it around his neck, get some string and tie the mike around his neck. It will still work.

What if the mike doesn't work? If you get no sound from a mike after having done all the things described in Chapter 4—pressing RECORD, and checking to see if your mike cable is plugged in—try the old standbys of wiggling the wire near the plug, wiggling the plug, or trying another microphone. Also check to see whether the mike itself has an ON/OFF switch on it that is turned to OFF.

In a pinch, you can even record sound without a microphone! Yep. Strangely, if you take your common stereo headphones and plug them into your MIC IN and talk into the tiny headphone speakers, the sound will be recorded. It may sound a bit tinny, may hum a bit, and may not even work at all, but for that emergency AUDIO DUB you need to make without a mike, this'll usually get you by.

These are the basic basics of audio. With them, you will be successful at recording the sounds you want most of the time. Audio is somewhat forgiving. If it is not done perfectly right, it is frequently still usable.

The rest of this chapter is dedicated to helping you make the sound perfectly right. If your sound is poor, it will distract the viewers from the message. If the sound is mediocre, the presentation will appear amateurish.

Many readers could stop right here and skip the rest of this chapter. (Hooray!) So why all the extra pages? Because true videophiles are sometimes perfectionists who want to do professional-looking jobs with their amateur equipment. Unfortunately, there aren't many simple books on the subject of audio, and nearly all video books gloss over the topic. Audio is complicated and involves a lot of connections and electronics terms. Audio, like brain surgery, is also a hard subject to tell just a little about. In order to really do something fancy, you have to know quite a bit. It is a subject as complex as video, and as you can see, one can write a whole book on video. Thus the choice facing the home TV producer is either:

1. Simply use the mike built into the camera, and except for an occasional AUDIO DUB, skip the rest of this audio absurdity, or,

2. Dive headlong into a sea of switches, segues, and shielded cables, learning how to manage mixers, microphones, music, and mouth noises.

If you selected #2 above, what you'll get next is a dose of professional audio edited for the amateur. With this taste of how the pros do audio, you can make the decision whether *you* want to get that involved yourself.

Professional-sounding audio is like paint on a car. The car drives okay without it, but that extra shine is what turns people's heads. How impressive do you want your shows to be?

Words to Know

ELECTRET CONDENSER MICROPHONE
This mike is usually built into portable TV cameras and many audiocassette recorders. It has good sensitivity to a wide range of sound volume and pitch, and it gives especially clear voice reproduction. It takes a small power supply to operate. When built into the camera, the mike gets its power from the VCR, just like the camera does. Stand-alone ELECTRET mikes cost $30 and up and require tiny batteries to operate.

DYNAMIC MICROPHONE
Lacks the fidelity and sensitivity of many of the ELECTRET CONDENSER MICROPHONES but is good enough for most VCR use. This commonly used microphone is rugged, relatively inexpensive, and quite trouble-free. Its name comes from the fact that its signal is generated by a moving (hence the DYNAMIC) coil of wire and a magnet.

CRYSTAL MICROPHONE
Makes its signal when sound vibrates a tiny crystal inside it. It is fragile and not very sensitive, but is very cheap.

The following terms describe "pickup patterns," the direction in which the microphone "hears" best:

OMNI-DIRECTIONAL MICROPHONE
Can hear in all directions regardless of where it is aimed.

CARDIOID MICROPHONE
Can hear very well in front of it, medium well to the side of it, and hardly at all in back of it.

DIRECTIONAL or UNI-DIRECTIONAL MICROPHONE
Can hear well in front of it and hardly at all anywhere else.

SHOTGUN MICROPHONE
A very UNI-DIRECTIONAL microphone that looks like a shotgun barrel. It hears very specifically in the direction in which it is aimed.

The following terms describe how the microphone's electrical signal travels to its destination:

HIGH IMPEDANCE (abbreviated HI Z where Z stands for IMPEDANCE)
An electronic term describing certain kinds of microphones. audio inputs, and audio outputs that should be used together. Microphones with 8-foot cables (or shorter) are probably HI Z. Inexpensive audio equipment and home TV setups are usually HI Z. CRYSTAL microphones are usually HI Z.

IMPEDANCE
IMPEDANCE is measured in OHMS, and 1000 OHMS (abbreviated 1000Ω or 1 kilo OHM, or 1 KOHM, or 1kΩ) or more makes something HI Z.

LOW IMPEDANCE (abbreviated LO Z)
An electronics term describing other kinds of mircophones, audio inputs, and audio outputs which should be used together. Microphones with 15-foot (or longer) cables are probably LO Z. Most professional TV studio equipment and all high equality audio equipment use LO Z. LOW IMPEDANCE is usually considered to be less than 500 OHMS.

IMPEDANCE MATCHER or IMPEDANCE MATCHING TRANSFORMER

A small device which changes something LO Z to HI Z or vice versa.

UNBALANCED LINE

An audio cable that has two conductors only. There is a center wire and a braided shield (the second wire) surrounding it. At the end of such a cable is a two-conductor plug, such as an RCA, MINI, PHONO, or PHONE plug (examples are shown in Figures 7–1 and 7–2). This method of carrying audio is usually found on inexpensive equipment and usually with HI Z microphones with short cables.

(continued on page 214)

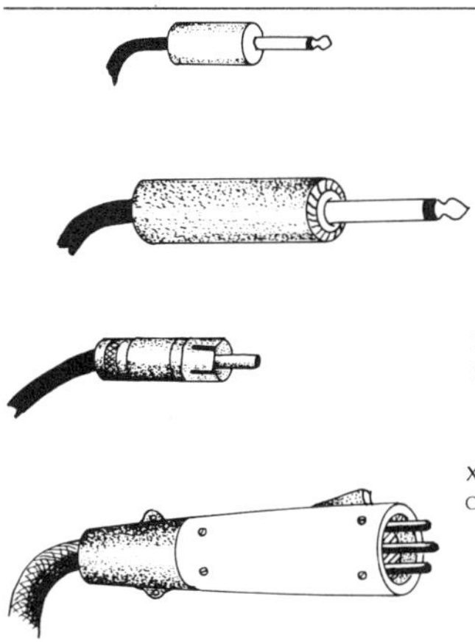

Name	Found on the end of a BALANCED or UNBALANCED LINE	Used with a HI Z or LO Z mike	Used with
MINI PLUG	UNBALANCED	usually HI Z	Audio cassette tape recorders; ½" VCRs; Small portable equipment
PHONE PLUG	UNBALANCED	usually HI Z	½" VCRs; Reel-to-reel audio tape recorders; Most school AV equipment
RCA or PHONO PLUG	UNBALANCED	usually HI Z	Some reel-to-reel audio tape recorders; Nearly all phono turntables
XLR or CANNON PLUG	BALANCED	usually LO Z	Most mike mixers and other audio equipment of high quality; Nearly all good microphones

Figure 7–1
Various plugs

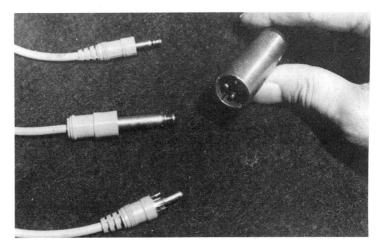

Figure 7–2
Audio plugs. Quiz yourself: Can you name each?

> BALANCED LINE
> An audio cable with three conductors (two center wires plus a braided shield). It terminates with a 3-pin plug such as an XLR or CANNON plug (shown in Figures 7-1 and 7-2). This kind of wire is usually found in expensive or professional equipment, usually with LO Z microphones. The BALANCED LINE is extremely impervious to extraneous electrical interference; or, put another way, unwanted buzzes and hums don't sneak onto your recording.
>
> *The following terms describe various popular audio plugs:*
>
> CANNON PLUG
> The Cannon company makes plugs, including what it calls the XLR plug shown in Figures 7-1 and 7-2. When Switchcraft makes this particular plug, it calls it an A3M plug. Similarly, other companies will call the plug other names. For some reason, the word "CANNON" has become synonymous with this particular kind of plug, much as the words *kleenex, xerox, scotch tape,* and *RCA plug,* though they are company or brand names, have taken on a generic meaning for particular items. This is a common phenomenon. This book uses the most common terms encountered in the field, so don't be surprised to see some items identified by company names.
>
> MINI, PHONO, RCA, PHONE
> Popular types of audio plugs used with UNBALANCED LINES. They all handle both HI and LO LEVEL signals, but the RCA and PHONO plugs are more often associated with HI, or LINE LEVEL signals. To confuse the issue, RCA plugs are also used on most home video equipment to carry video signals. PHONE plugs are frequently used to handle LO, or MIC LEVEL signals. They are also often used for headphones.

It must be becoming abundantly clear to the reader that audio has a language of its own. Although the vocabulary may be burdensome, how it all works is reasonably simple. That's why this chapter is so short.

The Microphone

HOW IT WORKS

Sound vibrates a mechanism inside the microphone, making a tiny electric signal that passes down the mike cable. Many things can happen to that signal before it gets to its destination; nearly all of these things are electrical in nature.

KINDS OF MICROPHONES

The words ELECTRET CONDENSER, DYNAMIC, and CRYSTAL refer to what's inside the microphone that makes it work. These components contribute to the microphone's characteristics: its fidelity, its sensitivity, its ruggedness, and its cost.

The words OMNI-DIRECTIONAL, CARDIOID, UNI-DIRECTIONAL, and SHOTGUN describe in which direction(s) the microphone is designed to hear.

You'll see later how the various kinds of microphones are selected for different situations.

IMPEDANCE

IMPEDANCE is an electronics term. LOW IMPEDANCE stuff is designed to work together, and HIGH IMPEDANCE stuff works together. Never the twain should meet except through an IMPEDANCE MATCHING TRANSFORMER which changes the one kind of electronic signal into the other. Generally, small, inexpensive audio devices are HI Z; large, expensive ones are LO Z. The IMPEDANCE of a microphone is usually stamped on it somewhere. The better microphones have a HI/LO Z switch enabling them to work with either HI Z or LO Z equipment. Sometimes the mike inputs on the equipment you are using will also specify the IMPEDANCE the equipment was designed for.

To work correctly, a HI Z microphone must be plugged into a HI Z microphone socket, and a LO Z mike must go into a LO Z socket. Some microphone mixers and amplifiers have switches near the sockets that will change the IMPEDANCE of the input, thus allowing either HI or LO Z mikes to be used. Behind those switches are IMPEDANCE MATCHING TRANSFORMERS (in case you were curious). Usually, home VCRs have microphone inputs which are a compromise between HI and LO Z, something one might call LOWER MIDDLE Z. Such machines have no HI/LO switches and will work with both kinds of mikes.

BALANCED AND UNBALANCED LINES

BALANCED and UNBALANCED LINES are two ways the mike cable can carry the signal to its destination. The higher quality BALANCED LINE has two conductors and a metal shield inside the cable. It can carry signals a long (over 50-foot) distance yet will pick up very little stray electrical interference (called "noise" by the experts). Most home video equipment is designed for UNBALANCED LINES. These cables have only one conductor and a shield inside them, making them thinner than their BALANCED counterparts. The mikes, the plugs, and the wires for UNBALANCED systems are inexpensive and simple to maintain. Since home VCR users generally work close to their machines, use the mediocre mikes sold in hobby shops, play back their sound through the tiny speakers on portable TVs, and usually aren't very discriminating about their sound anyway, UNBALANCED LINES are good enough for these users. There is one very strange problem some users encounter when using UNBALANCED LINES—radio interference. While recording, they may hear police calls, CB (Citizens' Band radio), nearby TV stations, or nearby radio broadcasts. These problems can often be ameliorated by using BALANCED LINES—if your mikes and VCRs have that capacity. Since most home video equipment does not work with BALANCED LINES, you may have to call for technical help to get rid of the unwanted interference.

ADAPTERS

Since there are so many different kinds of plugs and sockets, various IMPEDANCES, and BALANCED versus UNBALANCED LINES which have to mate, there must be adapters to make it possible to connect one thing to another.

Figure 7–3 diagrams some most-common audio adapters and names them. It wouldn't hurt to learn these names.

So what do you do if you're lucky enough to come across one of those expensive, high-quality professional mikes with a BALANCED XLR plug, but your VCR takes only a cheapie UNBALANCED MINI plug? There are two solutions to this problem, one quick and cheap, the other expensive but better. Let's examine the better first:

LINE MATCHING TRANSFORMERS An INLINE TRANSFORMER is simply a device which has (1) a socket (usually XLR) at one end,

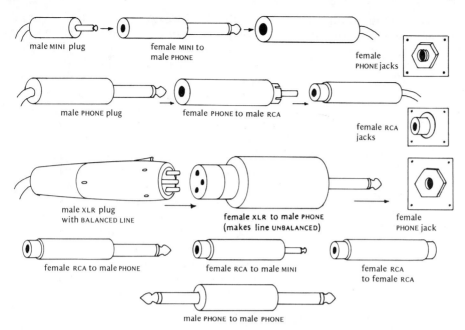

Figure 7–3
Audio adapters and their names

(2) a plug (usually MINI, PHONE, or PHONO) at the other, and (3) a circuit inside which changes LOW IMPEDANCE to HIGH or vice versa.

A BALANCING TRANSFORMER is a device which (1) has a socket for a BALANCED LINE (usually XLR) at one end, (2) has a plug using an UNBALANCED LINE (usually MINI, PHONO, or PHONO at the other, and (3) has a circuit inside which changes BALANCED LINES to UNBALANCED or vice versa.

Because people often have to interface a LOW IMPEDANCE, BALANCED LINE with an XLR connection, to their VCRs which have a HIGH IMPEDANCE, UNBALANCED LINE with a MINI, PHONE, or RCA connection, the INLINE and BALANCING TRANSFORMERS are often combined into one simple unit, often called a LINE MATCHING TRANSFORMER (shown in Figure 7–4). They can be purchased for $20–$25 from Comprehensive Video Supply as well as other electronics and audio-visual dealers. The LINE MATCHING TRANSFORMER will maintain the high fidelity of your microphone and take advantage of the "low-noise" qualities of the BALANCED LINES you're now free to use.

Rewiring If you have a mike with an UNBALANCED LINE terminated with an XLR plug, and you feel handy, you can lop off the XLR plug and solder on a MINI plug suitable for your VCR. Such a "quickie" fix *will negate* the advantages of the BALANCED LINE and will not guarantee a good IMPEDANCE MATCH, losing some fidelity, but it will usually work.

Here's how it's done (but skip the rest of this paragraph if you want nothing to do with electronics): The 3-pin XLR plug is removed from the end of the mike cable. One of the two center wires in the BALANCED cable gets soldered to its shield wire to make the BALANCED LINE, UNBALANCED. A PHONE or PHONO or MINI plug is now wired in the XLR's place. The shield wire attaches to the new plug's shield or "cold" terminal, while the remaining central wire attaches to the plug center or "hot" terminal. It's also possible to buy an adapter that has been similarly rigged (shown in Figure 7–3). It has an XLR plug on one end to service mikes and so on with BALANCED LINES, and a PHONE or some such plug on the other end to plug

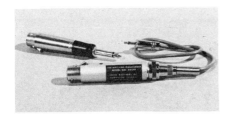

Figure 7–4
LINE MATCHING TRANSFORMERS, INLINE TRANSFORMERS, or BALANCING TRANSFORMERS

into cameras, etc., which take UNBALANCED LINES.

All the good things that a BALANCED LINE does are negated when you make it UNBALANCED. Even though the aforementioned adapter is attached to the *end* of the BALANCED mike cord, it makes the whole cable UNBALANCED and thus more subject to unwanted noises.

Choosing and Using the Proper Microphone for a Recording

GENERAL RULES

You want the best sound. You know that LO Z mikes with BALANCED LINES will introduce the least noise into your signal, so use them if you have them.

The closer the microphone is to the subject, the clearer the sound will be. However, placing the mike too close to the performer (say 3 inches) results in the mike's picking up unsavory snorts and lip noises. If the letter p has too much punch (audio experts call this "popping your p's") the mike may be too close. The farther the mike is from the talent, the more room echo there will be. The voices will also sound distant, while little background sounds will become exaggerated.

This is the situation that mitigates against using the camera's built-in mike. Picture your performer sounding 20 feet away—because she is 20 feet away—but *you're* zoomed into a tight close-up of her face. The distant sound and the close face just don't go together. A separate mike near the performer would be more appropriate.

ONE PERSON, ONE MICROPHONE

The talent (TV term meaning performer) is sitting or standing while talking. Unless aesthetically unsatisfactory, a LAVALIER (abbreviated LAV) or LAPEL CLIP mike is best. Figure 7-5 shows how one should be connected. Being close to the speaker, the mike rejects echoes and extraneous room noise in order to produce crisp presentations.

If the neck cord is too tight, the LAV mike will pick up too many throat sounds, making the speech hollow or muffled. If tied *really* too tightly, the LAV could choke the performer, resulting in a gagging, gasping sound. Connected too loosely, the LAV mike won't pick up as much sound from the speaker. It may drag on clothing and rattle against buttons. Good-quality LAV mikes are designed specifically for use as LAVALIERS and give poor sound when gripped in the hand. Some less-specialized mikes (less expensive with less quality) may be used as LAVS but are really HAND-HELD mikes which come with a string for hanging them around your neck.

If the performer expects to move around

Figure 7–5
LAVALIER and LAPEL mike placement

a lot, it is good to attach the wire to his body or have him hold some wire in his hand so that it will trail along easily. This is where a HAND-HELD/LAV mike may be handy; as the talent becomes active, he holds the mike. As he settles down, he rehangs it about his neck.

If the talent is too active, or if for aesthetic reasons the mike or cable must not show, the answer is a BOOM microphone (seen in Figure 7–6). The term BOOM microphone actually refers to the mechanism that holds the mike rather than the mike itself. Various mikes can be used on a BOOM, generally CARDIOIDS and UNI-DIRECTIONALS. The BOOM itself can be either a giant, wheeled vehicle or a simple "fish pole" with a microphone on the end. These mikes are used mostly for dramatic productions.

If a LAV mike must be hidden, its cable can be threaded up a pants leg or under a shirt. You can hang it just beneath the shirt, or, better yet, you can sneak the "head" of the mike out between the buttons on the shirt so that the sound is not muffled by the shirt. Mikes can be hidden at the cleavage in a bra, under a carnation, in lots of places. There are stories to tell about the places microphones have been hidden.

LAV and BOOM are usually the best ways to mike a single speaker. Next best is a CARDIOID or UNI-DIRECTIONAL mike placed on a stand and aimed toward the performer as shown in Figure 7-7. Though the room echoes will become significantly more noticeable, these mikes will reject much of the unwanted room noise.

If your desk mike must remain out of view of the camera, try hiding it in a prop, perhaps amongst the flowers in a centerpiece or camouflaged as part of some tabletop artwork.

One excellent way to mike a single speaker is with a SHOTGUN mike. These mikes can hear in one direction only, and they reject with a vengeance sounds coming from the sides or rear (see Figure 7–8). Because the SHOTGUN is so directional, it can be placed a little farther away from the speaker than the standard desk mikes, thus keeping it out of your picture.

The *least* desirable way to mike a performer is with an OMNI-DIRECTIONAL mike at any distance from the person. The OMNI-DIRECTIONAL mike will pick up echoes galore, along with the backstage sounds of people shuffling, kids squabbling, or neighborhood adolescents testing their car speedometers.

Remember, the mike on your camera is often an OMNI and is usually quite a distance from the talent, yielding a very echoey recording.

218 Recording Audio

Figure 7–6
BOOM
microphone

Figure 7–7
Stand and
desk micro-
phones

Figure 7–8
SHOTGUN mike

TWO PEOPLE, ONE MICROPHONE

It is better to have a separate mike for each person; that way you can adjust the volume of each source independently. If this option isn't available, then we try to get one mike to hear two people and the least room noise. A very close OMNI may work okay if the room isn't too echoey.

A BOOM mike or a SHOTGUN mike might also work if the mike has to stay out of the picture and if you have an operator to aim it.

As is done on many game shows and news interviews, the emcee or reporter can hold a mike (preferably a CARDIOID type) and direct it to each of the respondents to catch their replies.

SEVERAL PEOPLE, SEVERAL MICROPHONES

A LAV for each speaker is the best situation. This way, you can adjust each person's mike volume independently. But to put all these signals into your VCR's one input, you'll need a mike MIXER (to be discussed shortly).

If you run out of microphones and inputs for those microphones, on your MIXER, you'll have to compromise. Try grouping the people into threes or so and aiming a CARDIOID microphone toward each group. If you are severely limited in the number of microphones you can use, try planting an OMNI-DIRECTIONAL mike in the middle of the group with a LAV or CARDIOID mike delegated to the group leader or to the most important speaker.

For news conferences, where several individuals will be using a podium and others from the audience will be asking questions, place one mike at the podium and place another on a stand in the aisle for the audience. The intent is that the audience members will step up to the mike in the aisle when asking their questions. If a loudspeaker system is also in use for the conference, these microphones should be CARDIOID or UNI-DIRECTIONAL in order to reject as much of the loudspeaker sound as possible, thus avoiding FEEDBACK. Another way to mike a news conference audience is with a SHOTGUN microphone and an alert assistant to aim the microphone toward each person speaking.

If you have two mikes and two people speaking and don't want to bother with a MIXER, one cheapie trick is to get a Y ADAPTER (like Figure 4-10, except it has two *female* receptacles and one male). You plug the two mikes into the Y ADAPTER, which combines the signals and shoots them out one wire to your VCR's MIC IN. For this to work, both mikes should be of similar construction, both persons of about equal loudness, and all the plugs need to mate together right, which may require a few more adapters.

RECORDING MUSIC

If the performers are singing, fidelity is paramount. LAVALIER mikes are generally designed for speech and are therefore inappropriate. The best fidelity usually comes from the larger HAND-HELD mikes (which can also be used on a MIKE STAND or a BOOM). If possible, mike each singer separately for individualized volume control, and keep each performer 1 to 2 feet from the mike. If you run out of mikes, group the singers.

Musical recording is a science in itself. If it is necessary to group the musicians, do it so that the lead has a separate mike from the rest, the rhythm gets a mike, the bass gets a mike, the chorus shares a mike, and related instruments share microphones. This way you have independent control of the volumes for each *section* of the band.

Where do you place a mike for recording musical instruments? Stringed instruments such as violins, mandolins, banjos, guitars, cellos, and basses deflect their sound forward from their soundboards. Place the mike 2 feet in front of these instruments aiming toward them. Horns, drums, and electric guitar speakers also shoot the sound straight out but are louder. Place the mike 3 feet away or more, aiming toward the source. Loud rock bands can be miked from 50 to 100 feet away; closer will probably cause distortion. Woodwinds and flutes tend to make airy breathing noises near their mouthpieces. Mike them from behind the performer, aiming the mike over the talent's shoulder. Pianos are hard to do. Try placing your mike several feet from the high strings, aimed diagonally at the low strings.

BANISHING UNWANTED NOISE FROM A RECORDING

Wind Even a slight breeze over a microphone can cause a deep rumbling and rattling that sounds like a thunder storm in the background of your recording.

Solution #1: Stay out of the wind, and don't interview politicians.

Solution #2: Buy a windscreen, a foam boot that fits over the mike and protects it from breezes while letting other sound through.

Solution #3: In a pinch, take off your sock and put it over the mike to deflect the wind. Be prepared for wisecracks like "Your audio stinks."

Hand noise The shuffling and crackling of nervous hands holding a microphone can be avoided. Set the mike on a stand or hang the LAV around the talent's neck with the warning "Don't touch it, don't touch it, don't touch it." If the performers *must* handle the mikes, tell them merely to grip the mike and not to fidget with or fondle it.

Stand noise The mike is on a table stand and every time the talent bumps the table, it sounds like a kettle drum rolling down a stairwell.

Solution #1: Have the talent keep their hands and knees still.

Solution #2: Insulate the base of the mike stand from the table with something spongy like a piece of carpet, a pad, a tissue, or a pizza.

LAV noises Too tight a cord results in excessive throat sounds. A very loose cord results in a mike that swings like a pendulum and bumps things. Find the happy medium. *Before recording anything from a LAV, first check to see that there are no buttons or tie clasps for the mike to clank against.*

Mouth noises Peformers love to put their lips to the mike. Perhaps they don't trust the wizardry of electronics to sense their feeble sounds from a foot away and amplify them to spellbinding proportions. As a result, two things happen. When the performer speaks loudly, the sound distorts. When the performer pronounces the letters t, b, and especially p, it sounds like bombs bursting in air. **Solution:** Teach performers to trust the mike. Have them keep their distance. Cover the top of the mike with erect porcupine quills. **Another solution:** Buy the very expensive professional mikes designed for such abuse. They filter out most of the mouth noises and can also withstand the excessive volume found one-half inch from a rock singer's lips.

Room noise The closer the mike is to the performer and the louder the performer speaks, the less room noise will be heard. So what do you do if she starts "popping her p's," as described above? As shown in Figure 7-9, place the mike *at an angle to the side* rather than directly in front of her mouth. The offensive consonants will fly straight forward, hurting no one, missing the microphone.

Figure 7-9
Microphone placement to avoid "popping 'p'" at close range

As mentioned before, CARDIOID and UNIDIRECTIONAL mikes are best for rejecting extraneous room noise.

Try to place your performer in a quiet part of the room, away from windows, fans, and TV loudspeakers.

Feedback Sometimes known as "howl" or "back squeal," this familiar screech or whoop is very common when you monitor your recording with a TV set that has its sound turned up. It results when sound goes in the microphone, gets boosted by the TV, and comes out of the speakers only to be picked up by the microphone again and amplified. It goes out through the speaker and into the mike; 'round and 'round it goes, getting louder all the time. Figure 4-28 diagrammed the process.

Sometimes instead of getting a screech you get a hollow sound or a faint ring accompanying your sound. This again is FEEDBACK.

The immediate solution is to turn down the TV-set volume. Of course this negates the purpose of having it on in the first place, to hear how your recording is coming along. To get around the FEEDBACK problem yet still monitor your recording, you could:

1. Turn your TV so that its speaker doesn't aim toward your mike.
2. Turn your mike so it doesn't aim toward your TV speaker.
3. Use a CARDIOID mike which won't pick up as much of the unwanted sounds.
4. Keep the TV volume high enough for you to hear but low enough so your mike doesn't hear it.
5. Keep as much distance between the TV and the mike as possible.

A better solution than any of the above is to use an earphone to monitor your sound and turn off the TV sound.

TESTING A MICROPHONE

"Clunk, clunk, blow, blow, testing—1—2—3—testing—1—2—3—testing—testing. . . ." Such is the traditional prelude to every sound recording.

The ceremony of testing microphones has two steps. The first is to find out if it is working *at all*, and the second is to make it work well. The process is easiest if you have a helper at the controls to adjust volumes and to monitor the sound. After you have plugged the mike into its proper receptacle (MIXER, VCR, or whatever) and turned up the volume for that mike, you're ready for the tests.

First tap the mike with your fingernail (while listening for the *clunk-clunk* over the monitoring speaker). This determines if the mike is working *at all*. Next, speak *normally* into the mike while your assistant listens to the sound on a TV monitor or through an earphone, or better yet, over headphones (which cover both ears and block out distractions). This determines *sound quality*.

One way *not* to test a mike is to blow into it. The ritual of blowing into a mike is like squeezing eggs in a supermarket to test for freshness. Some mikes are too fragile to withstand the "blow" test.

SOUND COLORATION

Every mike has its distinctive sound qualities. You're hardly aware of these differences in timbre and fidelity until you use two *different* mikes at once. Suddenly you'll notice one person sounding tinny and the other sounding rich and full. To avoid this little distraction, use the same kind of mikes on everybody.

Similarly, sound recorded in one room will be slightly different from sound recorded in another. Different background noises or room acoustics or mike placement will also color the sound. Sound recorded on an audio tape and mixed with "new" sound will be different. Take this into account when planning your shooting. To avoid drawing attention to the fact that your tape may have been made in several segments, use the same mike in the same room when recording narration. Use the same mike on your audio recorder if prerecording bits of narration. Make AUDIO DUBS using the same mike you used during your original sound track. These steps will minimize the changes in sound coloration you get from recording session to recording session.

Automatic Volume Controls

This feature, sometimes abbreviated AVC or AGC (for Automatic Gain Control—*gain* is another word for *volume*) is present on all home VCRs and on some industrial units. The advantage of AGC is obvious: Recordings are made automatically at the right volume level—no muss, no fuss. It's all done with a circuit that "listens" to the audio and if it gets too loud, it turns the volume down. If it gets too low, it turns the volume up.

There are cases where AGC is not helpful at all. Say you were using an AGC machine to record an interview in a blacksmith's shop. The talent speaks, everything sounds fine, then somebody's hammer strikes an anvil. The AGC reacts to the loud sound by lowering the record volume drastically and then slowly raising it again to the level appropriate for speech. The recording could sound like this: "Under the spreading chestnut tree, the village smithy sta —WHANG! ......ty man....he....ith...and sinewy hands. And the muscles of —WHANG! ......rms...strong....iron bands." It would be better if the loud noise came and went in a flash, leaving most of the speech intact, like this: "Under the spreading chestnut tree, the village smithy stands. The smith, a mighty man is he with —WHANG! ...and sinewy hands. And the muscles of his —WHANG! ...arms —WHANG! ...strong as iron bands."

AGC is similarly troublesome in situations where long quiet pauses occur between sentences. When the talent stops speaking, the AGC circuit "hears" nothing and slowly turns up its volume. Still "hearing" nothing, it turns the volume up higher and higher. Turned way up, the machine records every little noise in the room, shuffling, sniffing, VCR-motor noise, some electronically caused hum or buzzing, automobiles outdoors, and the like. Then the first syllable out of the talent's mouth after this long pause is thunderously loud because the volume is far too high for speech and hasn't yet turned itself down.

In short, AGC is helpful when you expect a fairly constant level of sound. AGC doesn't like long silent pauses or short loud noises. Unless you switch to an industrial type VCR with manual controls, you'll be plagued by these AGC problems. Luckily, most sounds are quite steady and only rarely is this AGC difficulty noticeable.

Recording Stereo Simulcasts

FM radio is a high fidelity medium. TV is fairly low fidelity. It would be nice if you could record the picture part of a musical broadcast off your TV (through your VCR's tuner) and record the audio part from your FM radio. Some broadcasters are making this possible by presenting SIMULCASTS, the simultaneous broadcasting of a program on both FM radio and TV. To record it on your VCR, you simply tune the VCR to the TV station for the picture, tune your FM radio to the right station for the sound, and then find a way to get the radio's sound into your VCR.

For monaural VCRs and radios, you connect a PATCH CORD from the radio's TAPE OUT or AUX OUT to the VCR's AUDIO IN, like in Figure 4-5. If the radio only has an EARPHONE output (high level signal) or your VCR only has a MIC input (for low level signals), you will get raspy sound unless you run the signal through an audio ATTENUATOR (explained in Chapter 4).

If your FM receiver is stereo, and your VCR is monaural, then you have to combine the left and right channels before routing them to your VCR. Sometimes a switch on the radio will make it mono. In such cases, switch it to mono and then connect *either* of the two outputs to your VCR's AUDIO IN. If there is no switch, you can usually combine the two radio outputs with a Y ADAPTER (shown in Figure 4-10) and send the mixture to the VCR.

If you have a stereo FM receiver, stereo SIMULCAST, and a stereo VCR, you can record the show in living stereo. Using two PATCH CORDS, you'll connect the radio's left TAPE OUT to the VCR's left AUDIO IN, and the right to the right. (Unfortunately, the JCV HR-7650 and Akai VS-7U won't listen to their TUNERS

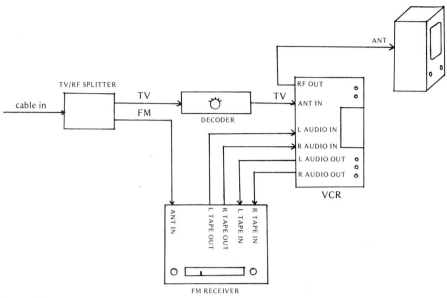

Recording stereo SIMULCASTS

and AUDIO INputs at the same time, making SIMULCAST recording tricky.)

How do you listen to your new stereo recording or monitor it while recording? You could plug earphones into the VCR and listen in private. Or, you could send the signal from your VCR back to your hi fi for everyone to hear. You need two PATCH CORDS again. One cable you connect between the VCR's left AUDIO OUT and the receiver's left TAPE IN. The other cable connects the right to the right. (For mono, you need only one cable.) Switch the FM receiver's SOURCE or INPUT SELECT button to TAPE while recording and here's what's likely to happen: The radio picks up the FM signal, sends stereo music out to the VCR, the VCR records it while sending a sample back to the radio which plays it in stereo through the speakers. Thus you hear what you're recording. Switch the VCR to PLAY, and the stereo tape will play back through the radio.

One more thing. Sometimes the cable TV companies use signals from far away—too far away for your FM receiver to pick up.

In this case, get a TV/FM SIGNAL SPLITTER (see Chapter 2) and connect your FM radio to the cable. Thus, the TV signals go to your VCR (or DECODER BOX) while the SIMULCAST FM signals go to your stereo receiver where you can tune them in.

Mixers

A MIXER accepts signals from various sources, allows each signal to be individually adjusted for loudness (even adjusted all the way down to no loudness at all), and sends this combination of adjusted signals to a VCR or some other recording device.

ADJUSTING AUDIO LEVELS ON A MIXER

Have each performer speak into his or her mike while you twiddle the volume controls and listen to some monitoring device, perhaps headphones. Label which performer's sound is controlled by which knob so you can find the right ones quickly during your show.

If the MIXER has a meter, adjust each volume control thusly:

1. Select one performer to test.
2. Have him speak *normally* into his mike (perhaps saying his A,B,Cs).
3. Turn that mike's volume control up until the meter wiggles just below the red area. If the meter doesn't wiggle check:
   (a) That the mike is plugged in.
   (b) That the mike is turned on.
   (c) That the MASTER volume control is turned up (say halfway, anyway).

The meter should wiggle when people speak, but it should rarely dip into the red area. "Rarely" doesn't mean never; loud outbursts are expected to sweep the needle into the red for a moment. If possible, listen to the sound on headphones or a speaker in order to judge the sound quality. Is the sound distorted? Is there a buzz, or a hum, or a hiss in the background? If so, something is wrong. Off to the troubleshooting section, Common Audio Ailments and Cures, at the end of this chapter.

Once the audio level is set (during checkout while you were setting up), the circumstances of the production will dictate whether it will need to be adjusted again. If the audio is from a professionally prerecorded source or if the audio is from a speech or dictation in which the loudness was relatively even, it is possible to make the entire recording without twiddling the audio knobs and with only occasionally checking the meter. If the sound source changes its volume frequently or drastically, you may have to "ride audio," twiddling the knobs and watching the meters intently.

How much do you twiddle? Answer: the least amount possible to keep the audio level about right. You're twiddling too much if every shriek or cough makes you turn the volume down. Brief noises are loud, yes. They sound distorted when recorded at high volume, yes. But they are gone in an instant and easily forgotten. Brief pauses and whispers are quiet, yes. But whispers are supposed to be quiet and silent pauses are not unnatural for us to hear and may even be a refreshing break from the monotony of constant chatter. While playing back a tape, it is irritating for the viewer to have to rise from his comfy seat to readjust rising and falling volume because of some over-zealous knob-twiddler who adjusted the record level too often. In short:

1. React quickly to sustained bursts of noise or substantial passages which would be lost if not adjusted for.
2. Don't react to momentary sounds or silences.
3. As conversations ebb and flow in volume, gradually make tiny adjustments in order to compensate. Do it so that no one will notice that the volume is being changed.

INPUTS TO THE MIXER

Microphones are generally plugged into the MIC INPUT sockets in the back of the MIXER, up to one mike for each corresponding mike volume knob on the front of the device. On the more professional models, the inputs are the XLR types and have a little pushbutton near each socket to release the plug so that it can be removed.

All the microphone inputs are LO LEVEL INPUTS, which means that they accept tiny signals only. It is possible to use (besides microphones) telephone pickup coils, guitar pickups (the signal comes on a wire directly from the guitar itself, not through an amplifier), tape heads (from unpreamplified audiotape decks), and phonograph turntables with these inputs because the signals are tiny and because the MIXER's inputs are very sensitive. Although turntables and tape heads have the right signal strength for use with MIC INPUTS, these sources may sound tinny or bassy when used this way because they require extra circuits to make the tone right (in a process called *equalization*). Some audio equipment accommodates these sources by having special inputs labeled TAPE HEAD or PHONO where the equalization is built in.

The MIC INPUTS are not designed for stronger signals like those coming from an FM tuner, a VCR AUDIO OUT, any earphone output, any speaker output, any preamplifier output, or just about anything which boosts a signal before sending it out. For these LINE LEVEL or HIGH LEVEL sources, one must use a different input on the MIXER (if it has such): the AUX (for AUXILIARY) IN or the LINE IN.

These inputs are less sensitive than LO LEVEL inputs and work well with stronger signals. Use AUX or LINE IN when you have some musical background or sound effects which you wish to mix with the voices on the microphones. The HI LEVEL or LINE outputs of tape decks, FM radios, record players, cassette players, or similar devices can be connected to the AUX IN for this purpose.

The microphone inputs may have switches next to each of them that say HI Z/LO Z. These switches switch the INPUT IMPEDANCE of each mike input. When using a LOW IMPEDANCE mike, set the corresponding switch at LO Z. Meanwhile, if another mike is HIGH IMPEDANCE, its input gets switched to HI Z. If you don't know the microphone IMPEDANCE try the mike and switch from one position to the other. One position probably won't work or will sound noticeably terrible.

Once everything is plugged into the MIXER and is working, *label each of the MIXER's knobs* to tell which source is controlled by which knob. For example, near each knob, stick a piece of masking tape with the words "Joe" or "RP" marked on it. This way, during a production, you don't have to be asking yourself, "Let's see, is Control #1 Joe or is it the record player?"

OUTPUTS FROM THE MIXER

The MIXER's output sends the combined signals to the VCR or other device. Just as each microphone's volume is adjustable with a knob on the MIXER, the volume of the signal the MIXER sends out is also adjustable with the MASTER volume control. Usually this knob is a different color, shape, or size from the rest. Turning the MASTER down turns down all the signals coming out of the MIXER. This is a very convenient feature, especially at the end of a program that uses multiple mikes—you'd have to be an octopus to turn down all the individual mike volume controls simultaneously.

The MIXER may have several outputs. Although one output from a MIXER is all that is needed to feed a VCR, the others are there to permit flexibility in setting up and using the audio system.

MIC OUT or MIC LEVEL OUT or LO LEVEL OUT This is an audio output that has a tiny signal, like a microphone; thus it can be plugged into the MIC IN of a VCR or other device. To this end, the audio cable and plugs are just like microphone cables and plugs. It is as if the MIXER were pretending to be a microphone: It is putting out a signal just like a mike and the signal goes to wherever a mike's signal could have gone.

LINE OUT or HI LEVEL OUT or AUX OUT A medium-sized signal emanates from this output and is destined to go to a VCR's AUDIO IN.

HEADPHONE This is another output for monitoring the audio signal over headphones. This output has a fairly strong signal, one inappropriate for feeding VCRs.

MONITORING AUDIO

The mikes and other things feed into the MIXER and get combined. This audio signal could be fed to the VCR and monitored either on headphones connected to the mixer, headphones plugged into the VCR, or on a TV receiver connected to the VCR. What happens if you get no sound on the TV set? Is it the VCR's fault, the MIXER'S fault, the mike's fault, or what? If the MIXER has a meter and the meter wiggles when someone speaks into a microphone, the MIXER and mike are most likely working. If you can plug headphones into the HEADPHONE jack on the MIXER, you can make doubly sure the MIXER and mike are working. If the phones and the meter are silent, then the problem is probably in the microphone or in the MIXER. Try another microphone: If it works where the first one didn't, the first mike (or its cable or plug) is bad.

Once you've proven that the signal is playing through the MIXER, you turn your investigation to the VCR. Is the MIXER's output plugged into the VCR's input? Is the VCR in the RECORD mode? Remember that many VCR's don't allow you to monitor their signals unless they are in RECORD. Is the VCR's TV monitor volume turned up? Are all the cables tight?

More Words to Know

DECIBEL
A decibel is a measure of sound volume. Zero db (0 db) means zero decibels. On professional sound meters (called VU meters: see Figure 7-10) the minus db numbers (like − 20 db) are very weak sounds, the slightly minus numbers (like − 3 db) are slightly weak sounds, 0 db is the "perfect volume," and the plus db numbers represent excessive volumes. The more "plus db" your meter goes, the more distorted the sound will become. More than +3 db is noticeably bad, while between 0 and +3 is not too irritating.

Figure 7–10
VU meter showing 0 db

PEAK LEVEL INDICATORS
Another device for measuring excessive volume levels. VU meters, though very useful for measuring overall volume levels, do not react fast enough to show brief bursts of sound. So in addition to the traditional VU meter, some new equipment also has an LED (Light Emitting Diode) that flashes when sound transients exceed an allowable level. When you see your PEAK LEVEL lamp flashing frequently, turn your volume down a little.

> **PAD**
> A small plug-in device (or it can be wired in by a technician) that decreases high volume audio levels to an acceptable range.
>
> **SEGUE**
> Changing from one sound source to another by turning up the volume of one source while simultaneously turning down the volume of the other source.

CONTROLLING EXCESSIVE VOLUME LEVELS

Sometimes, no matter what you do, you (1) can't keep the meter out of the red, or (2) can't get a volume control to give proper volume unless it is turned almost off, or (3) get a sound that is raspy and distorted. This is a case where a VCR or MIXER is receiving more sound than it can handle. It happens when you connect a LINE OUT or HI LEVEL OUT to a MIC IN or LO LEVEL IN. It also happens when you try to use the signal from an EARPHONE or HEADPHONE OUTPUT from a radio or cassette tape recorder or the SPEAKER or EXT SPEAKER output from a hi-fi or similar device. Such signals are usually too strong, even for HI LEVEL INPUTS to handle.

The solution for this is to buy an AUDIO ATTENUATOR or a PAD. No, this isn't the paper tablet you use for writing home to the folks twice a year, a week after Mother's Day or Father's Day. It is an inexpensive little box with a circuit in it that *throws away* most of an audio signal and *passes on* a tiny fraction of it. Plugging a powerful signal into it and then plugging it into your MIXER results in the MIXER's receiving an acceptable signal even though an excessive signal was being put out at the source. A PAD can also be plugged in between the MIXER and the VCR to cut down the MIXER's volume. If, for instance, the VCR had no HI LEVEL AUDIO IN socket and you were forced to use its more sensitive MIC IN socket to receive your MIXER's LINE OUT, the PAD comes to the rescue. You can buy one from most any electronics store.

OTHER GADGETS ON MIXERS

IMPEDANCE SWITCH Next to each microphone socket in the back of the MIXER there may be a HI/LO Z switch that allows you to use either type of microphone in that circuit.

LOW CUT FILTERS No, a LOW CUT FILTER is not a filter with a plunging neckline. This is a switch near each volume control which may allow you to remove deep bassy sounds from a signal.

This is especially useful when you use a room filled with deep echoes or when you wish to make speech sound especially clear. By switching the filters to the IN position (putting them IN the circuit), you cut out the low sounds and leave only the mediums and highs for maximum intelligibility. You can then turn the filters off (switching to OUT, taking them OUT of the circuit) when you wish to record music with full fidelity.

MIC/LINE input Some MIXERS have a switch on one or more microphone inputs that change the inputs from LO LEVEL to LINE LEVEL to accept stronger signals. By switching this switch to LINE you can now use one knob to control an audiocassette player, for instance, while the other knobs control microphones.

MIXING WITHOUT A MIXER

Sometimes you have two sound sources you'd love to mix together, but don't have a MIXER to combine them. Here are some alternatives which *may* work, but come with no money-back guarantee that they'll work well.

Acoustical mixing Picture yourself driving through Dodge City with the twang of country music on the car radio. A passenger is taping the sights out the car window to send back East to the folks. The camera mike picks up the sound of the radio and adds that Western flavor to the view, and when narration or comment seems appropriate, you just lower the radio volume.

Similarly, you can DUB narration and musical backgrounds over existing visuals. Simply find a record player and select appropriate music. Set up your mike (or the mike built into the camera—it can make AUDIO DUBS even though the camera isn't recording any new picture) close to you and the phonograph. Perhaps start with the music at normal volume, then fade it down while you speak. Then fade it back up when you're finished speaking. Although the sound fidelity won't be too great, the music-and-voice mix will add pizazz to your tape. Also, you're likely to enjoy the thrill of being a true DJ as you attempt to coordinate switching on the VCR, starting the record, fading the volume, and making sense as you try to speak at the same time.

Y ADAPTER Figure 4-10 showed how this little widget could take your VCR's one audio output and send it to your stereo's two inputs. A Y ADAPTER with the right plugs can work the same way in reverse, taking two sources and combining their signals to send to one VCR input.

For this to work well, both sources must have about the same strength signal, like two microphones or two similar audio outputs from VCR's or audiocassette tape players. Since the Y ADAPTER has no volume controls, you can't stop a strong source from overpowering a weaker source. Also, you'll probably notice a drop in volume from one source when you plug in the second. So if you're combining two microphones this way, each will appear to become half as sensitive as it was before.

Stereo amplifier or audiocassette deck
Most stereo hi-fis today have lots of inputs in the back for tape players, tuners, phonographs, and even microphones. By turning your amplifier's INPUT SELECT switch to one of these positions, your amp will "listen" to that pair of inputs. Often you can plug a microphone into one of the inputs (MIC input works best, but others may work too—experiment) and talk through your hi-fi. Since stereos have two inputs (left and right channels) for every source, you could plug a *second* mike into the twin input and talk through *two* microphones. By adjusting the left and right channel volume controls or by adjusting the BALANCE control on your amp, you can adjust the loudness of each mike.

Now to get the signal into your VCR: Find an output on your amplifier which you can connect to the AUDIO IN of your VCR. This output may be labled TAPE OUT, AUX OUT, LINE OUT, PREAMP OUT, or if there are no outputs, use the HEADPHONE socket (you may need to buy an adapter or special audio cable for this). Now you have a stereo output to send to your VCR, but most VCR's aren't stereo. To solve this, either switch the STEREO/MONO switch on the amp to MONO, which will mix the two signals, or use another Y ADAPTER to take the left and right signals, mix them, and run them down one wire to your VCR. Either way, you now control the volume of two sources using your amplifier's volume controls, and send the result to your VCR for recording.

If FEEDBACK from your amplifier's speakers is a problem, find a SPEAKER OFF switch to cut their sound. Alternatively, you may mute the speakers by plugging something into the amp's HEADPHONE input. Or you could switch on the REMOTE speakers if your hi-fi has them, sending the unwanted sound to another room. Put pillows in front of the speakers if nothing else works. *Do not* simply unplug the speakers from the amp in order to silence them. This may damage your amplifier's circuits!

Usually anything you do with the speakers will not affect the signal to your VCR. Monitor your signal to the VCR with the VCR's earphone or with a TV speaker. Now all your mike mixing can be done by twiddling your amplifier's volume controls as if it were a genuine mike MIXER.

Many stereo audiocassette decks have microphone and auxiliary inputs and independent volume controls. By plugging your mikes into the audio deck, twiddling the controls, and sending the deck's output to your VCR, you can mix audio sources like you could with your stereo amplifer. You need a Y ADAPTER to combine the deck's stereo outputs into a single signal.

Stereo VCR A few VCR's today record in stereo. It is possible to send one audio source to the left channel and another source to the right channel and either record a stereo presentation or to at least get two sources onto the tape at the same time. When playing back the tape, you adjust the volume of the two outputs to perform your sound mix. The disadvantage with this is that every time you play the tape, you have to adjust the controls manually to perform the right mix. It would be nicer if the mix were made first, *then* recorded on the tape correctly, so you wouldn't have to mess with the machine during playback.

Sound-Mixing Techniques

There is no substitute for creativity. There are some basics, however, that could help. In fact, your library probably has several whole books written solely on the basics of audio; it gets that involved.

SEGUE (pronounced SEG-way)

SEGUE is like a DISSOLVE in video, only you're working with sounds. One volume control is lowered at the same time a second is being raised. This is often done between two pieces of music. A more sophisticated SEGUE uses an intermediate sound when changing from one audio passage to another. For instance, to go from one scene in a play to another, after the last line in the first scene come a few bars of appropriate music. As the music fades out, the first line in the next scene is delivered. Briefer things like jokes or single statements may deserve a sound effect, laughter, applause, or a single note or chord of music (such a musical passage is called a "sting") between them. Some SEGUES prepare the listener for things to come, like faint machinery noise before we open the engine room door, or the sound of windshield wipers before the actors begin to speak in the car on a rainy night. A famous Hitchcock SEGUE is a woman's screaming suddenly changing to the scream of a train's whistle as it chugs into the next scene.

MUSIC UNDER,
SOUND MIX,
AND VOICE-OVER

Your production begins with a snappy musical selection. The title fades in and then cuts to the opening scene. Someone is about to speak. The music fades down just before the first words are heard. This is a MUSIC UNDER. The music became subordinate to the speech and is played *under* it.

Sometimes you have to decide whether to fade the music out entirely when the action starts, or to MUSIC UNDER, holding it in the background throughout the scene. If the music is needed for dramatic effect, either to create a mood or just to provide continuity through long gaps in action or conversation, just keep it in. If, however, the action or conversation is very important, then don't distract your audience with background music.

So how loud should the background music be? The answer, of course, depends on the particular situation; there's no hard rule. In general, keep in mind that background music is *background* music. Keep the volume low—lower than your natural inclinations would have you set it. How many amateur productions have you sat through straining to hear the dialogue through that "noise" in the background? One guide to proper volume setting can be your MIXER's VU meter. If your narration makes the needle hover, as it should, at around 0 db (the 100% mark on the scale), the background should wiggle the needle around −8 db (about 40%) on the scale.

Again, these are just generalities; some musical selections are inherently more obtrusive than others. For instance, while listening to a narration, the viewer may hardly be aware of instrumental music in the background. Conversely, a song with words competes with the narration for the viewer's attention. Because singing with words is so distracting, it's best to avoid it in favor of instrumentals.

Not all background sound is music. Street sounds, machines, sirens, motors, gunfire—all can be background to your dialogue. Some of these sounds may not be background at all but are interjected between dialogue, such as "thud," "crash," and the like. How all these sound effects and backgrounds are woven together is called the SOUND MIX, and it's quite a feat to mix them effectively.

Sometimes you wish to add narration to a tape. The original recording may be silent or may have sound on it already. Say you want the original sound on the tape to be kept, but only as background to the narration. This is what's called a VOICE-OVER. The voice you're adding is imposed over (and louder than) the original sounds. Although this will be covered more completely in Chapter 11, here, briefly, is what happens:

Mr. Expert brings in a tape showing his foundry in action. The tape shows the busy machines while you hear them foundering away in the background. Mr. Expert also brings a script that he wishes to read through parts of the recording. What you do is set up your VCR to record the video from his VCR as he plays his tape. The audio from his machine you send to a MIXER. Also you connect Mr. Expert's mike to the MIXER. The MIXER combines and regulates these two sources and feeds the combination to your VCR for recording. As his VCR plays, your VCR records, copying the picture and whatever original sound the MIXER lets through. Mr. Expert reads his script as he keeps one eye on the TV monitor screen. You adjust audio levels, sometimes favoring the background sounds (when the narrator is silent) and sometimes lowering them (when the narrator speaks). That's a VOICE-OVER. If he doesn't like the way the final tape comes out, you can erase it and do it over, since his original tape was not altered in the process. If, after repeated redoing, he still doesn't like it, you may choose to alter his original tape—over his head!

More sound handling techniques will be covered in Chapter 11 in the sections on AUDIO DUBBING.

SEVERAL PERFORMERS, EACH WITH A MICROPHONE

When all the microphones are turned on at the same time, you not only hear the sound of the one person speaking, but you hear the breathing and shuffling of the others in the background. You also get the hollow echo of the speaker as her voice is picked up on everybody else's microphones. In order to avoid this, turn down all inactive mikes, allowing only the speaker's mike to be alive. This is easy to do in a scripted production like a newscast or a play, but it is difficult to do in a free discussion or the like. In such cases, you have to suffer the disturbing background sounds resulting from leaving all the mikes live, because if you turn off the unused mikes and then turn them up after a new speaker has started speaking, you will lose the first few words. One partial solution may be to lower by a third or a half the unused microphone volumes, raising them after the person begins speaking. Although the first words may be weak, they will be audible and will soon be at full volume. Allow the dynamics of the discussion to be your guide. If the speech is coming in a crossfire from everyone, leave all mikes open. If someone seldom speaks, lower that mike. If the speakers render long monologues, take a chance on turning down the other mikes until the speech appears finished.

Good audio requires lightning-fast reactions coupled with a dose of anticipation. Always make a habit of marking the proper volume settings on the MIXER knobs so you know where to turn them during the show. This preproduction planning can save guesswork, confusion, and precious time during the actual shooting.

CUEING A RECORD OR TAPE

You wish to push a button and have that *boing* happen instantaneously, right in sync with the action of the performer as he winds his watch. Once you find the *boing* on the record disc, you need to get it backed up to just before the *boing* so that it will play the instant the switch is thrown. If you leave too much space before the *boing*, when you play it during the production, you'll get ... *boing*. That's too late. If you don't back up far enough you'll get *"ing,"* the tail end of the sound, or *woooing*, the sound of the turntable picking up speed while the effect is being played. Proper CUEING of a sound effect (assuming an appropriate turntable is being used) goes like this:

1. Before starting your recording, set up your record player and adjust your MIXER volumes so that the sound effect is at the right level.

2. Turn the turntable motor ON and put it in gear so the disc rotates.

3. Locate the *boing* on the disc by sampling various grooves with the needle. The record label may even tell the contents of various bands on the record.

4. Once you find the *boing*, pick up the needle and set it down two or three grooves earlier in the record. You can anticipate that the *boing* will come up in a few seconds. Be ready to stop the disc from turning.

5. Some disc jockeys prefer to stop the disc manually with a finger (touching the edge of the disc and allowing the turntable platter to revolve beneath it). Others shift the turntable to NEUTRAL and use a finger to brake both the disc and platter together. Others turn the motor switch off, brake the platter with a finger (along the edge) and put the turntable into NEUTRAL afterward. Whatever method you use, somehow you've got to stop the disc when you get to the *bo* ... in *boing*.

6. Once you have the *bo* in *boing*, rotate the disc in reverse—with the needle still playing—to find the beginning of the sound. Sometimes it is hard to tell exactly where a sound begins, so with your finger, you rotate the record backward and forward, trying to recognize the sound of the beginning of the *boing*.

7. Once you've got it, back up the record one eighth of a turn more—one sixteenth of a turn for good turntables. (The better the turntable, the faster it can pick up speed and the less backing up you have to do to assure it has time to pick up speed before it plays the sound.)

8. Put the turntable into NEUTRAL with the

motor turned ON, or leave the turntable in gear with the switch turned OFF.

9. Begin recording your show.

10. When you want the *boing*, just switch the gearshift to the desired speed or turn on the motor switch.

Two things to watch out for:

1. If you are rotating the disc itself back and forth with your finger while the turntable is still moving:

> **(a)** Make sure that the turntable's felt platter is clean so it doesn't scratch the record as it rubs against it. If your turntable has a rubber platter, cut up some flannel or soft cloth making a disc with a hole in it and lay that over the platter. This will cushion and protect your records.

> **(b)** Try not to shake the disc as you move it or else the needle may jump to another groove.

> **(c)** Once you are CUED UP, you may wish to hold the record by its edge with one hand and switch the turntable ON, waiting for the performer to wind his watch *while the turntable keeps turning*. When the time comes, just release the record and out comes *boing*. If, however, the watch-winding isn't for a while, you may prefer to stop the turntable's motion, release the disc, and, at the proper time, start the turntable going again for the *boing*.

2. Prior to your actual TV production, be sure to check the proper volume level for your sound effects.

The CUEING process sounds roughly like this:

Prerecording setup	. . .
Record plays forward in a search for the sound effect.	CLUCK, CLUCK . . . PLOP, PLOP . . . BOING, BOING. . . .
Lift needle and back up a couple of grooves.	. . .
Play the record.	—LOP, PLOP . . . .
Get ready to stop the disc.	. . .
Stop the disc at the first sound.	. . . BOIN—
Rotate the disc counterclockwise to find the beginning of the sound.	GNIOB. . . .
Rotate the disc clockwise to find the beginning of the sound, then backwards, then forwards again, eventually finding the starting "B" in "Boing."	. . . BOI— —IOB. . . . —BO— —B— —B—
Rotate the disc backwards an extra one eighth turn.	—B. . . .
Leave the turntable stopped in that position.	. . .
Start your recording.	. . .
At the appropriate time, start the turntable.	BOING

There are some record players which are not designed to be run backwards. Assumedly, the needle digs into the record groove when the record rotates backwards, damaging both record and needle. I have never seen this happen, but some manufacturers recommend against this practice. If in doubt, try a careful experiment while watching for possible damage, or use old records and old turntables for this purpose.

CUEING UP other devices requires a similar technique. Since some machines don't run backwards, you may be required to play forward to the desired sound, stop, rewind a speck, play again, and, by trial and error, find the beginning of the desired passage.

Some notes about rewinding a speck: Nearly all reel-to-reel audio tape machines permit you to turn both reels physically by hand. So if you wish to back up the tape a little ways, switch the machine to STOP and then manually roll the tape backward (or forward, if you wish) from one reel to the other to get to the exact spot on the tape that you want.

Audiocassette tape recorders cannot be rewound a speck by hand. Here you'll have to play the audio passage, memorize what sounds come after what, and then while playing the passage again, hit PAUSE on your audiocassette player just before the desired sound effect. When the right time in the video recording comes up, just UNPAUSE the audiocassette machine and pray your guess was accurate.

Special Audio Devices

Chapter 14 (paperback Volume II), Special Audio Devices, will explore some of the audio goodies you could hope to find under the Christmas tree. Meanwhile, here are the two most common audio options, DOLBY and GRAPHIC EQUALIZER.

DOLBY

Hi-fi enthusiasts use DOLBY to diminish background hum, hiss, and noises in recordings.

SOME AUDIOPHILES TAKE GREAT PRIDE IN THE SILENCE OF THEIR SYSTEMS DURING QUIET PASSAGES

By background noise, I don't mean the kind of sound that gets recorded as you interview someone on a busy street corner. DOLBY won't reduce the unwanted sounds that sneak into your microphone because of noisy surroundings. What it *will* reduce is the hum and hiss created *internally* by imperfections in your tape and recording equipment.

It would be nice if the sounds you *wanted* to hear were 100 times stronger than the noise which sneaked into your recordings. Often this is the case, but sometimes when your music or speech or whatever is very quiet, maybe 1/100 of its normal volume, the background noise is as loud as your music or speech. Thus, in quiet passages, the noise becomes quite noticeable.

DOLBY electronically raises the volume of the quiet musical passages during the recording in an attempt to keep the music many times louder than the noise. To make everything come out right, DOLBYIZED recordings must be *played back* through a DOLBY system to deemphasize the boosting which occurred in the recording process. The result: Quiet *and* loud passages sound normal through the DOLBY system, while background noise is almost eliminated.

DOLBY is built into a few models of VCRs like the Akai VP-7350US and JVC HR-7650U, making a marked improvement in their sound. Home VCRs are notorious for their poor quality sound because of their very slow tape speed. If your VCR doesn't have DOLBY but you wish it did and are willing to spend the bucks, you can buy a stand-alone DOLBY ENCODER/DECODER. You could build a Heathkit DOLBY or buy a unit from Integrex, Box 747, Havertown, PA, 19083. You would plug your mike (or other source) into the DOLBY ENCODER, then plug the ENCODER into your VCR's AUDIO IN. Now your VCR will be recording DOLBYIZED audio. Upon playback, the audio must be de-emphasized, so you run your VCR's audio output through the DOLBY unit on the way to a separate amplifier and speaker. You can't use your TV speaker as before because the RF going to your TV is carrying DOLBYIZED sound and that's what your TV will play. The sound will be tinny and shrill. You have to treat your VCR's audio separately to unDOLBY it (unless the DOLBY is built into the VCR).

Note that an off-air recording or a prerecorded tape that wasn't DOLBYIZED *during recording* is beyond help. DOLBY can't fix it. DOLBY only helps you when you can both record and play back through a DOLBY unit.

GRAPHIC EQUALIZER

Sound is made up of high, medium, and low tones. Bass and treble controls can boost or diminish these tones. Sometimes it would be nice to boost or diminish *one particular tone,* not messing up the rest of the highs and lows in the process.

A GRAPHIC EQUALIZER does this. It contains selective filters that allow you to pick a sound frequency you don't like and remove it, while passing the rest of the audio untouched.

A word about frequency: The higher a sound's frequency, the higher the pitch or tone. The lower the frequency, the deeper and more bassy the tone. Frequency is measured in Hertz, abbreviated Hz. Sixty Hz means 60 vibrations per second, a fairly low-sounding hum.

Healthy young people can hear in the range from 20 to 20,000 Hz. Dogs can hear even higher frequencies. Old people generally hear 100 to maybe 10,000 Hz. Good hi-fi systems produce the full range of 20–20,000 Hz. Good AM radios produce 40–8,000 Hz, while good portable radios work in the range of 50–4,000 Hz. TV stations broadcast their sound within the 50–15,000 Hz range. Telephones transmit 200–2,500 Hz. Most speech occurs in this range.

Thus if you want to improve speech sounds, you boost the 200–2,500 Hz frequencies while cutting down on the others.

Hum from a poorly grounded audio system can be diminished by filtering out the 60 Hz frequency. The rumble of wind can be diminished by filtering out the 30 Hz frequency. The high-pitched whine of an electric motor may require removal of the 10,000 Hz frequency. A dusty record may need removal of the 15,000 Hz frequency.

By filtering out the unwanted frequencies, and by boosting the wanted frequencies, you get a recording tailored to your needs.

Common Audio Ailments and Cures

Troubleshooting audio difficulties is like prospecting for gold nuggets. Putting them in your pocket is easy—it's finding them that is hard. Similarly, finding the source of an audio defect is most of the battle. Once the offending machine or cable is located, it is a simple task to check the obvious connections and switches, and then, if the device appears defective, to find a substitute and to send the dud off for repairs.

As you prospect for audio problems, keep in mind where the signal is going and where it may be interrupted. Figure 7-11 shows the progression of several audio signals through a system. Sound can be ruined at its source, in the cables, at the MIXER, in the MIXER's cables to the VCR, at the VCR, or in the cables from the VCR to an amplifier and loudspeaker.

One good place to start prospecting is at the MIXER. If the MIXER's meter doesn't wiggle and the MIXER headphones make no sound, the problem is *not* likely to be in the VCR or

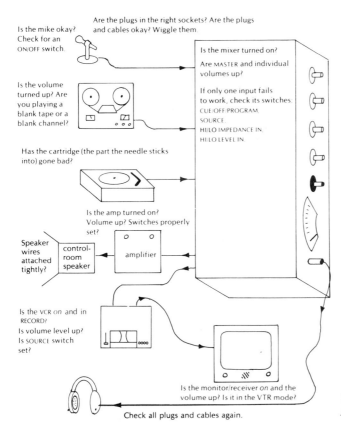

Figure 7-11
Audio troubleshooting

its monitors. The problem is either in the source or in the MIXER. To find out which it is, try two sources into the MIXER. If something plugged into one of the MIXER's inputs gives no sound while a second source plugged into another input does, the problem is most likely in the first source. It is also possible that a switch or knob controlling the first source's MIXER input is maladjusted.

If the MIXER's meter does wiggle or if there's a signal over the MIXER's headphones, you know the source being tested *and* the MIXER are okay; your problem, if you're having one, is downstream from the MIXER.

Downstream from the MIXER is the MIXER-to-VCR cable, the VCR, and the VCR's monitor. If there's no signal on any of these devices, suspect the MIXER-to-VCR cable or the VCR.

To test the VCR, plug an audio source directly into the VCR. If it yields sound, the VCR is working properly and your problem was in the MIXER-to-VCR cable. If the test yielded no sound, the VCR is either malfunctioning, or its RECORD button isn't pushed, or its controls are misadjusted. Check your INPUT SELECTOR.

MIXER PROBLEMS

MIXER passes a signal from all but one microphone The problem is in that microphone, cable, plug, or MIXER input. Does the mike have an ON/OFF switch that's OFF? Is the mike plugged into the right MIXER input? Do the wires look damaged near the mike or its plug? Are the MIXER's switches properly set? Does the MIXER's IMPEDANCE match the mike's? Try another mike in the same input. If the problem persists, check the MIXER once again, because that input is probably at fault. If the problem stops, the first mike or its cable is defective.

MIXER passes a good signal from all sources but one, which is very low in volume Turn up the individual source's volume. Check to see if a switch for that input is on HI LEVEL when it should be on LO LEVEL. Try switching the IMPEDANCE switch on the MIXER. If it doesn't have such a switch, check the microphone to see if *it* has an IMPEDANCE switch, then try switching it. Check the source for loose wires in the cable or plug, wiggling them to see if it makes a difference. Does the source have a volume control of its own? If so, turn it up. Substitute a duplicate source; if it works, your first source has a defect.

MIXER passes a good signal from all sources but one, which is loud and raspy Could it be that you are using a HI LEVEL output from the source going into a LO LEVEL or MIC input on the MIXER? Perhaps your source's signal is too strong for the input. Turn the source's volume down if you can. Are you using an earphone or speaker output as your source? That may even overburden a HI LEVEL or LINE or AUX input. If this is the case, put a PAD in the line (see the section on controlling excessive volume levels earlier in this chapter).

MIXER passes a good signal from all sources but one, which has a lot of hum, buzzing, and hiss and sometimes may also have a weak signal Hum, buzzing (which is just loud hum), hiss, and a weak signal are usually wiring problems. The troublesome input may be wired with:

1. A loose connection where the plug plugs into the socket. Disconnect it and plug it in again. Is it in *all* the way?
2. A broken wire in the cable. Replace the cable.
3. The wrong kind of audio cable (unshielded). Replace with a shielded cable.
4. A disconnected wire in the plug itself. (Sometimes you can unscrew the plug handle and see for yourself—each wire should be soldered tightly to its corresponding pin in the plug. No wire should have a loose or broken strand touching

another wire or another part of the plug.) Resolder the wires or replace the cable and the plug.

5. IMPEDANCE mismatch. Try switching the IMPEDANCE switch on the MIXER. The source's IMPEDANCE must match that of the MIXER.

6. Level mismatch. If a mike or other LO LEVEL source is plugged into a HI LEVEL input and the volume is turned all the way up, then hiss, hum, and a weak signal will be the result.

7. Proximity to noise sources. Sometimes, if the wire or the microphone is next to an electric device, the electronic field from the device is radiated into your sound system. Keep at least 2 feet away from fluorescent lamp fixtures, electric motors, light dimmers, high-power electric current cables, amplifiers, power supplies, TV sets, and things that use electricity.

If you encounter hum, hiss, buzz, or no sound at all, the quickest remedy is to find out if the culprit is the mike and its cable, the mike extension cable (if you're using one), or the MIXER input. Take the following steps to find out which is to blame:

1. First, disconnect the old mike (and extension cable) from the MIXER and substitute a similar mike. If this one doesn't work, suspect the MIXER input or associated MIXER controls. On the other hand, if this one works, the fault lies with the mike or its extension cable. On to step #2.

2. Second, reconnect the extension to the MIXER and plug the substitute mike into the extension. If this works, suspect the first mike or its cable or plug. If this doesn't work, suspect the extension cable. To confirm the extension cable faulty, do step #3.

3. Disconnect the extension cable from the MIXER and connect the first mike in its place. If the mike works now, the mike is good and the extension is bad. Off to the repair shop with the extension.

If the problem doesn't seem to be wiring, then it is likely to be the result of a sensitivity mismatch, somewhat like that described in step #6 discussed in the wiring problems earlier. If the source puts out a terribly weak signal and the MIXER must be turned *way* up to make it audible, then perhaps the source is just too weak for this kind of MIXER. It may need a boost from a preamplifier (turntables sometimes need this). Is it possible that the source needs power to make a hidden preamplifier in it work? Look for an AC plug. If it *does* have a stray plug, the device, when plugged in, will probably make enough signal to power a HI LEVEL input.

Inexpensive preamplifiers, if used on some equipment, may cause hum or hiss. If turning them off (and waiting a half a minute for them to "cool off") makes the noise go away, the noise is probably the preamp's fault. If the noise stays, it may be caused by the wiring.

A weak hum with adequate signal strength is often caused by correct but inadequate wiring between the source and the MIXER (the technicians call this a "ground loop" or "floating ground"). To check out this possibility, take a wire, touch it firmly to the bare metal chassis of the MIXER and simultaneously touch the other end firmly to the chassis of the source (or the metal body of the mike). Scrape it a few times to assure good contact. Does the addition of this wire decrease the hum? If so, you have a "ground loop." You'll have to beef up the cables or connect a wire between the chassis of the two devices.

MIXER passes a good signal from all sources but one, which has either flat or boomy sound Some mikes, some sources, and some MIXERS have controls which adjust their tone.

1. Check for a tiny switch on the mike. This may be a "roll-off" filter, which means it throws out certain tones and passes the

Common Audio Ailments and Cures 239

rest. Often, this filter throws out the low tones, thus making a thinner sound—which is good for speech but bad for music, where we like to hear deep bass notes.

2. Some sources have treble and bass controls. If the controls have a position called "flat," then put them in that position generally. This doesn't mean that the sound will be flat; the word on the device means that the device will give a true, even response to all sounds, both high and low. Some cheaper sources have a single "tone" control. Generally, turn it to its sharpest setting (HIGH or TREBLE), as the other settings may sound muffled.

Some turntables have "rumble" and "scratch" filters. Unless you hear rumbling and scratches, leave these switches off. The rumble switch filters out some of the low tones, while the scratch switch filters out some of the high.

3. Some MIXERS have LO CUT filters near each individual volume control. These switches cut the low tones, a move which may be beneficial for speech recordings but not for music.

If the problem is very bad, it may be a wiring difficulty. If the sound is tinny and weak, the IMPEDANCE may be mismatched, the BALANCED LINES might be connected wrong, or they might be connected incorrectly to an UNBALANCED input—or vice versa—or a wire may be broken somewhere. Try a substitute mike or a substitute source.

All audio from the MIXER has hum Test for a "ground loop" as described before by touching one end of a wire to the chassis of the MIXER while (this time) touching the other end to the chassis of the VCR (or wherever the signal is going). If grounding the two devices together reduces the hum, you may have to find a way to wire the two together. Incidentally, if you hear the same hum on the MIXER's headphone (if it has such an output), the problem may be in the MIXER itself. Try removing the AC plug, turning it one half-turn, and reinserting it. This sometimes cures hum.

Make sure the MIXER volume is not way up or way down. If it is, try to even it out. If the MIXER volume has to be way up in order to get enough sound, try connecting the MIXER to the VCR's MIC input, which is more sensitive than its regular AUX input.

OTHER PROBLEMS

All sound going into the VCR is too loud or is raspy Is your source's or the MIXER master volume up too high? Is the source's or the MIXER's HI LEVEL OUT feeding the AUDIO IN of the VCR as it should? Try a PAD to attenuate the signal going into the VCR.

The sound is good except when you hear radio stations, police calls, CB radios, the buzz of fluorescent lamps, or the tic-tic-tic of automobile ignitions in the background of your recording Presumably, you couldn't hear these sounds with your unaided ears; but you could through the audio devices. These signals are electromagnetic radiation inducing a signal in your wires. To keep this interference from sneaking in:

1. Try to keep a distance from interference sources.

2. Insure that you have good shielding on your source cables.

3. Assure that the cables from the source or MIXER to the VCR are also well shielded.

4. If you have a choice, run the MIXERS HI LEVEL OUT to the AUDIO IN of the VCR rather than using the MIC LEVEL OUT and the VCR's MIC IN. By using the stronger signals only, the interference will seem weak in comparison to the signal.

5. Ground the devices together with a wire connecting their chassis.

6. Use short cables when possible.

7. Contact the local FCC if you think someone might be broadcasting with too much power. If nothing else, the FCC may at least send you literature on how to modify your equipment to be less sensitive to this interference.

8. Use BALANCED LINES.

VCR has had "noisy" sound from the day you bought it Some VCR's have internal defects in the shielding of their wires. Radiated "noise" from other machine components "leak" into your VCR's sound circuits. Take the VCR to a repair service and have the technician perhaps beef up the shielding.

Perhaps the VCR needs a HEAD CLEANING or HEAD DEGAUSSING. Chapter 13 (paperback Volume II) tells how to do this.

At this point, the reader may ask, "Is this all worth it? Can't anything go right?" Usually, things do go right. The section on The Basic Basics in the beginning of this chapter covers the simple situations when everything goes right (admittedly this becomes rarer as your equipment becomes more complicated). Then we have complex setups where everything goes wrong.

Start simple. Enjoy the sweet sound of success before befuddling yourself with audio systems that are too sophisticated for you to handle. In time you'll grow, and your system will grow as well.

NEW DEVELOPMENTS

VCR audio fidelity is poorly suited for recording concerts, rock, or other musical productions (although Deborah Harry and Dolly Parton are fun to watch even with the sound turned off). Frequency response is a wretched 50–10,000 Hz as opposed to an ideal 20–20,000 Hz. Sony may change all this in 1984 with its Beta Hi Fi audio system claiming the full range of 20–20,000 Hz. Instead of recording the sound linearly on the standard audio track it will encode the sound into the picture as it is recorded. On playback, a circuit will separate the sound and play it out with full fidelity. To ensure compatability with other BETA VCRs and allow audio dubbing after the picture-and-sound have been recorded, the VCR will retain the standard audio tracks. Thus it will have the stereo hi fi audio signal combined with the picture *plus* the regular lo fi audio tracks. VHS makers plan similar systems.

Audio in a Nutshell

GETTING THE BEST
AUDIO SIGNAL

Room

1. Use a quiet room, one with thick walls and tight-fitting doors to seal out extraneous noise.

2. Keep the room quiet by turning off fans, air conditioners, and other machines while taping.

3. Use an anechoic room, if possible. Reduce echoes by hanging curtains, laying carpet, or draping blankets over the walls.

Microphone

1. If you're a perfectionist, use a LO Z mike with a BALANCED LINE to keep electronic "noise" from sneaking into your cable.

2. Wherever feasible, use DIRECTIONAL microphones to reject room noise, especially if the mikes cannot be kept close to the performers.

3. Keep mikes close (6 to 18 inches) to your performers.

4. When possible, keep the performers from handling mikes or cords during the recording.

5. If the mike has a switch for LO CUT filter, leave it OFF for music and ON for speech.

6. Use a windscreen if miking outdoors.

7. Place carpet or foam under the mike stand to keep floor or desk vibrations from being recorded.

8. For best fidelity, use a DYNAMIC or ELECTRET CONDENSER microphone in favor of a CRYSTAL microphone.

9. If a mike is to be hand-held, choose an OMNI-DIRECTIONAL one. CARDIOIDS and DIRECTIONALS are often too sensitive to hand noises and "booming" when held too close to the performer's mouth.

Cable

1. If using HI Z, UNBALANCED LINES, make sure you are using *shielded* cable only. Keep cable length to an absolute minimum. Let 30 feet be the maximum.

2. Keep people from tripping over your cables by tying them (the cables) to your mikestand base (as in Figure 7–12) and perhaps by taping them to the floor in heavy traffic areas.

3. Keep cable extensions from disconnecting by tying them in knots at their plugs (see Figure 7–12 again).

4. Loop the LAVALIER mike cable through the performer's belt so that as he or she moves, the cable isn't tugging directly on the microphone.

Inputs

1. Use LINE or HI LEVEL signals, whenever available, to feed AUX or HI LEVEL inputs (i.e., feed the MIXER's HI LEVEL OUT to your VCR's AUDIO IN). This reduces the electronic noise that can sneak into your cables.

2. If you must feed a HI LEVEL signal into a MIC INPUT, use a PAD.

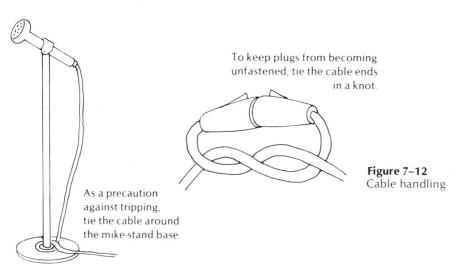

As a precaution against tripping, tie the cable around the mike-stand base.

To keep plugs from becoming unfastened, tie the cable ends in a knot.

Figure 7–12
Cable handling

3. Keep enough adapters handy so you can use *what's best,* not have to make do with what *fits.*

Mixers

1. To judge volume levels, trust the MIXER's meter. To judge fidelity, trust your ears. Monitor everything.

2. If the MIXER has filters, avoid using them when recording music. Use them primarily for speech or for adapting to poor room acoustics.

3. Don't run some volume controls high while others are low. Whenever possible, run them all nearest their middle settings. For example, don't run your audiocassette player's volume way up high while feeding the signal to the MIXER whose volume is way low. Split the difference.

VCR Fidelity

The VCR's sound quality improves as its tape speed increases. Record at the fastest speed for the best picture and sound. Exception: The newer BETA "Hi Fi" VCRs give excellent fidelity at any speed.

Also use the VCR's direct AUDIO OUT for the best audio signals. When audio is modulated by the RF generator and then demodulated by the TV tuner, to play through the TV speaker, it loses some of its clarity and gets a little noisy.

MIKES AND
SIGNAL SOURCES

The three basic kinds of microphones are the OMNI-DIRECTIONAL (hears in all directions), CARDIOID (hears mostly in one direction) and UNI-DIRECTIONAL or SHOTGUN (hears primarily in one direction). The farther the mike is from the performer, the more echoes and extraneous noise you pick up, and the less desirable the OMNI mike becomes. Cameras with built-in mikes usually have OMNIS.

The IMPEDANCE of the source (mikes, etc.) must match the IMPEDANCE of the VCR's MIC or AUDIO IN, for best fidelity. This information is usually listed among the equipment's specifications at the back of the instruction booklet.

Most home equipment uses UNBALANCED audio lines. If commingling professional equipment with home equipment, you may need adapters to match IMPEDANCES and BALANCED and UNBALANCED lines.

Don't test a mike by blowing into it. Tap it with your fingernail or have someone speak into it normally.

Send weak audio signals into the MIC input of your VCR. Send strong ones into the AUDIO IN jack of your VCR. If the sound distorts, try turning down the volume level of the source or inserting a PAD in the connection.

If hum, hiss, or distortion problems cannot be overcome, you can always put the VCR's mike (or the camera with its mike) near the source's speaker (i.e., in front of a record player) and pick up the sound that way. You'll lose some fidelity.

Chapter 8
Lighting

THROWING A LITTLE LIGHT ON THE SUBJECT

The human eye is an amazing thing. It can make wide-angle, crystal sharp images in color under the worst conditions. The eye is sensitive enough to see by candlelight and tough enough to perform in sunlight 20,000 times brighter than the light of a candle.

The television camera is frail in comparison. It needs plenty of light, but too much light can damage the vidicon tube. It can display only a two-dimensional image, which looks flat and dull compared with the 3-D panorama our eyes give us with each glance. Where the eye can discern 1,000 different levels of brightness, the best cameras under the best conditions can distinguish only 30 or fewer shades of gray.

Lighting serves two purposes. First, it illuminates the scene so that the camera can at least "see" it. Second, it enhances the scene to make up for television's visual shortcomings.

Color cameras need more light to yield a good picture than do black-and-white cameras. Also, with color cameras you have to account for the color and COLOR TEMPERATURE of your lights. Other than that, lighting for color is essentially the same as lighting for black-and-white. For this reason, the "rules" for black-and-white lighting will be covered first, and then we'll go back and examine the exceptions and special considerations concerning color shooting.

Sometimes the home TV producer simply wants a decent picture on the camera, really doesn't care about the nuances of lighting, and doesn't want to mess with a whole bunch of fixtures and stands and other lighting lunacy. That's fine. The Basic section of this chapter will cover the simplest setups, which require hardly any knowledge or effort at all. After the basics will come a gradual immersion into the world of professional lighting. Like in the audio chapter, you can stop with the basics—they'll get you by—or continue into more complex and advanced techniques so that you can light your TV shows "like the pros."

The Basic Basics

INDOOR LIGHTING

The biggest problem with indoor shooting is *not enough light*. The second biggest problem is *light in the wrong place*. The third biggest problem is *wrong color light for color TV*. Let's attack these challenges one at a time.

Enough light Turn on every light in the room. Take the shades off any lamps not appearing in your scene, just to boost the lighting more. Move lamps closer to the scene, if possible. Replace the light bulbs in lamps with the highest wattage rated for the lamp. (If using extension cords and lots of lamps, do not exceed the wattage maximums for your house wiring and extension cords. More on this at the end of the chapter under Power Requirements.) Use window light when possible. If you have a floodlight (perhaps stashed with your outdoor Christmas displays) aim that at your scene.

Every light that you add will add punch and contrast to your picture. Notice the scene differences in Figure 8–1 as light is added to the scene.

Lamp placement You want most of the light to come from behind you (the camera). Avoid light coming from behind the subject (bright windows, etc.) as that will silhouette your performer as shown back in Figure 6–4. Try not to have all your light coming from *too* near the camera or you'll lose all your shadows and everything will look flat and dull, like in Figure 8–2.

Figure 8–1
Ambient lighting vs extra lighting

Scene lit with regular incandescent home lighting

Scene lit by overhead office fluorescent lamps

Scene with just one extra 250-watt lamp added near camera

Scene with three, well placed, extra lamps

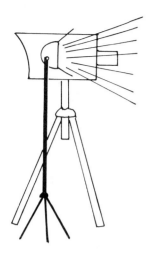

Figure 8–2
One light placed near the camera yields a flat picture with almost no shadows on the subject

Sometimes, because you're moving around from room to room, you have no choice but to carry your light with you. Try to mount the light above and to the side of the camera, about a foot away if possible. Figure 8–3 shows a lighting contraption which is inexpensive and easy to make which clips to your camera for traveling shots.

Gentle shadows are better than dark shadows, so if you have a deep shadow on the left of someone, place a lamp to the left of your camera to fill it in some.

Keep lights high because it's natural for shadows to fall below things.

Colors of light Fluorescent lights are bluish. The incandescent lights in your home are reddish. You don't see this difference with your naked eye, but your camera does. The COLOR TEMPERATURE and WHITE BALANCE controls on your color camera can make up for much of this variation, but it is better to use the right colored light to make excellent pictures to start with

If you have a movie light packed away (from your super-8 moviemaking days), use

If roving around inside buildings, clamp-on a fixture with a bare 250-watt bulb. You'll also need a long, heavy-duty, extension cord for it. All this can be found for under $6 at your hardware store. A special PHOTOFLOOD light bulb from a photo store would cost extra but would improve color shooting.

Figure 8–3
Cheapie roving camera light

it. You can also go to your photo store and buy some PHOTOFLOOD bulbs which screw into your regular lamp fixtures. You simply replace regular incandescent bulbs in the room with these PHOTOFLOOD bulbs to get the correct color light.

Movie and photo lamps are both specially made to give the same amount of red and blue (mixed in with the white) as old Mr. Sun does. Such light renders rich colors and nice flesh tones.

OUTDOOR LIGHTING

The big challenge with outdoor shooting is shadow control. Picture a bright, sunny day. The baby chases the family puss across the green lawn and under a tree. Suddenly the baby's rosy pink cheeks turn muddy gray. Your orange cat turns muddy gray. The green lawn turns muddy gray. And every once in a while you see a white flash as the kid's outfit is caught in a stray beam of sunlight piercing the leaves.

The trick here is to fill in the shadows:

1. Shoot on hazy days when shadows are soft.

2. Shoot with the sun mostly to your back so shadows are hidden.

3. Stop down your lens as far as possible to reduce excessive contrast.

4. Glue some tinfoil to a sheet of posterboard and "fill in" the shadows with reflected light.

5. Shoot with a bright light near the camera, even in broad daylight. Place the light in such a way as to "fill in" the shadows caused by the sun.

These are the basics. You'll see them again as we study light and shadow in more detail.

The Kind of Light a Camera Needs

ENOUGH LIGHT

Unless yours is a specially equipped industrial LOW LIGHT LEVEL CAMERA, your camera will need a fair amount of light to register any picture at all. For instance, if you're cuddled around a campfire at night, you can feel confident that none of your nosy friends will be spying on you from a hidden TV camera. To provide enough illumination for the camera, you'd have to be *in* the campfire, not next to it.

Normal home lighting is barely sufficient to yield a picture. Although faces and objects will be recognizable, the image will be rough, grainy, and very gray and flat-looking, like in Figure 8-1.

Office and classroom lighting is generally sufficient for shooting. Depending on the circumstances, you may even be able to "stop-down" your lens from its lowest f-number to its next lowest f-number, realizing a little better depth-of-field in the process. Office lighting, though it provides sufficient light to create a picture, doesn't create the shadows and contrast to yield a vivid picture; it will still look somewhat flat and lifeless, like in Figure 8-1.

On a cloudy day outdoors the light is adequate for shooting. You may be able to use f4 to f8 for good depth-of-field.

A slightly hazy day is perfect for shooting outdoors. Shiny objects won't be dangerously bright (thus not endangering the vidicon tube), and there will be plenty of contrast, even at f8, yet shadows won't be too pronounced.

Full sunny days are pretty good for shooting. Avoid highly reflective objects. Use f16 or so. The picture will be bright and vivid but may appear too contrasty. Shadows, especially, may look too dark, and anything lurking in them may be obliterated.

Views of the sun or its reflection off highly polished surfaces, a welder's torch in action, or direct views of an atomic-bomb blast constitute excessive light and must be avoided at all costs (unless special light filtering lens attachments are used).

LIGHTING RATIO

Place something very bright next to something very, very bright, next to something very dark, next to something very, very dark, and you will be able to distinguish one from another easily. A TV camera, on the other hand, will see only two white objects and two black objects. Although your eye can handle something 1000 times brighter than something else in the same scene, and although photographic film can distinguish between an object 100 times brighter than another in the same scene, a TV camera can accept a light ratio of only about 15. (Under the best circumstances and with the best equipment, this ratio can be as high as 30.) The brightest thing in the picture cannot be more than 15 times brighter than the darkest object in the same scene.

Here's what this means in practice. You wish to tape a person standing in front of an open window during the day. What your camera will see was shown in Figure 6-4. Since the light from the window is very bright, everything else looks black and silhouetted by comparison. The gradations of gray in the clothing and face are all lost. If you close the shade (see Figure 6-4 again), now the whitest thing in the picture is the

Excessive lighting ratio
Too bright a lamp, too close, causes excessive contrast and deep shadows

wall and some parts of the clothing. They are only about 10 times brighter than the hair and the other dark parts of the picture. As a result, everything between the blacks of the hair and the whites of the wall gets a chance to be seen as some gradation of gray, rather than appear black, as they did before.

In short, things that are super bright must be avoided. The brightest part of the scene should be about 15 times brighter than the darkest part of the scene. Shafts of light coming in the windows, shiny buttons, and chrome hardware (like mike stands) should all be avoided or subdued.

To dull objects that are too shiny (Uncle Frank's bald pate or neighbor Norton's chrome birdhouse) dust them with a little talcum powder. The talc washes off easily and will save you from a burned vidicon.

Words to know

KEY LIGHT
Provides the main illumination of the subject. It is similar to the sun in that it puts most of the light on the subject and creates the main shadows that will be seen.

FILL LIGHT
Partially fills in the shadows created by the KEY LIGHT and creates an overall brightness to the scene.

> **BACK LIGHT**
> Shines on the performers from high and behind, accenting their shoulders and the tops of their heads in order to give depth and dimension to the picture.
> **SET LIGHT**
> Illuminates the background behind the performers.
> **BARN DOORS**
> This is not the thing you close after the horses get loose. BARN DOORS are hinged flaps that can be moved to direct the light and to shade cameras from the light.
> **VARIABLE FOCUS**
> Allows you to direct the light into a small bright area or "spot" (hence the word *spotlight*), or it allows you to diffuse it to cover a wide area evenly, flooding it (hence the word *floodlight*).

Basic Lighting Techniques

EXISTING LIGHT ONLY

You're shooting on location and didn't bring lights. How do you illuminate your subject?

1. Place your subject where the existing illumination is best, like outdoors (in the daytime) or under office lighting.

2. If the camera, with its lens wide open (lowest f-setting), still shows a poor picture because of insufficient light, seek out other light sources such as desk lamps. Be aware, while placing such lamps, that the closer they are to your subject, the brighter your subject will be illuminated; however, the area covered by the light will be smaller. This can be a problem if your subject is a moving one, as she might slip out of the small bright area you have created. Also, moving subjects cause a brightness problem with close lights. If someone moves her head 6 inches closer to a lamp 4 feet away from her, the change in the illumination of her face will be unnoticeable. But if someone 2 feet away from a lamp moves 6 inches closer to the lamp, the illumination on her face increases sharply, causing a pronounced flare or shine on her forehead and cheeks.

3. Avoid bright windows or lights in the background of the shot. If you wish to use light from a window, get between it and the subject so that she, not you, is looking into the windowlight.

ONE LIGHT ONLY

You're shooting on location and you brought only one light (perhaps that's all that would fit in your saddlebags or bike basket). Where do you place it?

Don't place it next to the camera. Doing so gives a flat picture without shadows, as in Figure 8-2. In most cases, shadows are desirable because they create a sense of depth and texture in the image. Place the lamp at an angle 20°-45° to the right and 30°-45° above the subject, as shown in Figure 8-4.

TWO LIGHTS ONLY

This time, you could only bring two lights with you on location shooting (that's all your foreign sports car would hold). Where do you place them?

1. Moving it farther away from the subject.
2. Removing its reflector.
3. Aiming the light at something reflective nearby (a white posterboard, wall, or some aluminum foil). The diffuse reflected light will then fill in the shadows.

Individual taste and circumstances play a large role in setting up lights. There is no law that says a light must be 30° up and 30° over. No law says one lamp must be brighter than the other so that one is the KEY and the other the FILL; they could be equal. The ideas set forth here are generalities, not rules.

STUDIO LIGHTING

Perhaps you plan to set up a mini-studio in your basement or garage. Your biggest problem will be finding a way to get lights up high enough—the ceiling is in the way. Perhaps the garage or the attic or the chicken coop has a higher ceiling for your lights. Perhaps you can remove the "drop ceiling" in your office at work (your boss should love this) and use the extra space normally occupied only by ducts, pipes, and asbestos fibers. Let's assume you have found room to hang your lights (or you've put them on tall stands—but stands always seem to be in the way).

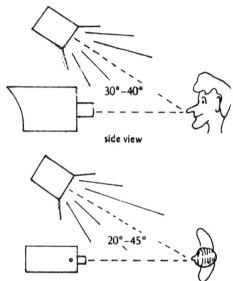

Figure 8–4
Optimal placement of single lamp to create depth through shadows

The first light you place 20°-45° to the side and 30°-45° up as described earlier and as shown in Figure 8-4. The second you place up and to the *other* side of the camera, as in Figure 8-5. The brighter of the two lights acts as the KEY LIGHT, providing most of the illumination of the subject, while the weaker lamp becomes the FILL LIGHT, filling in the shadows somewhat and softening the picture. If both lights are of equal brightness, one can be made into a FILL LIGHT by:

Figure 8–5
Using two lights

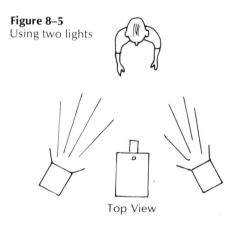

Top View

250   Lighting

Again, lighting serves two purposes:

1. To provide enough light so that the camera can "see."
2. To enhance the scene to overcome TV's inherent shortcomings, making the image appear sharp, vivid, and three-dimensional.

Figure 8-6 displays a TV lighting layout which should meet both objectives above.

KEY LIGHT This light illuminates the subject, creating the main shadows, as shown in Figure 8-4. These shadows help create depth and dimension in the scene. When a VARIABLE FOCUS fixture is used, like the movie light shown in Figure 8-7, it can often be adjusted either to FLOOD the area evenly with light, or to concentrate the light in a small SPOT.

Generally, if intense light is needed, adjust the instrument to SPOT. If the subject is large or moves around, SPOT may be un-

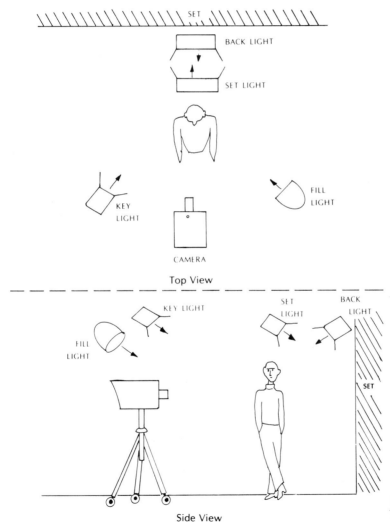

Figure 8–6
Typical lighting layout

Basic Lighting Techniques

Figure 8–7
Movie light used as a KEY LIGHT. Lever on top adjusts beam from FLOOD to SPOT

satisfactory because the area illuminated is so small. The solution is either to adjust the instrument—partially or all the way—to FLOOD (sacrificing some of the brightness as the light covers a larger area) or to obtain more instruments to cover the area.

FILL LIGHT Any light can be a FILL LIGHT if it fills in the shadows created by the KEY LIGHT.

You could make one out of a clamp-on utility lamp with a large shroud, as in Figure 8–8. You can further soften the light by covering the opening with layers of metal window screen. A PHOTOFLOOD bulb will give the proper color balance and plenty of brightness.

If you want all shadows to be harsh, deep, and noticeable and if you want textures to appear rough and super-three-

Figure 8–8
FILL LIGHT made from a utility light with a large reflector covered with metal window screen to further soften the light. The whole works should cost $10 from your hardware store

KEY LIGHT alone

FILL LIGHT alone

KEY and FILL together

SET LIGHT alone

BACK LIGHT alone

KEY, FILL, and BACK together

Figure 8–9
Various lighting effects in a darkened studio

KEY, FILL, BACK, and SET together

Basic Lighting Techniques

dimensional, you omit the FILL LIGHT. In most cases, however, you don't want *black* shadows, just the dark gentle hint of shadows. Figure 8-9 shows the image resulting from the use of a FILL LIGHT alone and in combination with the KEY LIGHT. The relative brightness of these two lights determines the depth of the shadows created.

While they're setting up lighting, some people use light meters and measure lighting ratios between KEY and FILL and the other lights. If that's not something you're inclined to do, a pretty good lighting job can be done "by eye" if you let the camera do some of the work for you.

1. Set up your lighting the way you think it should be.

2. Aim a TV camera at the subject to be recorded and look at a TV monitor to examine the image.

3. Readjust the lighting so that the image looks best *on the TV screen*.

Placement of the FILL LIGHT is generally 20°-45° to the side and 30°-45° above the camera-to-subject axis, just like the KEY LIGHT, only on the opposite side of the camera from the KEY LIGHT. This placement is flexible, however, and occasionally FILL LIGHTS may be found near the floor or near the camera.

BACK LIGHT The BACK LIGHT's placement is shown in Figure 8-6. The resulting image is shown in Figure 8-9. The BACK LIGHT is responsible for most of the dimensionality of the TV picture. Without it, the image is flat and dull; with it, the image stands out from its background and has punch.

The BACK LIGHT's job is to rim foreground subjects, separating them from the background. But don't make the BACK LIGHT too bright or it will light up the tops of actor's head and shoulders, distracting the TV viewer from the actor's face.

The BACK LIGHT shouldn't look straight down on the subject (as in Figure 8-10), for it will illuminate the nose (like Rudolph the

Figure 8-10
BACK LIGHT aimed straight down on subject

Red-Nosed Reindeer) every time the head tips back. The light should strike from above and behind at an angle 45°-75° up from the horizontal. The higher and farther back the lamp the better, because the light, being aimed *toward* the camera, has a tendency to shine into the lenses. This causes undesired optical effects (like those in Figure 6-5) and risks burn-ins when careless camera operators tilt too far up. Often BARN DOORS, little flaps that attach to the lighting instrument, are a help in shading the cameras from the lights while directing light only on the performers. Figure 8-11 shows a sample BACK LIGHT with BARN DOORS.

Figure 8–11
BACK LIGHT with BARN DOORS

that adjusting one lamp doesn't affect anything else (it will anyway, to some extent, but why make things worse?).

There are special SET LIGHTS designed for this purpose, but with a little fiddling around, you can get by without them. Of all the lights mentioned so far, the SET LIGHT is perhaps the least important for amateur use. Home studios are generally so small that the set gets illuminated just by the stray light from the KEY, FILL, and BACK lights.

Other kinds of lights There are enough specialized kinds of lights and lighting techniques to fill a catalog. Some are aimable spotlights designed for following an actor onstage. You could make your own by using a slide projector without a slide in it. Some project images on the background, such as the shadow of shutters, leaves, the image of stars, or various patterns. Again, a homemade slide in a projector can serve this purpose. Some instruments are de-

SET LIGHT Again refer to Figures 8–6 and 8–9. The SET LIGHT illuminates the set or background. Depending on the brightness of this light, a gray background can be made to look white, neutral, or black relative to the performer. It is best (usually) to have the background darker than the performers so that they stand out from the background, directing your attention to them. Light backgrounds tend to silhouette the performers.

Not just any light can do this job well. The set should be lit evenly. A regular light aimed down, near the set, will create a bright spot at the top of the set and will fade off to nothing at the bottom. Placing a lamp farther from the set will light it more evenly, but unless the subject is standing far in front of the set, much of the light will spill onto him. For good control, you want each light to illuminate one area only so

The woman in white fails to stand out from the white background. It is usually best to have the background darker than the subject (Courtesy of Imero Fiorentino Associates, Inc.)

Basic Lighting Techniques 255

signed for portable use; they are small and light, and they fold up compactly. Some are designed for battery use and others for use near explosive gases.

If you plan to do some serious video lighting, perhaps for a profit (a sidelight maybe?), you may want to buy a small lighting kit. Comprehensive Video Supply sells small kits with instruments, bulbs, stands, etc., for $200-$300.

If you prefer to assemble a lighting kit of your own, try to get all the components to match up. It's a pain stocking three different kinds of bulbs, three kinds of BARN DOORS, three kinds of light stands, etc.

Creative Lighting Strategies

If you arrange KEY, FILL, BACK, and SET lights as just described, your scene will have the "6 O'Clock News" look, stern and businesslike. To tone down the harshness and create a more casual, comfortable, homey look, get rid of the hard shadows by using diffuse FILL light only. Aim your KEY and FILL lights through wire screens, or bounce them off white posterboards. This also goes for folks with only a KEY light—to avoid the harsh "home movie" look, diffuse the KEY light (assuming you have a bright enough light to work with).

Chapter 6, under Creating Moods and Impressions with the Camera, described some lighting techniques to create a "night" look.

For the evil look, aim the lamp up from under the chin, as in Figure 8-12. For the soft, sexy bedroom look, use only reflected light by aiming several instruments at white boards. Figure 8-13 shows what soft, indirect lighting can do. Hard, direct lighting does just the opposite: It accents texture and flaws in smooth surfaces. To get hard lighting (as in Figure 8-14), avoid lamps with big reflectors. Use a typical KEY LIGHT as your FILL LIGHT. Have the lights hit the subject

Figure 8-12
Evil look with lamp from below

more from the side than from straight on in order to accent the shadows. The texture of a surface becomes more pronounced as the light skims along it from the edge.

With all the other lights in their normal position, adding a MODELING LIGHT, aimed from the side of the subject, can further accent the dimensionality and texture of the subject.

A small light placed near the camera's lens and shone into the talent's face will add a sparkle to his eyes. Be conservative; too much light will add tears to his eyes and complaints to your ears.

Glass, metal, or wet objects pose special lighting problems for television because of the reflection of the studio lights off shiny surfaces on the objects. In order to minimize these harsh shiny spots, use soft indirect lighting. If you have a white ceiling, try aiming all the lights at it for glare-free indirect lighting. Where appropriate, use screens or other semitransparent items to diffuse the light from the lamps. Make sure, however, that your diffusers don't melt or catch fire from the heat of the lamp.

Although this technique solves the shine problem and is appropriate for shooting mechanical objects up close, it is not likely to flatter performers. If both must occupy the same screen, compromises must be made. Can any of the shiny objects, like watches or bracelets or shiny buttons, be removed? Can the bows of a performer's

Figure 8–13
Soft, indirect lighting (Courtesy of Berkey Colortran, Inc., a Division of Berkey Photo, Inc.)

glasses be raised a half-inch to reflect studio lights downward? Can chrome mike stands be traded for ones with a dull finish? Can an actor's face or bald pate be powdered to reduce shine? If a shiny object plays an essential part in the scene, it may be dulled with soap, stale beer, milk, or even cloth tape. Close-up shots of super-shiny items, like silverware, may call for extraordinary dulling efforts. Here, one may erect a tent made of a white sheet over the objects. Lights aimed at the tent from the outside will softly illuminate the area inside the tent. The TV camera can poke its lens in through a hole somewhere to shoot the results.

Sets and backgrounds are usually too much trouble for the home video producer to bother with. Lighting can sometimes provide a quick-and-dirty way to create the illusion of something being there when it really isn't.

Are people supposed to be driving in a car? Park the car somewhere with a blank background (like in an open area), plant your camera on the hood aiming through the windshield, place a mike in the car, and have somebody wave a light across the car (from front to back) once in a while to imply movement.

Are folks chatting next to a fire? Wave some red or yellow lights around so flames dance across their faces.

For that "office look" in your home, make a 35mm slide out of cardboard that looks like venetian blinds and project that on a white wall behind the performer sitting at an office desk.

With imagination you can project plants, rain effects, explosions, rocket takeoffs, police car flashers, lighthouse beacons, colored spacecraft control panel reflection, green radar scope reflections, all kinds of things, on your performers and their backgrounds.

Figure 8–14
Hard, direct lighting

Creative Lighting Strategies

Lighting for Color

The principles of black-and-white lighting generally hold true for color. Color permits new lighting possibilities while exacting new constraints.

For color, you need more light than for black-and-white. Color cameras require this extra illumination because they are less sensitive than their black-and-white counterparts. With insufficient light, color cameras will produce grainy pictures with muddy color. So for color, you're stuck with erecting more instruments and consuming more power than for black-and-white.

Now that we're shooting color, we can get creative, illuminating our sets and performers with colored lights. Colored floodlights from your idle Halloween display may add some spice.

Blue lights can give the impression of nighttime, darkness, or cold; red lighting may convey warmth and happiness. Lighting an object with different colored lights from different angles offers dimensionality and visual appeal. We pay for this new creative freedom with new headaches. The camera doesn't "see" things exactly the way we do, so something that looks pleasing to the naked eye may look abominable on the TV screen. Lighting adjustments must be made with both eyes on a color monitor.

As the subject moves from one area of the room to another, the relative brightness of various lamps will change, thus changing substantially the vividness, the highlights, the darkness of the shadows, the overall contrast, and even the color of the subject. To accommodate these problems, don't position your lights too close to your performers, and plan—and light for—all moves in advance.

Background color can be used to add dimensionality to a scene. Where in black-and-white it was the BACK LIGHT's job to make a subject stand out from its background, now the complementary coloration of the background can help to emphasize the subject. Something to watch out for when choosing color backgrounds and props is the effect they may have on other colors in the scene. One colored surface may reflect light of its own color onto an adjacent object, such as an aqua dress casting a sickish blue-green tinge onto the neck of a TV performer or a yellow detergent box throwing a dingy hue on a nearby stack of clean white undershirts.

Faces are perhaps the hardest thing to illuminate correctly. No one will notice if a shirt, table, or backdrop appears more bluish or greenish than it's supposed to. But flesh tones that appear pale, greenish, or reddish-brown will not be tolerated (unless you're shooting Casper the Friendly Ghost, the Creature from the Black Lagoon, or Frankenstein's Monster). Not only are skin tones sensitive to outright color changes under various lighting and backgrounds, but these tones are also especially sensitive to changes in COLOR TEMPERATURE.

COLOR TEMPERATURE

COLOR TEMPERATURE is measured in degrees Kelvin. 3200°K (3200 degrees Kelvin) describes a lamp with a COLOR TEMPERATURE appropriate for color TV cameras. As this number goes up to about 6000°K, the light gets bluer and "colder" (this seems backwards, but as the COLOR TEMPERATURE increases, the scene looks "colder"). Fluorescent lights and foggy days exhibit such high COLOR TEMPERATURES. As the number drops down to about 2000°K, the light gets redder and "warmer." Incandescent lamps in the home create such low COLOR TEMPERATURES. A face, in order not to look too red or too pallid, should be illuminated by 3200°K lamps. This COLOR TEMPERATURE is available from QUARTZ-IODINE OR TUNGSTEN-HALOGEN lamps, or from the sun under certain conditions.

Much of what's been written about TV color relates more to artistic taste than to objective principles. Here are some of the more widely held "rules" of TV color:

1. Avoid pure whites. They will be too bright for most color cameras. Avoid pale yellow and light off-whites, as these may be too bright for the cameras. Light colors and light gray will probably all reproduce on TV as just "white." Medium-tone colors reproduce best. Dark colors, such as maroon, black, and purple, may all appear as "black" on TV.

2. Do not mix fluorescent lamps with tungsten or quartz lamps on your set. This creates COLOR TEMPERATURE problems.

3. The background for a colored object should be either gray or a complementary color. For instance, red looks best in front of a blue-green background, yellow in front of blue, green in front of magenta, orange in front of green, and flesh tones look best with a cyan background.

4. Bright multicolored subjects look best shown before a smooth, neutral background. Especially avoid "busy" backgrounds, as they distract the eye from the main subject.

5. Attention is attracted to items with saturated (pure or solid) color. Pastels attract less attention and are good for backgrounds.

6. Smooth objects appear brighter and their color appears more saturated than rough objects.

7. Colors appear brighter and more saturated when illuminated by hard light as opposed to those illuminated by soft, diffuse light.

8. Black backgrounds make both light and dark colors appear brighter.

9. Some colors become indistinguishable when shown on a color-TV screen. The colors between red-orange and magenta end up looking about the same on the screen. Similarly, blue and violet look about the same on the TV screen. You should therefore *avoid* highlighting a red apple with red-orange or trimming a blue robe with violet. These nuances in hue will not reproduce.

10. Things that look nice on a color TV sometimes turn to mud on a black-and-white receiver. Generally, your pictures will be *color compatible* (will look nice in both black-and-white and color) if the colors you are using differ in brightness. Where a pale green and a pale red will merge on a black-and-white TV to form a gray mush, a dark green and a pale red will create two easily distinguishable grays. Take a peek into your camera's black-and-white ELECTRONIC VIEWFINDER to preview the monochrome results of your color selection.

11. Yellow, gold, orange, red, and warm colors will appear lighter on camera than in real life. Greens look darker on TV than they really are.

TYPES OF BULBS

The home video producer faces three choices of lights in his or her price range: regular incandescent, PHOTOFLOOD, and QUARTZ-IODINE.

The regular incandescent light bulb gives off a reddish light (often 2800°K or less), doesn't direct the light without the help of a separate reflector (unless you borrow a spotlight from your patio or garage), generates a lot of heat, and blackens with age, cutting further its light output. But it's cheap at under a dollar.

PHOTOFLOOD lamps have standard bulb screw-in bases, many have reflectors, and they come in 3200°K and 3400°K, COLOR TEMPERATURES optimum for color shooting. As they age, their glass darkens and their COLOR TEMPERATURE drops. The bulbs cost about $5.

QUARTZ-IODINE bulbs usually require special fixtures equipped with special sockets. The fixture usually contains the reflector and often an adjustment for FLOOD or SPOT. They're designed to generate 3200°K light which remains the same throughout the lifetime of the bulb. They do not darken with age. Quartz lamps run cooler and their fixtures are smaller than their counterparts of the same wattage. The bulbs cost about $10 each. The special fixtures for these bulbs cost $30–$100.

Care of Lighting Instruments

FIXTURES GET HOT

And boy, do they! They make nearly as much heat as a toaster and can toast you if you don't watch out. Keep the instrument away from anything combustible or meltable. Make sure that the power cord for the fixture isn't draped over the instrument (the cord could melt). Watch where the lamp is aimed; you can feel the heat of a 600 w lamp from 5 feet away, so imagine how hot it is right in front of it. For instance, aiming the lamp at a wall or curtain less than 1 foot away or so could start a fire in a matter of minutes. When handling lights, let them cool before attempting to change bulbs (unless you go around wearing asbestos gloves). Do not attempt to store instruments until they have cooled adequately. When closing up after a shoot, do not lay lighting instruments on a carpet—not until they are *cold*. With new lights don't be surprised if the paint burns off the BARN DOORS sometimes; they are designed to take such punishment, but they smell awful in the process.

MOVING LAMPS

Do not jar, shake, bump, or attempt to move a lamp when it is lit. The filament in the light bulb is white hot and *extremely* fragile. When the lamp is not glowing, the filament is solid again and is fairly rugged. When you turn off a lamp, always let it cool for a few seconds before moving it.

FIXTURES GET HOT

CHANGING BULBS

You can assume that a lamp is burned out when it stops working. In order to confirm that it has expired, first *turn off the power to the instrument.* Take a close look at the bulb. If it has a big bulge, if it is blackened, if it is cloudy inside, or if the filament is clearly broken, the bulb is shot. If none of the above is true, perhaps the bulb is good and the fixture, switch, cable, or something else is defective.

Bulbs last between 10 and 500 hours, depending upon the manufacturer and type. When the bulb burns out in a professional or kit lighting fixture, be sure to replace it with exactly the same type of bulb or its equivalent. Some instruments can take bulbs of different power and brightness. *Do not exceed the power rating of the fixture.* Removing a 600 W bulb from a fixture designed for a 600 W bulb and putting in a 1000 W bulb will give you more light—until the fixture and its wires burn up.

Never touch a good bulb with your fingers or it won't be much good any more. This is especially true for quartz lights. Traces of oil from your fingers get on the glass bulb and cause a chemical change to occur when the bulb heats up. The glass devitrifies and fails right where the fingerprints were. Handle bulbs with a clean cloth or with the packing which came with the bulb in its box.

POWER REQUIREMENTS

Amps times volts equals watts. Homes and schools run on 120 volts, so if a circuit is good for 15 amps (as is typical in homes) then you may use up to $15 \times 120 = 1800$ watts of power on that circuit. If a circuit is rated at 30 amps (schools and offices usually are), then you can use $30 \times 120 = 3600$ watts. In short, the house current you get from a wall socket in your home is good for about 1800 watts. Institutional electrical outlets can sustain about 3600 watts. So how many 500 W lamps can you use at home without blowing a fuse or burning the house down?

Before turning on any light, check to see what else is on the same circuit and is also using power. Check also to make sure that you aren't running several lights off one extension cord. An extension cord rated for 15 amps (a label on it may say 15A at 120V, meaning that it can take 15 amps of electric current) can carry only 1800 watts of power. Even if you're working in a school whose outlets are rated 30 amps (3600 watts), your extension cord can safely handle only 1800 watts.

Once you're set up for a remote production and are satisfied that you aren't overburdening the wiring, you're ready to go. Switch the lights on one at a time rather than all at once, because during that moment when they are just lighting up they use abnormally high amounts of power. Switching all the lights on at once could cause a "surge" of power and blow a fuse. If you switch the lights on one at a time, the smaller surges are spaced out and are less likely to overburden the wiring.

Lighting in a Nutshell

When shooting indoors with a color camera, you're quite likely to need extra lights before you get a decent picture.

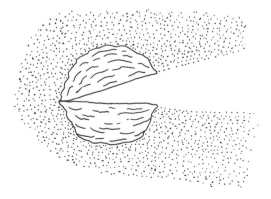

Place most of the lights behind you, aiming into the performer's face.

Use PHOTOFLOOD bulbs or 3200°K QUARTZ-IODINE lights for good color rendition. Fluorescent lights give a bluish tinge to everything, especially faces. Standard incandescent lights give a reddish hue to your talent and props. Adjusting your camera's COLOR BALANCE controls may counteract much of this miscoloration.

When shooting in bright sunlight, the image may appear too contrasty and the shadows too dark. Stop down the lens iris some and try to reflect some light into your talent's face to counteract the shadows. Hazy days are best for shooting.

Cameras abhor very bright and very dark things in a scene. A very bright sky will make the earth look dark. A dark dress worn by someone in a shadow will make the person disappear. Try to shoot things of moderate brightness and darkness for best results.

But don't carry the above too far; you'll end up with a bland picture. Shadows and ridges are needed to make figures pop out of the screen with reality and dimension. Encourage shadows by lighting scenes from an angle with a bright KEY LIGHT. Soften the deep shadows with a diffuse FILL LIGHT. Place a white rim over the head and shoulders with a carefully aimed BACK LIGHT. Use the SET LIGHT(s) to keep the background from being the same brightness as the foreground and merging with it.

If you enjoy the smell of burning flesh, then try changing a bulb in a fixture that's still hot.

Don't move lamps while they're on; they're filaments are super-fragile when glowing.

Unless you enjoy the smell of burning wire insulation and house fires, don't put a 600-watt bulb in an instrument designed for 300 watts—even if it fits. Don't plug too many lights into a wall socket or into an extension cord.

Chapter 9
Graphics

HIS MASTER'S ARTWORK

What is it that makes a professional TV production look "different" from the home TV epic? Smooth camerawork? Yes. Snappy lighting? That too. Background music and crisp sound? Definitely. But there's something that happens at the very beginning of a show that tells the viewers they're about to see something professional-looking. And that's a classy title. As your shows begin to include charts, tables, more titles, cartoons, and photos, they take on an air of authority and confidence. They become visually engaging. Consider the added flexibility available to you, the producer, as you now use all the other media available—16mm films, super-8 movies, 35mm slides, snapshots, computer displays, drawings, and commanding titles. With graphics you can do things you couldn't do before. You don't think they shot "All in the Family" in a house in Queens, New York, do you? Instead, they played a movie showing a rowhouse in Queens. This "sets the stage" for the interior shots produced on a stage in California. Your art and titles can similarly take your viewers to far-away places to see action tape recorded conveniently in your garage. Those far-away places could be shot from *National Geographic, Playboy,* or any number of magazines,* as well as from slides in your travel collection. In short, titles, art, and photos can add professionalism to your show while saving you effort obtaining difficult-to-get shots.

The Basic Basics

The essence of TV graphics boils down to three rules:

1. Make it fit the shape of a TV screen.
2. Keep it simple.
3. Make it bold.

*Note that 1978 copyright law forbids you from using someone else's published pictures or music without their permission. As a home video producer you should not be overly concerned with these restrictions unless you plan to sell or distribute your show.

263

Fitting Your Graphic to the TV Screen

ASPECT RATIO

A TV screen is a box a little wider than it is tall. If the screen were 16 inches wide, it would be 12 inches tall. If it were 4 inches wide, it would be 3 inches tall. However wide it is, it is three quarters as tall. This is called a 3:4 (three-by-four) ASPECT RATIO.

As a consequence, visuals for television should have a 3:4 ASPECT RATIO if they are to fill the screen evenly. Panoramas don't fit this ratio because they are too wide. Telephone poles don't fit because they are too tall. Strictly speaking, even a square box is too tall to fit perfectly on a TV screen.

When showing a panorama on a TV screen, one must either display a long, long shot of it, showing a lot of sky and foreground; or one must sacrifice some of the width of the panorama, getting just a fraction of it. In order to display the square box, one must decide whether to cut off its top and bottom in the TV picture or whether to get all of it, leaving an empty space on its left and right.

At least the composition of words, titles, and logos (a logo is a TV producer's symbol or trademark) is more flexible. One can arrange the words or whatever to fit the 3:4 dimensions. Figure 9-1 shows some good and bad graphic compositions.

Where good workmanship is important, great care must be taken when choosing

Figure 9-1
ASPECT RATIO

illustrative and printed materials so that they fit the 3:4 ASPECT RATIO. When such care is not warranted, one may simply "think boxes" when planning graphic composition. Things which are roughly box-shaped fit TV screens fairly well.

USE THE SAFE TITLE AREA

SAFE TITLE AREA

Two things you *don't* want to do are:

1. Show your audience the edge of your title sign board.

2. Have a piece of the title disappear behind the edge of the viewer's TV screen.

Both problems can be avoided by first leaving an adequate margin around the graphic to be shot and then shooting the graphic in such a way as to leave a little extra space around all sides on your TV monitor. This extra space allows for the fact that your camera may show the *whole* TV picture, while many home TVs cut off the edges. Some TV sets are poorly adjusted anyway, causing even further loss of the TV picture on the edges. To allow for this, the SAFE TITLE AREA is utilized, effectively confining all important matter to the middle portion of the TV screen.

For those who wish to be exact in this process, Figure 9-2 shows a template used by professionals in determining the SAFE TITLE AREA. The DEAD BORDER AREA is the camera operator's margin. It doesn't matter if the border is an inch or three inches in width, as long as it's enough to make it easy for the camera operator to shoot the picture without getting the edge of the title card in the shot.

The picture the camera operator usually sees is called the SCANNED AREA. Some call this the CAMERA FIELD, or the TRANSMITTED AREA. It comprises the whole title area plus some safety space around it, called the SUPPLEMENTARY AREA.

The picture seen by the viewer is contained in the SAFE TITLE AREA and is the only part of the picture which can really be relied upon to be visible on all TV sets, even the misadjusted ones.

To keep these areas in mind, some studios affix overlays or draw templates on their monitors and viewfinders. Figure 9-3 shows how these areas look on various TV monitors. What do you suppose would have happened to the home TV images in Figure 9-3 if the camera operator had zoomed in more and totally filled the viewfinder screen with the title?

When preparing TV graphics, it would be helpful to imagine a TV screen to guide you in judging size and ASPECT RATIO. The *total* size of your graphic doesn't matter (they all come out looking the same once you've zoomed in on them to fill the screen), just the ratio of height to width. Of course, if you make your graphics too small, you may need a special lens to get close enough to them. Also, the bigger the graphics, the less noticeable minor flaws in them become. On the other hand, when graphics become too large, people have a tendency to include too much detail ("Look at all that room, Martha, let's fill it up!") and not use bold enough lines.

Fitting Your Graphic to the TV Screen

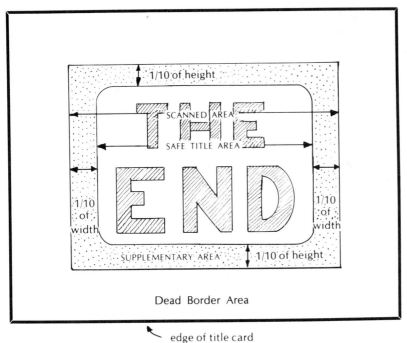

Figure 9–2
SAFE TITLE AREA

Figure 9–3
Picture areas compared

SLIDES

Thirty-five mm slides (also called *two-by-two slides* because the cardboard slide mounts are two inches square) are a staple of the video industry. A photographer with a simple lightweight slide camera can simply click off a few shots somewhere and come back (after processing) with pictures that look just as real as life when they completely fill the TV screen. Thirty-five mm slides of inanimate objects come out looking especially real on TV. Taking polaroid photos can even save the processing time.

With photos, the problem again is that of ASPECT RATIO. Slides, when mounted, make a picture 15/16-inch tall by 1-13/32-inch wide. That (if you're good at arithmetic) makes a ratio of 2:3. Even if you project the slide it still comes out with an ASPECT RATIO of 2:3, too wide for our 3:4 TV screen. This means that a piece of the left and right edges of the slide must be omitted in order to make it fit the TV screen. In addition, we mustn't forget Uncle Homeviewer with his misadjusted TV set. We must leave a SUPPLEMENTARY AREA around the edge of the slide just for safety.

Just keep in mind, while shooting the slides, that the most important material should appear in the center of the viewfinder and that something will be missing from the edges, especially the side edges.

Vertically mounted slides are abominable for TV use; they are just too tall and narrow. If you *must* use vertically oriented slides, then project them on a smooth, white, dull surface and aim a camera at the projected image. Zoom the camera in enough to fill the viewfinder screen with the picture, sacrificing the top and bottom of the projected picture.

Boldness and Simplicity

Figure 9-4 compares some examples of poor and good visuals for TV. Unlike cinema, slides, photos, and the printed page, TV is a fuzzy medium. Fine detail turns into blurry grays and hazy shadows. With your eyes alone, look at a newspaper from 3 feet away; you can probably read the entire page. Fill a TV screen with that same page and you can read only the main headlines, and even they don't jump out and grab you. Figure 9-5 shows an example.

Titling for TV needs to be brief, broad, and bold in order to have impact. Wordy subtitles that need to be small and unobtrusive should be limited to *no more than 25 to 30 characters per line* to remain legible. Remember, too, that something that looks pretty sharp in your TV viewfinder will lose a lot of oomph once it's recorded and played back. As you play your tape through RF into a casually adjusted TV set, it will lose even another layer of oomph. If you ever copy your tape, the copy will be degraded even further on the oomph scale. So, to survive the indignities of being recorded, RF'd, and perhaps copied, your graphics need to be simple and bold.

One experiment you may wish to try before you start drawing visuals is to letter something in pencil on a piece on paper. Next do the same with a felt pen. Next try it with a broad magic marker. Perhaps try a crayon too. Now shoot them all and compare the results. Use the medium that stands out the boldest.

One way to test for boldness without bothering with the TV camera is to step back from your proposed visual, squint your eyes, and look at it through your eyelashes. This is what it will look like when the user sees your visual. Give every title, drawing, and photograph the old "squint test" and two things are bound to happen: Your visuals will stand up to the rigors of the TV medium, and your friends will arrange optometrist appointments for you.

Gray Scale

Today's vidicon TV cameras are quite forgiving. Although all the books say "Never use white paper, never use black print,"

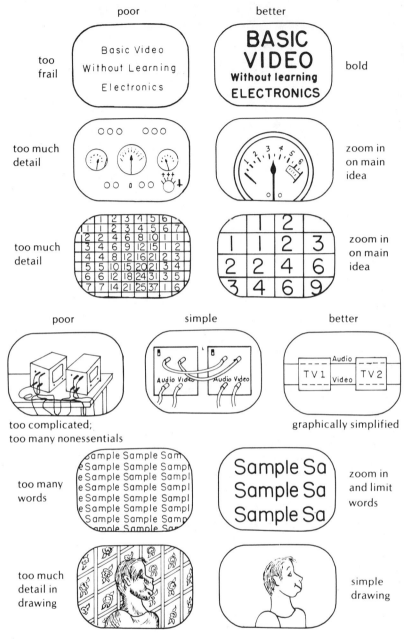

Figure 9–4
Boldness and simplicity in TV visuals

Figure 9–5
Comparison of a regular photograph with that same shot played back on a VCR.
Notice how TV loses a lot of detail

white paper and black print will look pretty good. Black print on off-white paper is even better, though, especially if a picture or half-tone drawing (one that contains various shades of gray) is involved.

The reason for this has to do with the camera's ability to see various shades of gray. If some part of the picture is blisteringly white, it will make all the light gray, medium gray, and darker parts of the picture look black in comparison. If, on the other hand, the whites weren't so terribly white when compared with the grays, all the tones would get a fighting chance to be seen.

So as a general rule, use off-white paper (yellow, light green, or buff) to allow the full range of grays to be seen by the camera.

Incidentally, those off-white backgrounds still look white when seen on the TV screen.

Typography

GENERAL RULES

Again, think boxes; keep it simple; make it bold.

1. Keep titles short. Use several title cards, if necessary.

2. For textual passages, put no more than about 25 characters on a line and up to 10 lines per graphic page.

Limit your long text passages to 10 lines, 25 characters each

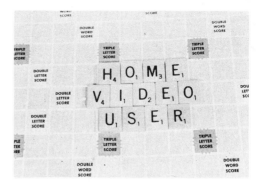

scrabble board letters

3. White or light letters over a dark background are easiest to read.

4. Avoid red or orange backgrounds. VCR circuits tend to make these colors smear and streak. Use blue or green for smoother color reproduction.

5. Use thick lines, not thin.

6. Outline things in black. Where colored masses meet, they tend to merge visually. The black edging sets them apart.

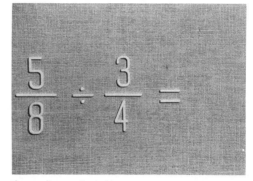

three-dimensional letters

What do you use to print your titles and subtitles? The answer depends on quality, budget, and purpose. Do you have a lot to say? Do you have a lot to spend? How good must it look?

HAND LETTERING

If you're blessed with a good hand for lettering, you can create some pretty snappy titles quickly with a felt pen, crayon, or other marker. Avoid longhand and fancy letters, again adhering to the principle of boldness and simplicity. Black letters on off-white backgrounds look fine, but with experimentation you may find other workable combinations.

Be aware of possible color compatibility problems on black-and-white TVs if you use letters and background of approximately the same grade of darkness.

rub-on letters

Lettering techniques

You will find that thin, colored lines will probably lose their color (a dark blue line on light paper will simply look like black on white). Broad, colored letters may tend to merge with their background unless you give them black borders. This edging is also necessary for setting colored parts of a picture apart. (For instance, draw a cartoon in black and *then* fill it in with color. The black borders will sharply define your illustration.)

SCRABBLE BOARD

Looking for some handy letters around the house? In the basement behind the furnace you'll probably find an old Scrabble game.

You can arrange the letters on the board for a true Scrabble look, or you can lay the tiles on a sheet of paper, or across a towel or carpet for varied backgrounds. The letters are easily moved and rearranged. Unless there's a Scrabble theme to your show, this type of lettering may look a little hokey. Maybe you can think of some other games around the house that come with little letter squares. How about kids' building blocks?

These methods are only appropriate for short titles, as you'll probably run out of letters (and patience) after just a few lines.

SPAGHETTI BOARD

This somewhat derogatory name hails from the days when this method of lettering was used primarily by restaurants displaying their menus on grooved felt boards with movable plastic letters. The method is fairly bold, fairly neat, and fairly rapid to set up. The letters are reusable. Care must be taken to adjust the lighting in a way that will not show the grooves from the lettering board. Opening the lens iris or *slight* defocusing may help some, too. This method is also appropriate only for short titles.

THREE-DIMENSIONAL LETTERS

These letters may be made of plastic and may have pins or magnets in them for attaching them to a lettering board. The method is simple, bold, and quick. The letters are reusable (until you lose them) and have the added advantage that lighting may be used to cast fancy shadows from the letters or to highlight their edges. Many such lettering sets look somewhat amateurish, however, especially if great care isn't taken in lining the letters up neatly. This method is appropriate for short titles only.

PRESS-ON OR RUB-ON LETTERS

These are wax-coated letters that come in a sheet and can be rubbed off onto a sheet of paper or posterboard. The process is quite simple and the letters are neat and bold. The wax sheets are inexpensive, and no equipment is necessary except for a stick with which to rub the letters. The process, though simple, is too slow for doing many words, but it is great for short titles, credits, short lists, and captions to be affixed to existing photographs, objects, or charts. The letters rub off, however, almost as easily as they rub on, so some care must be taken with the finished product.

Of all the methods, this one gives the most professional-looking results for the least bucks.

TYPING

Thrifty, but lacking somewhat in boldness, is our old friend the typewriter. Although it is much maligned as a TV titling device, when used creatively, type can be the budget TV producer's workhorse.

Typed titles

Given an IBM Selectric (or similar) typewriter and an assortment of type balls (about $20 each), you can type fairly nice-looking titles on off-white 3 × 5 cards in a matter of seconds. What your typed titles lack in "class" they may make up for in variety, particularly with the assorted lettering styles available.

Close-up lenses or inexpensive lens attachments permit the TV camera to take very tight close-ups of the typing. The resulting image is sharp, legible, and even bold when extremely tight close-ups of brief messages are taken. Just remember to zoom in to keep the lettering big, and don't try to put too many words on the TV screen at once.

Typing on various paper surfaces yields an assortment of background possibilities. Making clean photocopies of the cards (perhaps with a Xerox machine) increases the density of the letters, making them bolder. Making copies of copies of copies—as many as twelve generations down—increases the density even further, while adding a unique character to the letters.

The 3 × 5 cards are easy to handle, organize, and store. They take up very little room, unlike larger title signs. Illuminating them is simple with a desk lamp or two. Special effects, such as a spinning title, can be handled by attaching a card to a phonograph turntable with the TV camera viewing from overhead. Once a production is finished, the used 3 × 5 title card makes a good label for the master tape when affixed to an appropriate spot on the tape box.

One disadvantage of the typed-title method, however, is the likelihood that a minor flaw in the card's preparation will be exaggerated when the small picture area is magnified to large TV-screen size. The author remembers once when his associate sneezed near one of the little title cards during a production. The damage went unnoticed until the tape was played back on a 21-inch monitor. The flaw then became grossly apparent. Very grossly.

As one might guess, typing is useful for both short and long textual passages. It is perhaps more appropriate for the long ones, while the short, frequently reused and infrequently changed titles are relegated to the slower, bolder methods of typography.

HOME COMPUTER

Professionals often use a CHARACTER GENERATOR, an electronic keyboard that allows you to type material directly onto the TV screen. You often see it used in news subtitling for the hearing-impaired or for flashing players' names across the screen in sports.

Except for Comprehensive Video's $795 VideoScribe VS–100, CHARACTER GENERATORS, at about $3000 up, are expensive for displaying the occasional "Our Trip to Milwaukee" title, but if you're a home computer hobbyest, you may have something almost as good. Simply type your titles into your home computer and play the result into the RF input of your VCR and record them. The Atari 800, for instance, can generate a four-color message similar to the moving light display in New York City's Times Square. The computer program that does this is called "Marquee." Chapter 17 (paperback Volume II) tells much more about this and other things you can do with your home computer.

If you find that your VCR won't record the computer's signal directly (sometimes the computer's signal is not compatible with standard video), you can always try throwing the display onto your TV screen and with your camera, recording it from there. If the picture appears to "pulse" slightly in your electronic viewfinder (your naked eye won't see this flicker, only the camera can), try adjusting your TV set's contrast and brightness to minimize it.

IN-CAMERA TITLER

How would you like a handy CHARACTER GENERATOR, built-into your color TV camera? Canon, GE, Magnavox, Panasonic, and Quasar now offer just that in their $1100–$1400 cameras.

To work them you simply throw a switch on the camera to the TITLE position. The letter A will appear in the upper left corner of your viewfinder. If you want the letter C, then press a CHARACTER FORWARD button which changes the A to a B, then to a C. To go to the next letter in your word, press POSITION FORWARD and now instead of dealing with the upper leftmost corner of the screen, your typer will be dealing with the next position over. By pressing the CHARACTER FORWARD, CHARACTER REVERSE, POSITION FORWARD and POSITION REVERSE buttons, you can "print" any sequence of numbers, letters, and spaces, anywhere on the screen. Once the viewfinder screen displays the desired title, you record it by:

1. Setting your VCR in RECORD/PAUSE
2. Making sure the INPUT SELECT is on CAMERA
3. UNPAUSING the VCR for the length of time you want the title to show.

The IN-CAMERA TITLER isn't quite as fast as touch typing, but at least it won't dry out like felt pens do three letters before you finish printing your title.

Displaying Graphics

LIGHTING GRAPHICS

A nice-looking title is generally one that leaves no hint as to how it was constructed. Curly edges on letters and grainy paper fibers in the background make titles look amateurish. Even flat, smooth titles have minor scratches, ridges, and lumps in them which remain hidden until they are revealed by the all-seeing TV camera.

Some of these flaws can be deemphasized by the use of flat, shadowless lighting. As shown in Figure 9–6, lamps are set up to the left and right of the camera and aimed at the graphic. Each light "washes out" some of the shadows created by the other light, making the image fairly shadowless. The "softer" (more diffuse) the lights, the better.

Two 100-watt desk lamps with frosted bulbs and large reflector shades may work very well. The lighting will easily be bright enough since the lamps are so close to the object being illuminated.

Sometimes it may seem impossible to rid a visual of reflections, especially if the visual happens to be a glossy photograph or the like. It may help somewhat to take extra effort to flatten out the photo or graphic item. Mounting it on a posterboard may help. A slight curl will reflect light from many directions, whereas a flat object will reflect light mostly in one direction, like a mirror. Thinking of the flat visual as a mirror, position it (or the lights) so as to reflect the light away from the camera lens.

FOCUSING ON GRAPHICS

All of the focusing procedures in Chapter 7 also apply to focusing on graphics. However, small things like graphics are harder to focus on than larger things. Although a foot may make little difference 20 feet away, an inch makes a big difference 6 inches away from the camera lens.

One way to minimize the focusing problem is first to assure that the graphic is exactly perpendicular to the camera's line of sight, as shown in Figure 9–7. This way, all parts of the graphics are equidistant (almost) from the camera lens and are therefore all in focus at the same time.

Another way to minimize the focusing problem is to flood the visual with light and to "stop-down" the camera lens to a high f-number for maximum depth-of-field. In very tight close-ups, it may be impossible to focus all parts of the visual accurately because the edges of the visual are a shade farther away from the camera lens than the center of the visual. In such cases, stopping-down the lens may be your only recourse

Figure 9–6
Lighting graphics (top view)

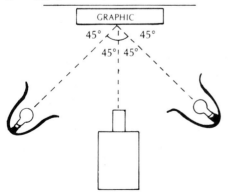

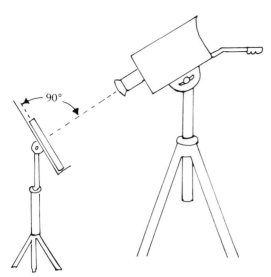

Figure 9–7
Graphic kept perpendicular to camera's line of sight

for getting an all-around sharp picture. Be aware that the heat from the intense lights, if they are too close to the visual for too long, may curl your visual, especially if it is an unmounted photograph.

Another reason for making sure your graphics are perpendicular to your camera's line of sight has to do with a phenomenon called KEYSTONING. The closer part of the graphic will look bigger than the farther part of the graphic. Figure 9-8 shows the results of aiming a camera off-center at several rectangular graphics.

CHANGING GRAPHICS

The easiest way to go from one graphic to the other is to PAUSE your recording, change graphics, then UNPAUSE. You could spice up the process by fading out your camera or defocusing between shots. Some of the camera tricks in Chapter 6 may be helpful.

No law says you can't see graphics being changed. A hand could turn pages like in a book. Pages could also slide off the screen, revealing more pages beneath.

Could the changing of the graphics somehow be related to the content of the program? Could beach waves obliterate titles scratched in the sand? Could a thin spray of water make your letters bleed down the page, ushering in a rainy-day setting? Perhaps a Father's Day show could have the titles and credits attached to his favorite places—his chair, his pipe rack, his *Playboy* collection. A pan around the room (to his favorite music) would locate each title in turn.

Changing graphics may seem like the most mundane of video production nuisances, but if handled creatively this simple aspect of your show can add pizazz and visual interest.

CRAWL and ROLL When words slide across the screen from right to left, it's called a CRAWL. You can CRAWL words across a screen by typing them on a wide sheet of paper, setting the paper in front of the camera and either panning the camera or smoothly moving the words. The process is mechanically hard to do and is best left to electronically generated text.

A ROLL is the vertical movement of text, sliding upward through the screen like the credits at the end of a movie or TV show. These are a little easier to handle with home equipment. First you need to prepare your text in a long, narrow column, no more than 25 characters wide. As is shown in Figure 9–9, the TV camera is wheeled up to a wall on which are mounted the titles, lists, or whatever on a long strip of paper. The strip should be lit evenly from top to bottom. A close-up lens may be necessary if the list is typed or reproduced in small print. Align the camera and list so that the camera may be aimed at the top of the list (or even above the top), and, on cue, its pedestal may be slowly, evenly (and carefully) cranked down, lowering the camera. The words will appear to rise from the bottom of the screen and slide to the top and out of sight. Leaving blank spaces at the bottom of the

Displaying Graphics 275

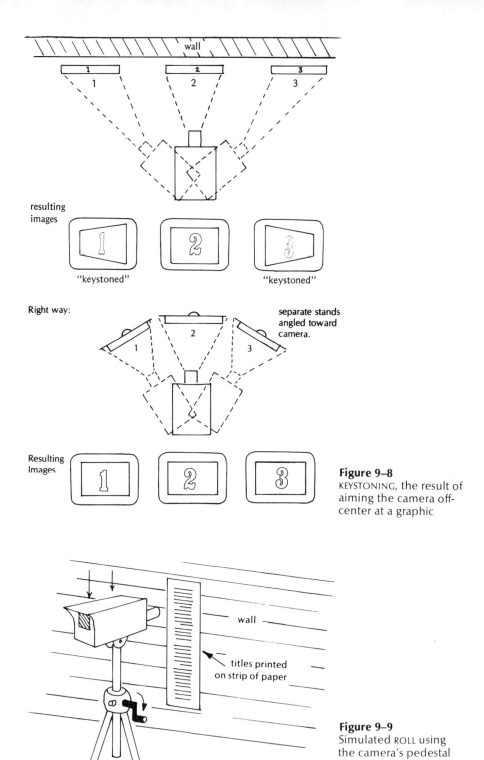

Figure 9–8
KEYSTONING, the result of aiming the camera off-center at a graphic

Figure 9–9
Simulated ROLL using the camera's pedestal elevator

strip allows the last words to be cranked right off the screen, leaving it blank.

Simply tilting the camera will not—in this case—achieve the desired effect. First, it's very hard to tilt a camera slowly and smoothly. Second, since the top and bottom of the strip are farther away from the lens than the middle, they will be out of focus. Third, for strips of any length, the KEYSTONE effect would be quite noticeable.

If typing your titles or credits, an easy ROLL technique awaits. Simply type your text in a narrow column, leaving white space at the top. Reinsert your text into the typewriter and disengage the ratchet so that the paper moves smoothly as you turn the roller. Stretch the paper or support it so it stays flat as it exits the roller. Using a CLOSE-UP lens and a tripod, aim the camera at a flat place and "roll credits."

Backgrounds for Titles

There is no reason why titles have to be shown over plain backgrounds. Black lettering over a simple picture (see Figure 9–10) is far more interesting. Just make sure the picture doesn't distract the viewer from the title (unless the picture is of paramount importance while the words are only incidental).

There may be occasions when the title remains the same but you'd like to change the background for the title. Rather than remaking the title several times on various backgrounds, one may make the title on a clear acetate sheet and lay it over the various backgrounds each time it's used.

Another trick is to use two projectors. One projects a background slide onto a small white screen or posterboard. The other projects your titles, photographed so that the letters are white and the background is black. By aiming both projectors at the same spot on the screen and shooting the results with your TV camera, you end up with white words SUPERIMPOSED over your background slide. By changing slides on your "title" projector, you can record the titles changing while the background remains the same. You could similarly change the backgrounds under the static title.

Another trick is to show the title through

Figure 9–10
Title backgrounds (Drawings by Jose Marjolin and Sheela Teeluck, courtesy of Joseph Sauder)

one camera and the background through another and KEY or SUPERIMPOSE the words over the picture. The process requires a special camera and another device called a SPECIAL EFFECTS GENERATOR, both described in Chapter 14 (paperback Volume II).

Making Graphics Come Alive

Those of us blessed with creativity will have no trouble making graphics come alive. Good artists and photographers can find ways to make almost anything look real or interesting. Although art, set, and graphic design are really subjects in their own right, inappropriate for discussion in a basic video text, a few related considerations should be mentioned.

There is no law that says that a visual must remain stationary. A camera can pan, tilt, and zoom over a photograph or a painting as if it were shooting something "live." Quick cutting and active movements can make the pictures themselves seem to be moving. Still photographs become movies. Battle scenes become the actual battles (with the help of sound effects). Ships roll back and forth while earthquake scenes shake up and down. Amusement park rides streak by, while the lights of the Midway grow blurry and dissolve to the next scene.

Cutouts can be placed over visuals and moved, simulating animation. Scenes can have holes cut in them with movement behind the holes simulating running water, snow, vehicles passing by, or whatever. Lighting changes can make three-dimensional objects seem to move. Strong BACKLIGHT with very little KEY or FILL light simulates night scenes. Raising the KEY and FILL lights ushers in the day. Puppets, when not sharing the scene with real people, may begin to look like full-sized people themselves. Visuals may be burned on camera. Title lettering may be blown away. Delicate hands may enter the scene and turn over tarot cards, revealing titles or credits. Smoke in the foreground can lead to the final FADE OUT. Blood (ketchup, unless you're a stickler for authenticity) may drip onto a title. The camera zooms in on the drip, filling the screen, and fades out as the next visual is revealed in the blackness of the drip. Titles can be made to disintegrate into a mushy cloud as the camera shooting the title goes out of focus.

Creativity—it's the fun part of television. Use it. At the same time, however, keep the objectives of your production in mind and don't let the fancy stuff carry you away. How appropriate would it be for the ending credits on Grandpa's Birthday tape to change from one to the other with holes burning in them or blood dripping on them?

Film-to-Tape Transfer

ADVANTAGES AND DISADVANTAGES

So you have slides or movies which you'd like to transfer to video. Perhaps the inconvenience of dragging the projection screen out of the hall closet is the reason. Perhaps smashing your shins on the coffee table while crossing your darkened "theatre" has something to do with it. Maybe you'd prefer the sound of background music and polished narration to the cement-truck sound of your projector fan. Or maybe your family album is composed of 20 different media, some slides, some snapshots, some regular-8 movies, some super-8, a few audiocassette tapes, a reel-to-reel audio tape, one of those old records you recorded in a booth at the amusement park, some cherished postcards, and Fibber-McGee-knows-what-else. And you'd like to organize and consolidate this mess into a single, neat videocassette, to be the family's illustrated autobiography.

Excellent idea. But before you start blowing dust off the film cans, take another peek at Figure 9-5. Look at the detail you'll lose. "Is that Baby Bootie in Mumma's arms or is she hugging Tinkle the cat? Or was that the bellhop and Mumma is the one bending over—no, that's the luggage. Wait—I see the cat now, he's coming in the door, but Papa, do you remember him having one big eye in the middle of his forehead? . . . He's what? Oh, he's going *out* the door."

In short, decide whether the advantages of a well-organized, convenient family documentary are more important than the sharpness and detail of the original photographic media. If after thinking it over, you still opt for video, at least consider the many methods you've seen so far for making the photos come alive. By zooming in and panning across pictures at close range, you can preserve much of the important detail.

DO-IT-YOURSELF METHODS

There are several methods for teaming your video camera up with a projector. The three which you can probably afford are: FRONT SCREEN PROJECTION, REAR SCREEN PROJECTION, and TELECINE ADAPTER.

FRONT SCREEN PROJECTION This method is cheap, flexible, and nearly as good as any other method. You simply aim your projector at a screen and aim your camera at an image to record it as shown in Figure 9–11.

Use a flat white posterboard as your screen. Any smooth, matte-finished (not shiny), very white, flat surface will also do. Regular projection screens tend to reflect "hot spots," overly bright areas on the screen, while allowing the edges and corners to look dim.

Position the projector so that the projected image is about 18 inches tall. Larger images start to become too dim. Smaller images accentuate the minor imperfections in the screen surface. Experiment. Be sure to focus the projector *very well*.

Set up your camera on a sturdy tripod. Position it as near the projector as possible so that you don't get KEYSTONING of your image. Turn out *all* lights (or close curtains) so that *no* light seeps in to "wash out" your projected image.

Focus the TV camera. Adjust your camera's iris, color temperature, and white balance controls as usual. The "indoor" setting on the color temperature control may work best, but experiment.

The CONTRAST RATIOS of slides and movie film are up to 100:1, while your camera can handle only up to 30:1. If you notice blooming whites, lag, or burn-ins starting to occur, try to reduce the projector's brightness. Perhaps stop-down the camera's lens iris or add a neutral density filter to it.

Adjust your camera's zoom lens so that the projected image fills the TV screen. Because film pictures have a wider ASPECT RATIO

Figure 9–11
FRONT SCREEN PROJECTION method for transferring film to tape

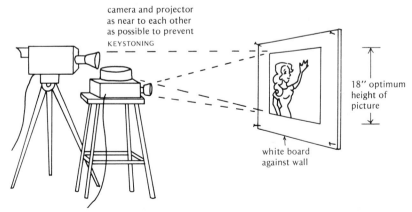

than video, this will mean the loss of a little of the left and right borders of your scene. Feel free, however, to zoom, pan, tilt, fade, or defocus your camera in order to eliminate dull parts of a scene or to focus attention on the main event, or to add movement when needed, or to dress up the transitions between scenes.

By PAUSING your VCR, you can omit that blank or black screen you get between slides on most projectors, or while changing movie reels.

Rehearse your narration while recording if you want, just to get a sense of timing, but save for a later audio dub (assuming your VCR has this feature) your actual sound track. You have plenty to handle right now, whereas later, giving audio your full attention, you're more likely to do a smoother, more creative job. Also, you won't have projector fans roaring in the background.

REAR SCREEN PROJECTION You've seen them in shopping centers, travel bureaus, and schools, these boxes with a gray, translucent screen and a projector inside displaying a sound film (usually scratched and dirty with unintelligible sound) or slides. You can buy just the rear projection material and make your own projection setup, like in Figure 9-12. The material is available from hobby and photo stores, or you can buy a sheet of Polacoat lensscreen for about $5 a square foot from Raven Screen Corp., 124 E. 124th Street, New York, NY 10035. Staple it around a foot-square frame. In a pinch you could also use translucent or vellum paper to project onto.

The projector shoots the image *through* the screen while the TV camera, on the other side, registers the image. You get more room to work than with FRONT PROJECTION, and there are no KEYSTONE problems if you line everything up perpendicular to the screen's surface.

The image is usually plenty bright for the camera, and depending on the circumstances, you *might* be able to work with a dim light on somewhere (your shins will thank you for that).

There are disadvantages too. Good REAR SCREEN material is expensive and hard to find. Poor material causes "hot spots" at certain angles and also yields slightly fuzzy pictures. Since your camera views the projected images from the rear, all pictures come out reversed left-to-right. Slides can be put into the projector backwards to solve this problem, but movies can't be flipped over and run through the projector (unless you like making new sprocket holes in your heirloom film).

TELECINE ADAPTER This is a little box with a mirror and a tiny, good quality, REAR PROJECTION SCREEN in it. You project your film

Figure 9–12
REAR PROJECTION

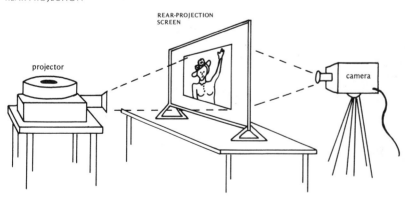

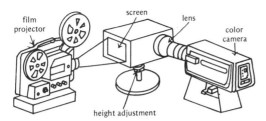

Figure 9–13
TELECINE ADAPTER

image into one hole in the box and aim your camera lens into another hole in the box to "see" the image, as shown in Figure 9–13. The image comes out correctly (not reversed) and the box shields enough light so that you don't need a darkened room.

They cost $100 to $250 and do a pretty fair job. Some models may give you a little trouble focusing (you may need a macro lens) and some don't accept square or vertically oriented slides without cutting off a piece of the picture.

OTHER TRICKS

Say you were using FRONT or REAR SCREEN PROJECTION and you had *two* projectors. Can you think of some way to switch creatively from slide to slide?

For about $25 a day, you can rent (from a photo shop) a DISSOLVE CONTROL for two slide projectors. The device will fade down one projector's picture while simultaneously fading up the other's. The result is a smooth, classy-looking transition from one slide to another. And since you get the effect without pausing your VCR between slides, your tape ends up glitch-free.

Here's another idea. Buy a Sony HVS-2000 video switcher ($200), and its HVM-100 black-and-white accessory camera, and you'll be able electronically to insert titles over your slides. But let's save that for Chapter 14 (paperback Volume II) when we play with Options and Accessories.

TELECINE ADAPTERS Left, Magnavox TeleCine converter for video camera, Model 9057 (Courtesy of Magnavox) Right, Quasar TeleCine converter for TV camera, Model KT 502QF (Courtesy of Quasar)

SHUTTER BAR

Horizontal bar in picture resulting from mis synchronization between the light from your movie projector and the scanning of your camera's electron beam

MOVIES

There's one annoying problem faced by anyone trying to transfer a movie to video tape, and that's SHUTTER BAR. Your video image flickers or has a dark horizontal band running through it, perhaps moving, sometimes stationary. Here's the cause:

Movie projectors work by beaming light through still, transparent pictures and focusing that beam on a screen. A shutter mechanism in the projector beams the light through one picture, then blocks off the light while changing to the next picture, then opens to beam the light through the next picture. The process happens so fast that you see a continuous moving image. Super-8 projectors display 18 pictures per second (24 per second for sound), and industrial 16 mm projectors show 24 frames per second.

Your TV camera scans its sensitive vidicon tube at the rate of 60 pictures per second. Because the numbers don't match, there are times when the TV camera is "looking" for a picture but your projector isn't making one—it's busy changing pictures and the screen is black. Thus the projected image, which looks fine to your eyes, looks bad to the TV camera.

Professional TV setups, like studios and commercial film-to-tape transfer shops, have special motors to resolve this problem. There's not a lot you can do at home to banish the SHUTTER BAR from your recordings. One thing which may help, if you have it, is a variable speed control on your projector. If you can increase your projector's frame rate to 20 or slow it down to 15 frames per second (both are multiples of 60), the bar may be minimized somewhat.

COMMERCIAL SERVICES

If you want to save the hassle, time, and expense of transferring your slides and movies to video, you can always have someone else do it for you (for a price, of course). A commercial service will employ better cameras and projectors than you have, and usually will give better results.

At last look, Fotomat, which promises 10-day service, charged $16 to transfer 400 feet (about 27 minutes) of super-8 movie film to BETA or VHS film (the tape is extra). For 120 35mm slides, the price was $10. Fotomats are available nationwide for drive-up service.

Or you could mail your stuff to Concept Video Productions, located at 3125 W. Burbank Boulevard, Burbank, CA 91505, and supposedly get a three-day turnaround. They charged (at last look) $14.50 per 400 feet of super-8 (plus tape) and $8.50 to transfer 120 35mm slides to BETA or VHS. One question to think about before mailing out your film for transfers: Do you trust your photo treasures to the U.S. Postal system? The U.S., I'm told, stands for Usually Smashed.

Of course you would have to assemble your materials in the right order for these companies to produce the tape the way you wanted it. Unfortunately, you lose nearly all "creative control" over the result. There would be no panning, zooming, fancy transitions, or mixture of media (slide, movie, then snapshot) like you could do at home.

Don't forget that AUDIO DUB feature on your VCR. Even though you had Fotomat copy your silent home movies onto tape, they don't have to stay that way. Add narration, or at least light music. You'll be amazed at how much "class" this adds to your slides or movies.

ORGANIZING YOUR PROGRAM

Now that you've learned the *mechanics* of putting pictures on tape, it's time to consider your message and how you present it. Presumably, the aim of your program is not to induce sleep. The two keys to visual entertainment are to (1) tell a story, and (2) keep it short.

Organize your slides or movies so that a theme or a story emerges. Design a beginning, a middle, and an end to your presentation. This way your show will have direction and purpose and an excuse to hang the pictures on.

Your running monologue should tie things together, not merely verbalize what is shown. For instance, instead of "(picture) And here's Uncle Frank wearing a hula skirt . . . (next picture) And here's another shot of Uncle Frank wearing a hula skirt," say, "So what did Uncle Frank decide to wear to this Luau?" (show picture) "Of course, the same outfit he wears to every yearly Luau." (more pictures of Frank, Hawaiian music in background).

Brevity is the soul of a good production. Who wants to see 166 slides of Baby Bootie's third birthday? Pick the best 10 or 15 slides and make a short story from them. Choose the action and emotion shots like "Babe eating the birthday candles—while still lit" or "Babe smearing frosting all over Grandpa's hearing aid," and of course, "The Little Angel falls asleep, surrounded by wrapping paper and mangled toys."

Graphics in a Nutshell

Titles, pictures, and other media add professionalism to the amateur production. Also, photos make it possible to create settings for your video dramas without the expense or trouble of actually being there.

The three rules of graphics are:

1. Make it fit the shape of a TV screen.
2. Keep it simple.
3. Make it bold.

Leave enough margin around titles and around important elements of graphics so that nothing gets lost behind the edge of a misadjusted TV screen.

Slides have a wider ASPECT RATIO than a TV screen. When transferring them to video, their left and right edges will be lost.

Avoid graphics with very white or very dark places; these will not reproduce well on TV. The whitest part of a picture should not exceed 30 times the brightness of the darkest part.

Keep titles brief and bold. Text should not exceed 25 characters per line or 10 lines per screen page.

Shadows make three-dimensional letters stand out well, but they make flat lettering look amateurish. Light graphics equally from both sides to cancel shadows.

For sharp focusing on graphics, keep the graphic perpendicular to the camera's line-of-sight. Close-up lenses or attachments are sometimes necessary, but if they make the image fuzzy, flood the graphic with more light and stop the lens down to higher f-numbers.

Spice up your titles with appropriate backgrounds. Feel free to zoom, tilt, and pan across pictures to make them come alive.

Slides and movies can be transferred to video for ease in display and for adding music and sound effects. The TV medium, however, will lose much of the detail of your movies and photos.

Although it is possible to videotape the image from a movie projector, your home projector will probably not synchronize with your TV camera. The resulting tape is likely to have a grayish SHUTTER BAR running through it or will appear to pulse in brightness. Better transfers are available from commercial concerns like Fotomat, but then you lose the flexibility of editing together differing media in creative ways.

The entertainment value of your video/slide/movie/graphics program is enhanced if you:

1. Tell a story.
2. Keep it short.

Chapter 10
Copying a Video Tape

HIS MASTER'S COPY

I hate to tell you this, but home-format video was never meant to be copied.

If you record an excellent signal off-the-air or from a camera, some of the signal quality is always lost in the recording process. This happens to an imperceptible degree with $40,000 broadcast recorders; it happens to a small degree with $2000 industrial VCRs; this degradation is quite noticeable on $900 home-type VCRs. Even with its slightly fuzzy video, shaky sync, and its impure colors, the miracle of home-recorded television is still satisfying. Make a copy of that marginal tape, however, and you record the flaws of the original tape while adding new flaws in the process of recording. The copy plays back with twice the imperfections of the original. You can clearly (or unclearly) see the degradation in the picture and hear it in the copy's sound.

This doesn't mean home tapes *can't* be copied. They can, and are. But they lack oomph. If the subject matter warrants copying a tape, even though the copy may be oomphless, then do it. You'll see how to in a moment.

If you expect to make a recording which you *know* is going to be copied, and it's important to have *good* copies, consider this: Rent, or borrow from a school or business, a 3/4U videocassette recorder (and perhaps a better camera), and make your MASTER (your original recording) on that. Later, you can copy from that higher-quality medium onto your home format and get much more satisfying results.

The Basic Basics

Here's how to copy a tape: Get two VCRs. Using separate audio and video cables, connect the audio and video *outputs* of the player (called the MASTER machine, or the VCP, for videocassette player) to the audio and video *inputs* of the recorder (called a SLAVE in video circles). Connect your TV set to the SLAVE's RF output so you can monitor the results as shown in Figure 10-1. Chuck your MASTER cassette into the MASTER player and a blank videocassette into the SLAVE. Adjust the INPUT SELECT switch on the SLAVE so

A) Original
As seen by camera

B) First generation
Same scene recorded and played back on VHS machine at 2-hour speed

C) First generation
... and at 6-hour speed

D) Second generation
Copy of example B made at 2-hour mode

E) Third generation
Copy of example D made at 2-hour mode

Picture degradation with copying

Making the Best Possible Signal for Copying

YOUR BEST MASTER

The tape you record can be no better than the tape you're copying from. If your original tape is fuzzy, grainy, or jittery, your copy of this tape will be fuzzier, grainier, and jitterier. So the first step is to start with good tape. Use your highest-quality tape for MASTERING. Perhaps high grade (marked HG on the boxes) tape would be best.

Virgin (fresh from the box) tape is good, and will also be free of the residual magnetism sometimes left on used tapes which haven't been completely erased.

Strangely, the *best* tape for MASTERING is

it "listens" to the MASTER. Select the desired tape speed; hit PLAY on the MASTER and RECORD on the SLAVE. Let the machines do the rest.

HG tape which has been played about 10 times. Such tape will have the *least* dropouts (tiny dots of snow). This is because when the tape is played, the spinning video heads "scour" the tape, cleaning off tiny contaminants.

After 30–100 playbacks, this scouring process begins to work to your detriment as the head starts to wear off some of the magnetic oxide in places. The tape also will have had opportunity to pick up dust and scratches.

In short, ten-times-used tape is best, virgin tape second best, and old tape worst for MASTERING.

The second step to making a good MASTER is to record your MASTER at your VCR's *fastest speed*. This ensures the highest-quality picture and sound to start with. Record your MASTER from a good, strong antenna signal, meticulously fine-tuned, or with a well-lit, carefully miked setup. Again, the better your original signal, the better your copy will be.

YOUR PLAYER AND RECORDER

North American Standard (NTSC) VCR's are all compatible as far as their audio and video signals are concerned. Whether one machine is a BETA and the other is a VHS or CVC or whatever, the audio and video outputs should feed a satisfactory signal to any other machine.

The machines you pick to play the MASTER and record the copy should be in tiptop shape. Any aberration in their reproduction will appear even worse as you play your final copy. Maintain your machines and clean the heads as per the manufacturer's instructions.

If possible, play the MASTER on the same VCR that made the MASTER. Like your own mother, the machine that originated a tape is more "blind" to the recording's minor imperfections and idiosyncrasies and will play it at its best. If you can't play the MASTER on the machine that recorded it, then pay particular attention to TRACKING as you play your MASTER. This may have to be adjusted to yield a smooth, steady picture.

If you're copying someone else's tape using your own two machines and you have a choice which to use as MASTER and SLAVE, then try the tape on both VCRs to see which plays it best. Often you will find that the machine equipped with the most visual effects (like NOISELESS SEARCH, REVERSE SCAN, STILL FRAME, etc.) will give the steadiest picture. The machine with the widest video heads (check the VCR's specifications for this) will probably give the sharpest, smoothest picture. Whichever VCR plays the tape best should be the MASTER.

If the MASTER tape has been stored at a different temperature than your VCR, allow the temperature to stabilize before playing it. If the tape hasn't been played for a while, wind and rewind it once to "relax" it and also to dry it out.

Your SLAVE recorder should be set at its fastest speed for making as stable a copy as possible. Again, use good tape.

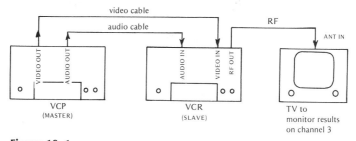

Figure 10–1
Simple hookup for copying a videocassette

Connecting Up Your Equipment 287

SIGNAL PROCESSING
EQUIPMENT

Although a MASTER SOW's ear can't be copied into a silk purse, there are devices which can accentuate desirable attributes of your picture and diminish the negative aspects. There are PROCESSING AMPLIFIERS, which stabilize your video signal, and IMAGE ENHANCERS which crispen the picture. These devices, which are connected in the path of the signal as it goes from the MASTER to the SLAVE, are covered in Chapter 14 (paperback volume II).

Connecting Up Your Equipment

Basically, you want to get the signal out of the player and into the recorder. One way is to send the RF OUT from the MASTER to the ANTENNA IN of the SLAVE. But we are loath to do this because we know straight video and audio have better fidelity than fuzzy RF. So VIDEO OUT and AUDIO OUT are fed from the VCP to the VIDEO IN and AUDIO IN of the SLAVE VCR.

Next we monitor our results. If you have two TVs, you could hook one up to the RF OUT from each machine. This way you could tell if the signal leaving the VCP was good and if the signal entering the SLAVE VCR was good. If something went wrong, the two monitors would tip you off as to where the problem was. No picture on both would imply something was wrong at the VCP end. No picture on only the SLAVE monitor indicates a cabling problem or something wrong with the SLAVE VCR. If you have only one monitor, hitch it to the SLAVE VCR as in Figure 10-1. This will tell you when the signal is being *received* okay (which implies that it was sent all right) and is also handy for playing back and checking your copy afterward.

Figure 10-2 compares several more ways to connect VCRs and TVs for tape copying. The connections in methods A and B have the added flexibility of allowing you to view off-air TV while the VCRs are busy copying a tape. The TVs can also monitor either VCR's signal or can display a playback from either VCR. The main difference between A and B is that A uses two TVs, where B only needs one but needs an added antenna switch. Method C in Figure 10-2 is not the best way to wire your VCRs but will work in a pinch.

Making the Copy

Before you start, once you've decided which machine is to *play* and which is to *record*, take this one precaution, which will save you apologies and a pint of stomach acid. *Place masking tape (or some other reminder) over the RECORD button on the VCR which will do the playing so that you don't accidentally push the button and erase something from the MASTER.* You don't want to forget which machine is which and suddenly find yourself with a Rosemary Woods gap in the MASTER tape.

Actually making a copy of a tape is straightforward:

1. Connect up the VCRs and monitors as shown in Figure 10-1 or 10-2.

2. Switch the SLAVE VCR's INPUT SELECT to "listen" to the other VCR. Select the desired speed, preferably a fast speed.

3. Play a sample of tape from the MASTER and adjust the TRACKING if necessary. Once everything looks good, rewind the cassette to the program's start.

4. Press RECORD/PLAY on the SLAVE and let it run for about three seconds, to get past the threading-caused dropouts at the beginning of the tape. Then hit PAUSE.

5. Press PLAY on the MASTER.

6. Watch your TV monitor. At first sign of clear program, UNPAUSE the SLAVE.

7. If you have any reason to believe the process might not work, stop copying after a minute and view the results. If okay, rewind both VCRs and make the copy for real.

8. When finished making a copy, view a little of the *end* of it. If your video heads clogged during the process, it's the end that is likely to be bad.

Now that you've successfully copied that video tape, what do you do next? Sit down and have a smoke, right? *Wrong!* You *label the copy* and make sure to put the word *copy* on that label. Copies and MASTERS look too much alike to be easily sorted out—but they're not *really* alike and should not be treated as equals. So label them and keep them separate.

Time for a riddle: A teacher shot some

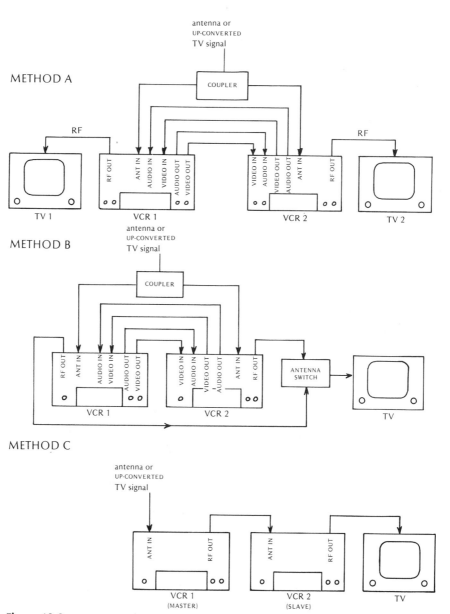

Figure 10-2
Ways to wire system for copying

video tapes at a school and gave them to her AV Department to edit together electronically into one tape. The tape came out so well that the AV Director immediately made a SAFE COPY of it in case the precious original ever got damaged. When the school board asked for copies, the AV Director wisely used the SAFE COPY as a WORKING MASTER (the MASTER you usually make copies from) to avoid risking the original MASTER. One schoolboard member copied his copy of the tape and sent it to another school. The teachers liked it and had their librarian make a copy of it for classroom viewing. A student liked it and copied the teacher's copy to play to his folks on his home videocassette recorder. His folks liked it, copied it, and sent the copy off to Grandma, who also had a VCR. The question now is: What did Grandma see? Answer: *crud.*

Why? Every time a tape is duplicated, it loses quality. The more generations you go down, the worse the picture and stability become. The moral of the story is: *Try to stay as close to the original as possible.* Make copies from the original whenever possible. Make copies from a WORKING MASTER only when it is really necessary to protect the original and when using professional equipment which affords the luxury of going down one more generation.

Copying Commercially Prerecorded Programs

A WORD
ABOUT COPYRIGHT

Commercially prerecorded videocassettes generally cost $6 and up to rent and $50 and up to buy. Wouldn't it be cheaper just to rent a tape and copy it for yourself, or borrow a friend's tape and copy that? Yes, it would—but: When the program is copyrighted (most recent ones are), duplicating it without permission from the copyright holder is illegal. It is also immoral, considering the copyright holder spent money producing his show to *sell* to you, not to have you simply take it.

The copyright laws and the doctrine of Fair Use are pretty complex. It takes a lawyer to unravel it all, but if you laid all the copyright lawyers end to end, they'd point nowhere. So, given vague laws and diversity of legal opinion, you, the VCR owner, have to decide what you think is safe to do and what isn't.

I'd like to help you (oh-oh, you know what's coming already) but the author is not a lawyer and cannot give legal advice. I *can* render my *opinion,* based on significant research and the legal opinions of others.

First a brief history: In 1976 MCA/Universal and Walt Disney Studios sued the Sony Corporation, four Los Angeles retailers, and a Los Angeles VCR owner for copyright infringement. Disney claimed that video recording TV broadcasts in the home violated copyright laws and decreased the market for Disney's shows. In 1979, U.S. District Court Judge Warren Ferguson ruled in favor of Sony, noting "Noncommercial home use recording of material broadcast over the public airwaves does not constitute copyright infringement. Such recording is permissible under the Copyright Acts of 1909 and 1976 and is fair use of the copyrighted works." In 1981, the 9th Circuit Court of Appeals overturned Ferguson's decision, ruling that home taping *did* violate the Copyright Laws. The Appellate Court then told the District Court to set damages. If it does so, it is likely that a royalty will be imposed on the sale of VCRs and/or blank tape. The proceeds would be pooled and divided up among the copyright holders according to a formula. The argument is now on the way to the Supreme Court, if Congress doesn't act to amend the laws sooner. Representative Stan Parris (R-Virginia) introduced Bill HR 4808 which would allow video recording of broadcasts to be made in the home for private use.

What is likely to happen? If HR 4808 passes, you're free to record. If Congress goes another direction, or if the Appellate Court ruling stands, you'll be paying for the right to record whenever you buy VCRs or tapes.

The bottom line: Is it okay to record TV shows for use at home today? Technically, it is against the law (as of this writing). However, damages have not been set, making the law somewhat inert. Disney and others have also announced they have no intentions of chasing down video tapers in their homes. In short, it appears pretty safe to go ahead and do it.

It is *not* safe to copy prerecorded videodiscs or tapes or pay-cable broadcasts *and distribute them for a profit*. That's playing Russian roulette with the law. If, however, you record a copyrighted work and play it *in the privacy of your home*, it's technically illegal (now) but unlikely to bring the FBI down on you.

ANTI-PIRACY SIGNALS

When you copy a copyrighted tape without permission, the producers call it "piracy." In order to thwart the unlicensed duplication of their wares, many distributors do something to the video signal to make it uncopyable. These antipiracy signals go under the names "Copyguard," "Videoguard," "MV Guard," "Stopcopy," "Copyguard II," and others.

They usually involve the reduction of the tape's video sync signal enough so that your VCR can *play* the tape, your TV can *show* it, but another VCR can't "lock on" to the signal to record it. Unfortunately for legitimate tape purchasers, the sync is sometimes too far out of whack for their VCRs to play, or their TVs to show. They get diagonal lines, vertical rolling, jittering, mistracking, picture pulsing, or a black line crawling through the picture. Sometimes these problems are surmountable by adjusting the vertical hold on the TV set. Old TVs and VCRs are especially susceptible to the "Copyguard Blues," while newer VCRs find it impossible to employ their "special effects" like fast search or still frame. In such cases, one can often bring the unplayable tape back to the dealer in exchange for an "unguarded" tape. If this isn't possible, reputable dealers will usually refund your money (one good reason for knowing who you're buying from).

While some equipment can't play guarded tapes, other VCRs have such sophisticated circuits that they can record guarded tapes with ease. The Sony SL5400, 5600, 5800, and the Zenith equivalents are such animals. Also the VHS Panasonic PVI1640, Quasar VH5155, RCA VDT625, and Hitachi VT8500 seem to record guarded tapes without difficulty. How these machines do it is unclear, but it may have something to do with their "special effects" circuits which are designed to "lock on" to the lousy sync the machine endures when scanning or still-framing normally recorded tapes.

So what do you do if (a) you can't even get your guarded tapes to *play* on your VCR, or (b) you wish to duplicate a guarded tape (copyright holders' wishes notwithstanding)? You buy yourself a VIDEO STABILIZER for between $80 and $250, a little box which goes between your video player and your recorder (or TV). You simply run your MASTER'S VIDEO OUT to the STABILIZER's input, and run the STABILIZER's output to the SLAVE's VIDEO IN (or ANT IN for some models). The STABILIZER strips off the guarded sync and replaces it with good sync suitable for solid TV pictures, uninhibited "special effects," and (ahem) illicit copying. More on this and other video gadgets in Chapter 14 (paperback volume II).

Scan Conversion

What if you wish to copy just a segment of a guarded tape to send to Grandma (perhaps the shark attack scenes from *Jaws* or the cannibalism scenes from *Night of the Living Dead*? You don't want to buy a STABILIZER just for that. Or what if you recorded a tape on a "sick" VCR which makes a picture good enough to watch but not stable enough to copy?

Or what if a friend from Copenhagen pops in with an armload of X-rated cassettes and her portable *PAL standard* VCR and TV? You watch as she plays her exotic tapes and lust for the chance to copy them, but her video signals are incompatible with your American NTSC-standard VCR. (And you can just taste the inheritance now, thinking of how Grandma will react to copies of these tapes!)

There is one process of making a video copy which isn't phased by sync problems, standards problems, or decoding problems, and that's SCAN CONVERSION. The process is shown in Figure 10-3 and involves aiming your TV camera at a TV screen and recording the image that's being played from another VCR. This process will result in a video copy that is stable, has good sync, and displays no picture breakup. *The quality of the overall picture will suffer noticeably,* however. The image will end up fuzzier, and it will have limited contrast and imperfect color. The process is a compromise between sharp pictures with unstable sync and fuzzy pictures that are stable.

Here's why the process works: Even though the signal being played from the MASTER VCR has imperfect sync, the TV can often still "lock on" to it. Most newer TVs are very forgiving that way. The fact that the sync is bad doesn't show *to your eye*. But that same signal is inadequate if sent electrically to another VCR. The SLAVE VCR chokes on it and you get a useless copy. Similarly, foreign-standard VCRs and TVs will make a perfectly *watchable* picture from foreign-standard tapes, even though their video signals are unpalatable to your U.S.A. equipment. So what SCAN CONVERSION does is re-record the picture *optically* from the TV screen. Your camera records what it "sees," which looked good. The camera manufactures its *own* sync internally, which is "standard" and good. Since audio is pretty much compatible everywhere, you can record that directly. Your VCR thus records the whole works with fresh sync, but because of the

Figure 10-3
SCAN CONVERSION

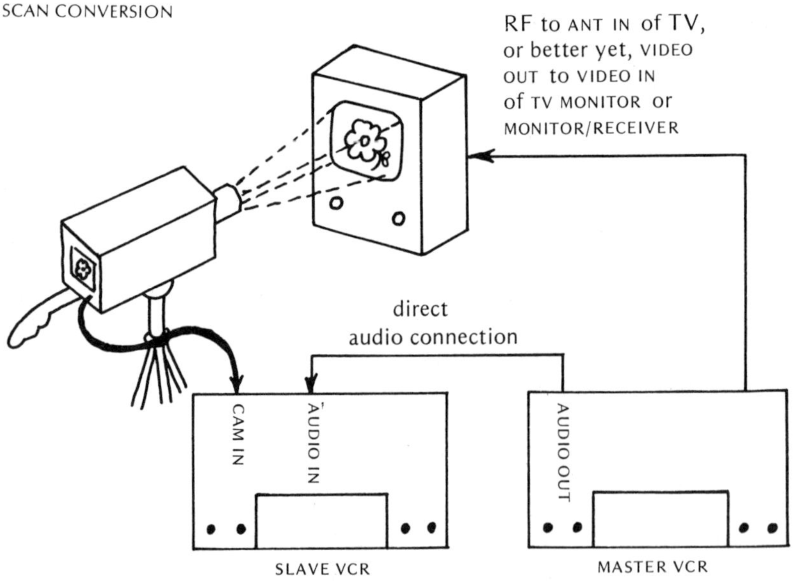

inadequacies of the TV screen and your camera, your picture quality decreases.

Here's how to carry out a SCAN CONVERSION.

1. Prepare your tape for playback by finding a good VCP and TV. Use straight video from the VCP to a monitor/receiver, if possible. Otherwise, use RF.

2. Set up a VCR and camera as in Figure 10-3. The camera faces the TV screen and is zoomed to a FULL SHOT. Audio is run directly from the MASTER VCP's AUDIO OUT to the SLAVE VCR's AUDIO IN. If possible, set up a VCR monitor from which to judge your final picture quality.

3. Play some of the program and adjust the TV picture controls. Note that the TV-screen image that looks best to your eye does not necessarily look best to the TV camera. Adjust the TV screen's brightness and contrast to make a good picture *for the camera*, regardless of how it looks to your eye. Use the VCR monitor as a guide to the optimal picture settings. There's a lot of experimentation in this process.

If you notice faint diagonal or curved lines on the image from the camera, try tilting the camera tripod slightly (lengthening a leg on the left or on the right). The faint lines are called MOIRÉ and the effect is hard to avoid. Experiment more.

If you notice the picture flashing or pulsing in brightness on your VCR monitor as it shows the picture from the VCP, try plugging all the equipment into the same outlet (or into each other) for power. If using a foreign-standard VCP and TV, *they* may have to be run on battery power if they don't take U.S.A.-type power. Still, the picture may pulse some. If it does, try adjusting the TV's brightness and contrast, and the camera's iris to minimize it.

4. When ready, turn out the lights (in order to avoid reflections in the TV screen face), start the VCR recording, start the VCP playing, and cross your fingers.

Copying Other Video Signals

Video tapes aren't the only thing you can copy. Anything that can be displayed on your TV set should be recordable (copyright regulations notwithstanding). TV games, computers, and videodisc machines, all have RF outputs which normally feed to your TV's antenna input. Simply feed the signal to your VCR's antenna input and switch to the proper channel to record.

Some computer equipment uses a nonstandard sync signal which your TV set may accept but your VCR won't. In such cases, you could find out if the computer manufacturer makes a TV STANDARDS CONVERTER (another name for TV STANDARDS is RS-170 STANDARD), which will translate the signals for you. Or you could use the SCAN CONVERSION technique just described.

Some videodisc machines have a VIDEO OUT which may be used in preference over the fuzzier RF OUT when copying.

If you're considering copying a videodisc program (and the copyright laws don't inhibit you), consider this: The blank tape you use probably costs more than the prerecorded disc. And is your time worth anything? How about wear and tear on a $1000 VCR versus a $500 disc player? Just a thought.

Photographic Copying

You were in the audience during the taping of "Family Feud" and the camera panned right by you jumping up and down and waving your hands hysterically in your typical accounting executive behavior. You taped the episode at home and want to show everybody that you were on network television. Only nobody wants to watch 30 minutes of Richard Dawson just to see you making a coast-to-coast spectacle of yourself. They might look at a snapshot, though.

Or you produced a tape of your own

which you'd like to distribute, but sending sample cassettes to prospective clients or employers is too expensive. Even the postal charges are prohibitive. Photos, however, are inexpensive to copy and easy to mail.

Maybe you wish to show Grandma a few scenes of her little granddaughter in the sandbox. Only she's in the hospital recuperating from the Copenhagen tapes you sent her and won't go *near* a videocassette tape recorder. Or, perhaps you would like to share a computer program or some VIDEOTEX displays with a friend and don't have a printer or graphics plotter, or other mechanism for storing the data. Photos of the TV-screen images could do the trick.

Or maybe you're writing a book, like me, and wish to show authentic examples of TV-screen images to your reading public.

Maybe the convenience of carrying around slips of paper appeals to you more than lugging heavy VCRs, TVs, and bulky cassettes. Whatever your reasons for wanting to photograph a TV screen, the process is not hard to do. But first a science lesson (oh, no, not again!) to explain why things have to be done a certain way.

MORE ABOUT HOW TVs AND VCRs WORK

The electron gun in your TV scans a picture onto the screen once every 1/60 of a second—sort of. Actually, it makes a picture, but *skips every other line as it goes.* This every-other-line picture is called a FIELD. During the next 1/60 of a second the gun goes back and zaps the lines it missed. The TV word for this is INTERLACE; it's the weaving together of the odd and even lines of the TV picture. This crazy procedure results in a TV picture which has almost no noticeable flicker *to your eyes.* Every 1/30 of a second, a totally new, complete picture appears. This is called a FRAME.

If a photo camera were to take a picture of a TV screen at, say, 1/120 of a second (a good speed for stopping motion and not requiring a tripod), you'd get a picture of a half-scanned TV FIELD. The TV's electron gun would have gotten only half the way down the screen before the shutter closed. The resulting picture (see Figure 10-4) would

Figure 10-4
Photos of a TV screen

Photo of TV screen taken at 1/120 of a second

SHUTTER BAR could appear anywhere on screen

Photo at 1/60 second

Photo at 1/30 second

Figure 10-4 *(Cont.)*

Photo at 1/15 second

Photo at 1/4 second

Photo at 1/2 second

Photo of STILL FRAME at 1/4 second

show a dark place where the TV screen hadn't been zapped yet. The resulting diagonal line is called a SHUTTER BAR. To avoid the bar, you have to shoot the TV screen at 1/30 of a second or longer, thus getting a complete, fully-scanned screen. Longer is even better, but if things are moving on the screen, a long exposure will make them blur.

How about pausing the playback for a STILL FRAME and recording that? The image (once TRACKING NOISE is adjusted out of the picture) will remain still long enough to shoot with your photo camera on a tripod. There's one problem with this, albeit a small one. The VCR's spinning video heads are timed so that one head swipes across the tape in 1/60 of a second and plays all the odd lines in a TV FIELD. The second head swipes across and makes the even lines in a

TV FIELD. As the tape moves, the machine plays the odd, even, odd, even parts alternately, creating a complete FRAME every 1/30 of a second. But if the tape stops, both heads swipe across the same place on the tape and play back the same FIELD, over and over. You may not notice, but you're seeing only the odd or maybe only the even lines reproduced on your TV. You're getting only half of the picture detail. Thus STILL FRAMES (which technically aren't FRAMES at all, but are FIELDS) are twice as fuzzy as when the tape is moving.

This may not be awfully noticeable, so go ahead and shoot STILL FRAMES. But if your scene is *not moving anyway*, why lose the sharpness? Wait until a still moment comes up and snap the picture then. You'll get a sharper picture—so sharp you'll cut yourself.

PHOTOGRAPHING THE TV SCREEN

Film For black-and-white prints, a medium-speed film such as Plus-X gives good results. High-speed films will work too.

For color prints, use daylight-balanced film. The same goes for slides. Use an ASA rating of 50 or higher. Since the color on a TV screen is slightly bluish, perfectionists can compensate by using a CC40R color-compensating filter on their photo camera lens. Open your photo camera's iris one stop extra to make up for the additional darkening caused by the attachment.

TV screen Avoid small TV screens. They're often not as sharp as bigger ones. Twelve- to 17-inch screens are fine.

First adjust the TV picture so it looks its best to you. Blacks should be black, whites white, and intermediate tones not too contrasty or overcolored. Next, turn the TV's brightness down a shade. This will sharpen the picture more.

If shooting black-and-white photos, use a black-and-white TV to get a sharper picture. Also, turn the contrast down a notch because black-and-white film tends to exaggerate contrast.

Turn out the room lights to avoid catching reflections in the TV screen.

Photo camera Place your camera on a tripod or other firm surface. Aim it straight into the TV. Almost fill the camera's viewfinder with the TV image. If the TV screen is small, you may need a close-up lens attachment on your camera to get close enough. Focus very carefully, perhaps trying to see the TV's scan lines on the screen.

Exposure is tricky. Use your camera's built-in exposure control, or use an exposure meter aimed at the TV screen. For critical pictures, bracket your exposures by one stop either way. Set your shutter for $1/8$ of a second or longer when possible. Professional photographer Jann Zlotkin gets her best results by shooting $1/5$ of a second at f5.6 using 400 ASA Ektachrome color slide film in her Nikon FM.

Use a timer or shutter release so that you don't wiggle the camera while pressing the button.

Cameras with LEAF SHUTTERS make less obtrusive SHUTTER BARS than FOCAL PLANE types do. If you *have* to shoot at speeds faster than 1/30 of a second, try to use a LEAF-SHUTTERED camera.

Polaroid shots of the TV screen are also possible. Through-the-lens reflex models are best for accurate aiming. The SX70 and Time Zero films are good for color, while the type 107, 667, and 87 films are good for black-and-white. Because most Polaroid cameras are automated, you may have to "cheat" the shutter into working at 1/30 of a second or slower. To do this, mount a used flashcube in the camera's socket (this disengages the shutter's "electric eye" mechanism). Recent Polaroid cameras have a "darker-lighter" control which will slow down the shutter speed of the camera. You may have to experiment. Don't forget to say "cheese."

For best results, pick recorded scenes which are close-ups or medium-length shots. The detail in long shots gets lost in the fuzz.

On the other hand, you can get some offbeat and creative effects by disregarding these rules. Super close-ups of the TV screen give you a mosaic effect from the TV screen. Boosting the color or shooting the screen from oblique angles can provide neat distortions to the image.

Retrieving text information stored on tape can get rather expensive if you try to photograph numerous screen pages. Here's a possibly cheaper way out: Buy some high-contrast 35mm film. Have the film developed, but not printed or turned into slides. Next waltz into your college or public library and sit down at their microfilm reader/printer, but instead of sliding a 35mm New York Times microfilm into the machine, thread *your* 35mm "microfilm" into the printer. Drop in your dime and press PRINT to get a cheapie copy of your screen shot.

If reading fine print doesn't bother you (and the print is unlikely to end up too fine, considering how few words can appear at once on a TV screen) you can always have your roll of 35mm film developed and *contact printed.* Then you get 36 TV pages on a single 8 × 10 sheet. Don't forget your magnifying glass.

Taking super-8 movies of your TV screen is quite unmanageable. You can't control most movie cameras' shutter speeds to get long enough exposures to avoid the SHUTTER BARS.

For more information on photographing a TV screen, consult Kodak's publication AC-10, *Photographing Television Images.*

Copying in a Nutshell

You need two VCRs, one to play the MASTER tape, the other (the SLAVE) to record your copy. Run the MASTER's video and audio signals directly to the SLAVE's inputs.

To make good copies you need good MASTERS. Use premium-quality tape recorded at fast speed on a well-maintained deck.

It is generally illegal to copy professionally recorded tapes, but many people do it anyway. There seems to be no grand cam-

paign to "crack down" on videophiles who copy programs for their own use in their own homes.

Anti-piracy signals make many professionally recorded tapes uncopyable. Sometimes these signals make the tapes downright unplayable. VIDEO STABILIZERS are electronic devices which convert guarded signals back to normal signals for copying or viewing.

Another way to copy a tape having sync abnormalities is to SCAN CONVERT. You play the nonstandard tape through a TV set and using a camera aimed at the TV, copy the picture off the screen. The audio can be connected directly between the player and recorder. The copy will be stable but somewhat fuzzy.

Most signals which can be displayed on a TV can be copied on a VCR by routing their outputs to the VCR's inputs.

To photograph a TV screen, use a shutter speed slower than 1/30 of a second, preferably 1/15 or 1/8 of a second.

Chapter 11
Editing a Video Tape

HIS MASTER EDITED

In the old days, if you wanted to delete something from a video tape, you had to slice it out with scissors and join the loose tape ends together. The method is still in use today with movie film and audio tape. What made scissor-editing difficult with video tape was that to make a smooth edit (no "glitch"), one had to cut *between* the magnetic video pictures, and we all know that magnetism is invisible. Even if the slice was properly made, the "bump" caused by the adhesive at the slice point often caused a glitch anyway. And there was always a risk that the tiny ridge between the tape ends would snag on the spinning video head, damaging it. For these reasons, scissor-editing is not done with video tape (except in dire cases, like salvaging a mangled tape by amputating the bad part—a process described in the Maintenance chapter).

In the professional video world, video editing is carried out on specially designed EDITING VIDEOCASSETTE RECORDERS capable of erasing *precisely* certain video "pictures" and replacing them *exactly* with new ones, all without a glitch. Sometimes these more-expensive VCRs are teamed up with computer-controlled BACKSPACERS, EDITING CONTROLLERS, or EDITING CONSOLES which make sure exactly the *selected* pictures are deleted and the chosen substitutions reinserted.

The home video user, with PAUSE button atwitter, is equipped to perform quasi-edits by stopping and starting the tape during the recording process. The resulting program may look satisfactory in the home to friends and relatives, but it should be understood that "PAUSE button edits" are not true edits. They leave glitches, sometimes tiny, but the glitches are still there. Professionals and broadcasters look down their noses at these "home-style edits" and don't consider them to be *true* edits. Professionals and broadcasters are forbidden to play glitchy edits over the air (FCC rules). In fact, much of their highfalutin equipment chokes on glitchy edits, yielding a few seconds of visual vomit every time one comes along. In short, if you have professional or broadcast plans for your video tape, edit it on professional editing equipment. Also, remember that home video equipment doesn't produce a very high-quality picture, making it ill-suited for editing, copying, and distribution. Use pro-cameras and VCRs if you want a sellable or broadcastable product.

Now for the good news: PAUSE edits are easy to do, look pretty good for home use, and add a whole world of new flexibility in video recording. Let's see how.

Words to know

ELECTRONIC EDITING
This is a method of editing on a specially equipped VCR through the mere press of a button. There are no glitches.

DUB
To copy something in its entirety.

AUDIO DUB
The sound portion of a tape is erased and replaced with a new sound track.

EDIT
To copy or record *parts* of something, perhaps changing the sequence of events or deleting events.

ASSEMBLE EDIT
With each ASSEMBLE edit, parts of a program are pieced together serially, like building blocks. Each new edit is added to the end of the existing recorded material.

INSERT EDIT
This kind of edit replaces a piece of existing program. It replaces a chunk in the middle rather than adding onto the end, as ASSEMBLE does.

RUSH or RAW FOOTAGE
Segment of video tape or film, perhaps shot on location, which will become part of a final production.

CREDITS
A list of the producer, director, audio director, camera operators, performers, and others who contributed to the production.

BACKSPACE
The act of rewinding video tape a measured distance from a desired edit point. This provides time for it to get a "running start" before it reaches the edit point.

Assemble Editing

You're ASSEMBLE editing when you first record one segment of a program, stop, then continue recording another segment, stop, and so on until you've assembled all the components into a complete show. There are two ways for you to stop your recording each time. Let's call one way a STOP edit, and the other a PAUSE edit.

Both ways allow you to record your segments minutes, days, or miles apart. One video excerpt may be a title, the next the opening scene in a long, establishing shot, and the next a close-up of your talent. After a series of such shots from different angles you may finish with a "The End" sign and a list of CREDITS ("Directed by—Alan Pevar, Camera—Alan Pevar, Audio—Alan Pevar, Screenplay—Alan Pevar, Conceived by—Mom and Dad Pevar").

STOP EDIT

Here you record a sequence and then hit STOP on the VCR. You practice the next sequence and when ready to tape it, press RECORD/PLAY and proceed with the recording. When the scene is finished STOP the machine again. Continue the process until you have assembled your whole show.

This is not a great way to edit. It's the *only* way if you have an old VCR without a PAUSE function. It's perhaps appropriate if there will be a long wait between recording sessions (you can't leave a VCR on PAUSE for more than a couple minutes without damaging the tape). The big disadvantage is that when you press STOP on VHS VCRs, the machine unthreads itself, losing your exact place on the tape. (BETA VCRs stay threaded until EJECT is pressed, so this is less of a problem with BETAS.) On a VHS machine, when you press RECORD/PLAY the tape rethreads itself, but to a point about *13 seconds earlier* than where you pressed STOP. Recording from this point will erase the last 13 seconds of your previous segment.

To avoid this tape pullback problem, you could try pressing PLAY instead of RECORD/PLAY, and waiting 13 seconds while watching your tape come up to your edit-out point. When you get there, press PAUSE, then press RECORD and PLAY, then UNPAUSE to begin recording. If your VCR has no PAUSE feature (or it doesn't work), try pressing RECORD and PLAY together when you reach the edit-in point.

STOP edits are cumbersome and inaccurate compared to PAUSE edits.

PAUSE EDIT

Here you record a sequence, hit PAUSE, set up the next sequence, UNPAUSE to continue recording, and so on until the end of the show. This is probably the method you've been using to edit-out the commercials from TV broadcasts.

Unlike the STOP edits, with PAUSE edits the tape doesn't unthread. You stay put, right where you left off—almost. There are exceptions. Some newer VHS VCRs automatically BACKSPACE (run the tape backward) about half a second when you hit PAUSE. They do this to ensure a smooth, almost glitchless edit. That's nice for smoothness, but a little rough on planning. On such machines you have to learn to hit PAUSE about a half second late, so when the machine trims off that half second, you'll be where you want to be.

One problem with PAUSE edits is that you can't leave your VCR in PAUSE for very long without risking your tape and video head (and some machines shut their PAUSES off automatically after a couple minutes). This limits the amount of time you have to arrange your next scene, plan your next shot, or get your next camera angle.

Some machines allow you to switch the VCR's POWER off (thus stopping the spinning video head) while in PAUSE. This allows you to take your time setting up between segments. It also saves battery power on portable VCRs. Hitachi calls this feature "power saver" while JVC calls it "record lock."

PAUSE edits are "cleaner" than STOP edits, and they display a barely perceptible blink on the TV screen. The edits are usually better when made at the VCR's fastest speed. If your VCR is so equipped, use the remote (solenoid operated) PAUSE in favor of the mechanical pushbutton on the machine for less glitch. JVC's HR-2200 and HR-7650, Hitachi's VT 6500A and VT 6800A, RCA's VGP-170 and Sony's SL-2000 make excellent glitch-free ASSEMBLE EDITS.

DROPOUTS AND ADS

When editing anything of consequence, don't start at the tape's very beginning, where the threading dropouts are. Record 30 seconds of black using your capped camera before beginning your show.

If you're cutting the ads out of broadcast TV shows, you will soon discover that a half-hour program is about 6 minutes shorter without the commercials. Likewise, an hour show takes about 48 minutes of tape, and 5 half-hour programs will just fit on a 2-hour videocassette. Figure most shows to be 20% ads.

Recording Something Over

You're recording the Memorial Day Parade to send to Aunt Blanche. You assemble fascinating shots of the flags, the bands, and now the horses, trotting majestically by . . . uh . . . whoop! You didn't really want to test Blanche's heart pacer with a giant color close-up of dobbin doin' a dandy on the pavement. PAUSING at this point won't help; the toothpaste is already out of the tube, so to speak. Your only choice is to REWIND a ways and replace the unsavory scene with something else, perhaps Girl Scout Troop 106.

So you REWIND, PLAY, and watch your viewfinder for a good breaking off point and hit PAUSE. Next hit RECORD/PLAY. UNPAUSE when Troop 106 gets in range.

You may notice later that such edits aren't very pretty (and it's not the fault of the girls in Troop 106). There may be a few seconds of herringbone lines wiggling through the picture, looking much like Figure 2–20 (paperback volume I). There may also be a rainbow or smear of colors lasting a few seconds. Why?

If you look back at Figure 4–21 (paperback volume I), you'll see that the ERASE HEAD, which "cleans" the signal off the tape before the VIDEO RECORD HEADS put a new signal on, is upstream of the video heads a ways. If you start recording in the middle of an already-recorded tape, there will be some tape that hasn't been erased yet, already past the ERASE HEAD and on its way to the VIDEO RECORD HEADS. Thus the VIDEO RECORD HEADS are putting a new signal on top of an old one for a while until the erased tape gets to the VIDEO RECORD HEADS. When you play back this spot with two video signals, the machinery and TV get confused, yielding a confused picture. The only reason you get *any* comprehensible picture at all is because the new signal overpowers the old one, so *its* picture dominates the TV screen.

Such is the disadvantage of recording something over; you get a glitchy edit. (Perhaps the horse poo would have been better.) There is no cure for this problem when using present home video equipment although this may change shortly. Professional ELECTRONIC EDITING VCRs have what's called a FLYING ERASE HEAD spinning just a shade ahead of the spinning VIDEO RECORD HEADS. It whisks off the old video just a millisecond before the VIDEO RECORD HEADS put new video on, making a "clean" edit.

But back to the affordable world. It's best to avoid recording over already-recorded tape when possible. This is another reason for avoiding STOP edits, too, because to get up to your edit-out point, as described earlier, you have to PLAY the 13 seconds of tape that unthreaded when you pressed STOP. If you start recording before the 13 seconds are up, you start recording over some old, remaining signal, giving you herringbone and Rainbow City. If, incidentally, you let more than 13 seconds go by, getting past the old video, you get into unrecorded tape. Then you'll get snow, and that looks as bad as a glitch.

If you *must* record over, two things you can do to minimize the glitch are:

1. Use the fastest tape speed for recording. This way the double-recorded tape will move through the machine faster, displaying a shorter glitch.

2. Make your edits on pauses in conversation or lapses in action. Then nobody feels like they're missing something important.

Insert Editing

You've taken great pains to produce a full-length home video version of "War and Peace." Upon playback, you discover an error in the middle of your presentation and wish to delete it. You want the material before and after the error to remain intact and you just want to replace a segment of the recording with new picture and sound.

This is what's called an INSERT edit, where you exchange an existing scene for a new one of the same length. The order of the remaining scenes on the tape doesn't change, they're untouched. The whole production remains the same length too. You're simply recording a new scene over an old one (erasing it). Figure 11-1 diagrams the concept. Here's the process:

HOW TO INSERT EDIT

First you play the tape to find the error. Next you back up the tape a ways, searching for a good point to "edit-in" or begin the edit. A point where activity pauses is best. Remember this spot and jot down the tape counter number so that you can come back to the spot easily.

Now, play ahead to find an appropriate place to "edit-out," that is, to terminate the new recording and go back to the original presentation. Again, a pause in action and conversation is usually a good place to come back to the old material from your edit. Once you find the place, note the number from your counter.

Essentially, you'll want to record the new passage, starting with the first index number and ending with the second number. This means that the performance must be timed to last *exactly* the length of tape you wish to delete. If the replacement scene is

Figure 11-1
Difference between ASSEMBLE and INSERT edits

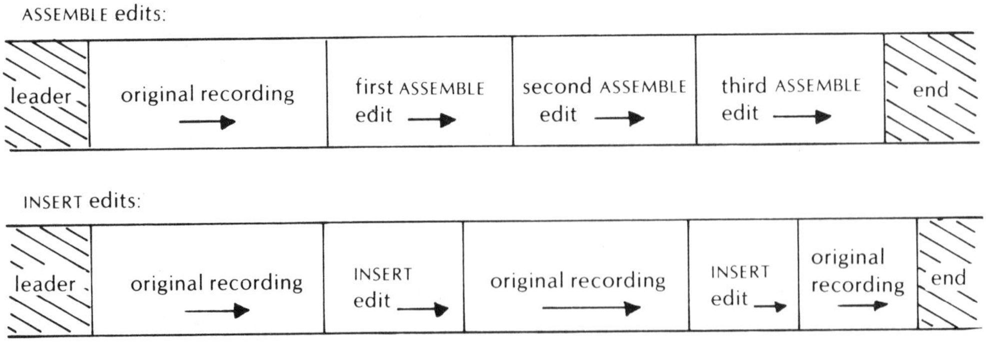

ASSEMBLE edits:

| leader | original recording → | first ASSEMBLE edit → | second ASSEMBLE edit → | third ASSEMBLE edit → | end |

INSERT edits:

| leader | original recording → | INSERT edit → | original recording → | INSERT edit → | original recording → | end |

302   Editing a Video Tape

too short, you end up with a long pregnant stare at somebody's smiling face while you wait for the index number to come up. You *must* wait for the number because if you terminate the edit too soon, you'll end up not deleting the tail end of the segment you want removed. If the replacement scene is too long, you end up erasing your way into the following material, which you wanted to keep.

This process is not easy! Besides being mechanically difficult, it requires precision timing from the performers and the VCR operator alike. To make things worse, the tape footage counter isn't all that accurate and will throw you off by a second or so anyway. Perhaps a stopwatch would be helpful if you're handy at running two devices at once. Instead of marking the footage for the edit-out point, mark the elapsed time from the edit-in to the edit-out point. Then you can let the timepiece be your guide for when to edit-out.

The aspect of INSERT editing that makes it so difficult is the fact that once you've pressed the RECORD button, you're flying blind. You can't see what you're erasing; you only see what you're recording. You have no visual cue for when to stop other than your index counter or your timepiece. Add to this the fact that if you make a mistake and edit-in too long a passage, you'll irrevocably erase the next scene as you record over it. For this reason, it is worthwhile to rehearse the edit several times in order to get the timing exact. You may wish to play the tape; *pretend* to edit (as it plays); have the performers dress-rehearse the scene; then *pretend* to stop the edit; and, by looking at the screen at this point, determine how far off you were and what should be done about it.

It should be noted that to be a *true* INSERT edit, the edit must be glitch-free. Perfectly "clean" INSERT edits are possible only on specially designed professional VCRs. What we're doing with the home VCR is *simulating* the true INSERT edit and calling it an INSERT,

much as we did with our quasi-ASSEMBLE edit earlier.

Now that the cautions are over, here's how you make the edits happen. Editing-in is done as described before in the Recording Something Over section of this chapter. To review: Find the edit place. PAUSE the tape. Push RECORD. When ready, UNPAUSE. Near the end of the replacement passage, get ready to edit-out by placing your hand over the PAUSE button. Watch your index counter or stopwatch, being aware that you may have to delay or hasten your action for a second in order to accommodate your performer. When the time comes, hit PAUSE, then STOP. You're done.

What does the edit look like? The edit-in has squiggly lines, double audio, and some picture breakup like in Figure 2-20 (paperback volume I). The edit-out will look even worse. The picture will break into snow which may last about 8 seconds for VHS at the 2-hour mode, or proportionately more at the other speeds. For BETA, the breakup lasts about 5 seconds at the fast speed, longer at the slower speeds. The snow will gradually slide off the screen during this period followed by a clean picture once again. Figure 11-2 shows an example.

Considering the disadvantages of this kind of edit, it would seem warranted by only the most dire circumstances. It would, for instance, be counterproductive to make

Figure 11-2
Snow fills screen, then slides off in about eight seconds

Insert Editing 303

an INSERT edit 10 seconds long, considering that editing-out would mangle 8 of those seconds, leaving you with perhaps 2 seconds of substituted programming and 8 seconds of garbage. *Short* INSERT edits aren't alone at being hard to handle. *Long* INSERT edits are difficult to time accurately and have a way of encroaching on existing material if the edit is not stopped in time.

One last reminder about INSERT *edits*: They cannot be used merely to delete something. Something must be recorded in place of the part being removed. The program doesn't become shorter when you make an INSERT edit, it just has a new segment substituted for an existing segment.

MODIFYING YOUR VCR
FOR BETTER INSERT EDITS

The Hitachi VT-6500A and VT6800A VHS recorders are designed to give "clean" INSERT edits, both in and out. It is possible to modify your regular VCR to do approximately the same thing.

If you're not mechanically inclined, if your warranty is still in effect, or if you don't want to mess with your machine, skip this section entirely. If you or a friend can do electronic repairs, there's an easy way to improve how your VCR handles those snowy edit-out points. First let's look at what causes the problem.

The cause Figure 4-21 (paperback volume I) showed how new video goes onto the tape at the VIDEO RECORD HEADS, which are some distance from the ERASE HEAD. When you edit-out, the ERASE HEAD has "cleaned" the tape but the VIDEO RECORD HEADS haven't gotten around to recording it anew. This leaves the gap of snow (unrecorded tape) seen on the TV screen in Figure 11-2. Figure 11-3 illustrates the gap on the tape.

Wouldn't it be nice if, when you finished your INSERT edit, there wasn't any gap of snow? One way to eliminate this gap would be to have the ERASE HEAD stop erasing a few seconds before the end of the edit. This way the old video would "catch up" to the new video. There might be some overlap between the old and new video, causing the color moiré and squiggly lines described in the Recording Something Over section, but even *this* is nicer than the snow. Compare the overlapping video of Figure 2-20 (paperback volume I) to the mess in Figure 11-2.

The modification Inside your VCR, two wires go to your VIDEO ERASE HEAD. Cutting one will disable the head. Putting in a momentary pushbutton switch which is normally "closed" (breaks "open" the circuit only when pressed) into the circuit at this point allows you to "turn off" the ERASE HEAD at the push of the button. Route the wires from the head, where you make the disconnection, to somewhere on the chassis

Figure 11-3
Gap of unrecorded tape left at end of home-style INSERT EDIT

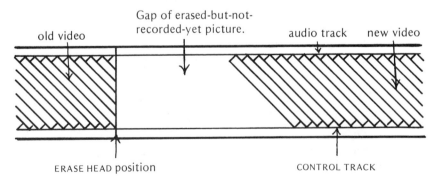

where you can mount the pushbutton. Pick a convenient mounting spot and label the button ERASE BYPASS. Make sure the wires won't drag on any moving parts inside the deck.

Using the ERASE BYPASS

1. Determine your edit-in and edit-out points.

2. Determine another point 8 seconds before your edit-out point. You could use the counter or a stopwatch for the above two steps.

3. Get ready for recording. Back up your tape and start it playing. At the edit-in point hit PAUSE.

4. Press RECORD/PLAY and when ready UN-PAUSE. Start timing if necessary.

5. Eight seconds before the edit-out point, press your ERASE BYPASS button and hold it down.

6. When you reach the edit-out point hit PAUSE. Then hit STOP. Then release the ERASE BYPASS button.

7. UNPAUSE, REWIND, and PLAY to see how it all came out.

Experiment. Perhaps 8 seconds is wrong for your particular machine. Again, I warn that this modification is only for the technically brave whose VCR warranties have expired.

Using Scan Conversion for Glitch-Free Edits

Your heart was in the right place. You produced a content-oriented video tape with varied scenes, unrestricted camera angles, and numerous edits; all this was done without expensive or elaborate editing equipment. Upon review of your tape, you find the picture breakup between each edited scene excessively distracting and wish you could remove the offensive picture problems from your otherwise excellent tape.

There is a simple way to hide the editing problems, but the method will cost you something in picture quality. The method involves using the SCAN CONVERSION technique described in the previous chapter. It amounts to playing the tape over a television, while with another VCR and a camera aimed at the television you rerecord the tape. Before you reach each edit point on the tape you're playing, you close the camera's iris or turn the TV screen brightness down (to black). After the edit point, you open the iris or turn the brightness back up again. In essence, you are making a copy of the tape with FADE-OUTS at each edit point. Here is the process in more detail:

1. Prepare your tape for playback by finding a good VCP and TV monitor. Also, log all the edit points (unless you trust your memory), either by tape footage on the player's index counter, or by a script. This will guide you as to when to FADE.

2. Set up a VCR and camera as in Figure 10-3 (paperback volume I). If you do your FADE-OUTS by turning down the TV's brightness control, you may wish to connect a second monitor to the VCP so you can see when the glitch is over and when it's safe to FADE UP the other monitor. The camera faces the TV screen and is zoomed to a FULL SHOT. Audio is run directly from the VCP's AUDIO OUT to the VCR's AUDIO IN.

3. Play some of the program and adjust the TV screen picture for best copying. Note that the TV screen image that looks best to your eye does not necessarily look best to the TV camera. Adjust the TV screen's brightness and contrast to make a good picture *for the camera*, regardless of how it looks to your eye. Use the camera's electronic viewfinder or a separate VCR monitor as a guide to the optimal picture settings.

If you notice faint diagonal or curved lines on the image from the camera, try tilting the camera slightly.

4. When ready, turn out the lights, close the camera iris or dim the brightness on the TV screen, start the VCR recording, start the VCP playing, and then FADE IN.

5. As you approach each edit point, get ready to close the iris or dim the brightness control. Close or dim to FADE OUT during the picture breakup, while turning your attention to the second VCP monitor. As soon as you see the picture in the second monitor restabilize, FADE back IN.

6. Continue this process to the end, where you perform the final FADE-OUT.

As described earlier, this process gives you excellent sync but fuzzier pictures. You have to make the tough choice over what's most important.

SCAN CONVERSION editing does provide one unique opportunity for graphics or special effects. First of all, you can play around with the TV's contrast and colors, creating unusual fantasy or science fiction scenes.

With careful lighting which can illuminate something held in front of the TV screen but not reflected from it, you can mix "live" and taped action. For instance, a title poster could be held in front of the screen, then slid off to reveal a picture. Also, one could affix white letters to a clear acetate sheet and lay it over the screen, creating an effect called KEY.

Hands and faces can be introduced into the taped picture, especially if you're using a large-enough TV screen. I remember entertaining many with a video tape of a pompous college president giving a graduation speech. A close-up of him reading at the podium was livened up each time he raised his hand—to pick his nose! He didn't really pick his proboscus; that was my hand reaching up and digging around on the right spot on the TV screen. But it looked authentic. And there were all kinds of things this hand could do to the face on the screen. It was even possible to bring in a second face in front of the TV screen and give the Mighty Leader a kiss on the cheek. All it takes is a large-enough TV and close-enough shot to make the scale match, and some careful lighting. The skin tones, incidentally, are easier to match in black-and-white than in color.

Again, experiment. This process has a lot more possibilities than I've mentioned here (or would dare to mention anywhere).

Audio Dubbing

Most VCRs, whether they can edit electronically or not, have an AUDIO DUB capability. This feature allows you to erase an old sound track while substituting a new one.

Say you wanted to dump your existing sound track and substitute music for the entire duration of the tape. The process would go something like this:

1. Find the beginning of the tape.

2. Press the AUDIO DUB button.

3. Check the music's audio level.

4. CUE UP the music.

5. When ready, hit PLAY while holding down the AUDIO DUB button. Then start the music playing.

6. Perhaps FADE OUT the music at the end of the recording where the video also fades to black.

If you want to replace only a portion of the sound track, you use the same method, but with more concern for the timing and placement of the DUB. The process is much like the INSERT edit in that you need to determine not only where to *start* the DUB, but also where to *stop* the process lest you erase part of the audio in the following scene. To narrate a particular scene, for instance, the process would go like this:

1. Find the beginning of the scene, noting the tape index number.

2. Play the scene, perhaps timing it as it plays. At the scene's end, note the tape index or the timing, and familiarize yourself with the TV picture at the exact place where you wish to stop.

3. Prepare a narrative which runs for the allotted time.

4. Practice the narration with the tape playing, as sort of a dress rehearsal.

5. When ready, again find the place where the DUB should start.

6. Press the AUDIO DUB button and get an audio level check. On some machines you have to hit AUDIO DUB and PLAY to make a sound check. To avoid recording this practice session, press PAUSE. Thereafter, UNPAUSE to start your AUDIO DUBS.

7. To start DUBBING, hit PLAY while holding the AUDIO DUB button down (or UNPAUSE as just described). Start the narration.

8. Prepare yourself to switch the VCR to STOP as you approach the appropriate index number.

9. Using the TV screen as a more accurate guide to the exact place to stop (assuming that the scene is one which offers visual cues you could use as a guide) and praying that the narration will end on time, stop the VCR at the end of the DUBBING sequence.

You'll remember how home VCRs have their erase heads quite a distance in advance of their record heads. AUDIO DUBBING on such machines should be done with the expectation that the first and last few moments of the DUB will be of dubious (or DUBious?) value. The beginning of the DUB will have the new sound recorded on top of the old (and not-yet-erased) sound for a moment. The end of the DUB will be followed by a few moments of silence as the erased-but-not-yet recorded tape comes around. For this reason, avoid complicated DUBBING. Never try to delete just a word or two. Try to start and finish the AUDIO DUB during pauses in conversations. Experiment to see what your machine will do.

Don't forget, AUDIO DUBS, like INSERT EDITS, *replace* material, they don't add or subtract it. The total program always remains the same length. Also, you lose the old sound as you record the new sound. There are exceptions to this rule: (1) A few newer VCRs on the market today like the Hitachi VT6500A and VT6800A feature SOUND-ON-SOUND (by pushing a button, the VCR records new sound *over*—and mixed with—the old sound). (2) Akai has stereo VCRs which will replace *one* sound track as you DUB, but leave the other alone, allowing you to play back the old sound, the new sounds, or a mix of both. (3) The ERASE HEAD BYPASS modification may also allow you to do SOUND-ON-SOUND while AUDIO DUBBING.

One of the great strengths of AUDIO DUBBING is its ability to tie scenes together. A montage of unrelated pictures all becomes related by a single strain of music, sound effect, or conversation.

Editing scenes with divergent sound levels can be smoothed with an OVERDUB. Say you shoot a birthday party. The long shot of the whole gang has a lot of kids screaming in the background. The close-up of the cake's candles being blown out is fairly quiet. The close-up of someone eating the cake has many voices in the background. The visual scenes here are fine. The disparate sound levels are jarring. So, bring an audiocassette tape recorder with you to the party and record a whole sound track. Except where important words are spoken and must be heard synchronously with the lips speaking them (called LIP SYNC), you can simply DUB your prerecorded background sound onto your edited tape to create a smooth din of authentic party noise. No one may ever notice that the cacophony of sound doesn't exactly match the bedlam of visual activity.

Many of the TV ads you see have had their audio DUBBED. Sometimes it's because the best-looking models don't have the best voices, or because better sound control is available with someone standing exactly one foot in front of a mike in a sound studio, rather than moving around a noisy set. Although you would have difficulty performing an exact LIP SYNC like the pros, you can often use these techniques in many scenes where a close, well-aimed mike is impossible. Redoing the audio later permits you to remove unwanted sound effects (e.g., jet planes in an old-time western, or the roar of electric fans in your "windy" scene). In the studio, you get only the sounds you want. And while you're at it, perhaps you can *add* a few sounds of your own for authenticity (clippity-clop in the western, thunder and gale winds in your storm scene).

One thing is likely to happen to you as you assemble together scenes taken in different places, and replace old voices with DUBBED-IN ones. You'll discover that each room has its own characteristic "room tone," a color to the sound created by room echoes, faint background noises, and the kind of microphone used. As your edited and AUDIO-DUBBED tape plays, you'll suddenly hear tonal changes from scene to scene, and from "live" to DUBBED portions. To minimize this, it is best to use the same microphone for all the scenes, including any AUDIO DUBS, and to perform the DUBS in the same room where the scene was shot (to capture the same sound coloration).

The most common use of AUDIO DUBBING is in a "show and tell" sequence. Here, the first scene shows a reporter telling the viewer (the camera) that a parade is coming. Next we switch to the parade, the band, etc. A few scenes later we come back to the reporter. After the sequence has been recorded you go back, DUBBING in the reporter's narration during just the parade shots. The result is a sequence with a reporter speaking (LIP SYNCED) to the audience, followed by some parade shots (same voice, narrating), followed by the reporter again (LIP SYNCED). The process is deceptively easy to do considering how professional-looking the results are.

There are many hi-fi enthusiasts out there who are adept at splicing and rerecording audio tapes together to make action-packed home radio shows. If there's a frustrated disc jockey within you somewhere, yearning to create something really fancy, then here's your chance. There's no law against dreaming up a hard-hitting audio drama just like the golden days of radio. Make a complete audio-taped program. Then go back and put visuals to it, either performing the drama to the sound of your prerecorded tape (fed into your VCR's AUDIO IN) or by DUBBING the audio in after you've completed the visuals. A superb sound program unleashes a whole new dimension to videography, one which can make an average "video" show into a dazzling audio/video production. The home video user tends to "forget" the power of audio, relegating it to the nether world of secondary importance. Judicious use of the AUDIO DUB feature offers you the opportunity to exploit the full capability of the television medium.

Editing from Another Video Tape

ADVANTAGES AND DISADVANTAGES

The problem with ASSEMBLE editing is that everything must be done in sequence. You progress through your shots in chronological order, unable to shoot the end scene first, then shoot all the airplane shots, then go to all the bedroom scenes. If you could shoot all the similar scenes in one sitting you could save a lot of running around locating your performers and setting up lights, audio, and so forth.

The problem with INSERT editing is that if you wish to *add* something to an existing tape, you have to *take something out* to make room for it. If you could alter the content of a show by simply removing scenes or putting in additional scenes, you could exercise uninhibited control over your production.

For these reasons, most professional teleproducers shoot RAW FOOTAGE, or as the film industry calls them, RUSHES. These are recordings made (on separate tapes, usually) at different locations, at different times, sometimes by different people. These tapes are brought back to the video tape editor, who assembles the best scenes together to make a final tape. One nice thing about shooting RAW FOOTAGE rather than the real thing is that you can take chances shooting scenes which might not work out in the final production. If they don't, well, just don't use them; you're not stuck with them. Also, you can shoot the same scene over and over, perhaps from different angles, until you get it right. Afterwards, you can select the best of the RUSHES to incorporate into your final production.

The disadvantage of editing from RAW FOOTAGE is that your final tape is already one generation down in quality. You know home video doesn't copy well, and you're building this deficiency into your program.

It's a very difficult decision, deciding which way to go—high picture and sound quality with limited editing, or unlimited editorial control but second-rate technical quality. And to further complicate the decision, you have another alternative: You could ASSEMBLE edit most of your production (perhaps because it's all shot at one time in one place under easily controlled conditions) yielding a sharp final tape. Then for a few difficult shots (waiting for Junior's "perfect" dive, or recording a whole movie just to get the juicy car crash) you record them separately and edit them into your final production. This way, most of your show is crisp-looking, and you were able to get the "hard" part also, although it's a bit grainy.

HOW TO DO IT

The mechanical process of editing from another video tape is the same as for copying a tape, only you're doing little pieces at a time rather than a whole tape. The procedure would go something like this:

1. Connect your MASTER and SLAVE VCRs for copying a video tape. It's advisable to put masking tape over the MASTER machine's RECORD button so you don't accidentally press it during the confusion of editing.

2. Try a test recording to make sure the SLAVE is getting a good signal from the VCP (videocassette player) playing the RAW FOOTAGE.

3. Play your SLAVE up to where you want the next edit to begin and hit PAUSE then press RECORD/PLAY.

4. Play your MASTER up to where the first scene to be copied begins and hit PAUSE *about 5 seconds before that point.*

5. Take a deep breath. Then UNPAUSE the MASTER player. Immediately position your finger over the SLAVE's PAUSE button.

6. When you see the desired scene come up, UNPAUSE the SLAVE.

7. When the scene ends, PAUSE the SLAVE.

8. If it won't take you long to set up the next scene, just leave the SLAVE in PAUSE. This will yield the "cleanest" edit.

If it will be more than 3 minutes to prepare the next segment, then switch POWER OFF (with PLAY/RECORD and PAUSE still ON). With many machines, you can later switch the POWER back ON without upsetting anything. This will yield a pretty "clean" edit too.

If yours is the kind of machine that switches itself to STOP when turned to OFF, your tape will unthread (on VHS machines). In such cases, when you're ready to record the next segment start back at step #3 and go from there. On BETA recorders, you may press STOP without having the VCR unthread itself.

MAKING GOOD RAW FOOTAGE FOR EDITING

You can't make chicken salad out of chicken feathers. Your RAW FOOTAGE should be of the highest quality for your edits to come out looking good. To do this, you follow the rules for Making the Best Possible Signal for Copying set forth in the beginning of the last chapter. Use good tape, plenty of light, and your fastest tape speed when recording. In addition, follow these extra practices:

1. Start your VCR recording a few moments before the scene begins. Whenever your VCR starts and stops, it leaves a glitch on the tape. This "startup time" allows you to get past the glitch and have a very stable signal before you begin the important parts of the scenes. You don't want to edit glitchy scenes.

How many moments head start is enough? If you're PAUSING your VCR between shots, leaving it in the RECORD/PLAY mode, a 3-second wait will do. If you have pressed STOP, you'll have to allow 8-10 seconds before you can trust your recording. (This applies to VHS, not to BETA. The BETA wait here will still be 3 seconds.) If you ejected, played, or skipped around on the tape before starting to record the scene, allow 8-10 seconds before you trust your recording. (This applies to both VHS and BETA.)

2. Call out your "take" number, recording it, before the beginning of each scene. Even the professionals do this while using a CLAPSTICK. No, this isn't something Suzie Chaffie tells us to put on our lips in the winter. It's a handy device (see Figure 11-4) for announcing the beginning of each scene to assist in locating and identifying the scenes later. Some "takes" look so much alike that unless they're numbered, audibly or visibly, you can't tell them apart when editing.

3. Cue your talent silently. Otherwise each edit will begin with the echo of your voice hollering "Action." However, hand signals aren't always good if your performers aren't supposed to be looking up at the beginning of the scene. The best way to cue them is to give a countdown with the last two counts

Figure 11-4
CLAPSTICK used by professionals to mark the beginning of each "take"

silent, like "five—four—three—...—...,"
and the action starts. Of course, let your
talent know this is how you plan to cue
them so they can count along with you.

Putting the above three steps together,
here's what you, the director/equipment
operator, might say and do:

A. "Everybody ready? Okay, quiet. (Start tape recording) Restaurant, take 1. Five—four—three—...—..." (scene begins).

B. (Scene ends. Director records a moment longer before stopping or PAUSING the recording). "Thank you. Let's try it one more time, a little faster. Ready? Quiet. (Starts tape) Restaurant take 2. Five—four—three—...—..." (scene begins).

And so the process goes, into the wee hours sometimes.

PUTTING IT ALL TOGETHER

GETTING ORGANIZED

You've been shooting the RAW FOOTAGE of your little teledrama for three days. Your lights and cables have cluttered the living room, dominated the kitchen, and enslaved your every thought throughout this time. Your family is sick of "acting," and if they hear you say "... just one more time," just one more time, they're likely to terminate you along with your video lines.

And yet the fun's just begun! Editing this mess together will test your patience and your stomach lining. The mechanics of editing aren't too hard; it's mostly a matter of pushing buttons at the right time. The planning and decision-making of editing is the hard part. Planning, viewing the RAW FOOTAGE, logging the scenes, and charting the edits are all tedious, unglamorous ordeals. Perhaps this is why most folks skip these steps and end up making uninspiring productions.

Here's what you can do to get organized:

1. Before you start, decide on how your story will unfold and determine the scenes necessary to tell the story (more on this in the next chapter). Then shoot your RAW FOOTAGE.

2. View all your RAW FOOTAGE and log it with index numbers as shown in Figure 11-5. In this example, NG stands for "no good," CU for "close-up," and MS for "medium shot." This sheet will make it easy for you to find quickly the desired "take" on the tape when you get around to editing this scene into your final production. These SHOT SHEETS are especially helpful when some time goes by between when the scenes were recorded and when the editing is to occur.

3. Plan your editing strategy using an EDITING SHEET like that shown in Figure 11-6. You lay out the sequence of scenes you wish to assemble, enter the counter numbers of these scenes from your SHOT SHEETS, and you're ready to edit.

Notice that the beginning and end titles can be shot "live" with a camera (as could other scenes, for that matter) and be assembled along with taped scenes.

If you find your "edit-in" and "edit-out" points untrustworthy because of the inaccuracy of the tape counter, then play the scenes a couple times to yourself to memorize them before editing. You'll find it easy to "edit-by-ear" as you observe the edits in progress.

SHOT SHEET — Cassette #3

Project: Dining Out **Date:** 12/7/82

Take	Counter	Action	Comments
1	0–23	Leader	NG
2	24–44	Testing	NG
3	45–65	C.U. sandwich hits floor	excellent
4	66–88	C.U. sandwich hits floor	dark
5	89–100	C.U. sandwich hits floor	OK
6	101–152	M.S. sandwich hits floor	fuzzy
7	153–208	Al sits at table	OK
8	209–250	Al sits at table	glitch
.	.	.	.
.	.	.	.
.	.	.	.

Figure 11-5
SHOT SHEET for logging "takes"

EDITING SHEET — Cassette #27

Project: Dining Out **Date:** 12/15/82

Segment	Action	Cassette	Edit-in	Edit-out
1	Intro	Camera		
2	boy meets girl	5	275	301
3	invites her to dinner	5	870	880
4	takes her to restaurant	1	40	65
5	sits at table	3	153	208
6	orders jelly sandwich	2	59	75
7	sandwich arrives	2	422	435
8	boy spills sandwich	2	501	503
9	sandwich hits floor	3	61	63
10	girl leaves	1	70	75
11	boy stunned	1	93	94
12	end	Camera		

Figure 11-6
EDITING SHEET for organizing sequence of scenes

Now that we've handled most of the *mechanics* of editing, let's explore the aesthetics and strategies of editing.

Don'ts and Do's of Editing

Here we go again with a list of "rules." And again I stress that, although these editing principles apply most of the time, they all have exceptions.

Refer to Figure 11-7 for examples of some of these editing faux pas.

JUMP CUT

If you mounted your camera on a tripod and shot an interview, and then edited together parts of the interview, you would see the person's head magically "snap" from position to position with each edited sequence. If the person was looking to the left at one time in the interview and looking to the right at another and these segments were edited together, the "snap" of his head from one position to the other would make it very obvious that the tape had been

DON'T

JUMP CUT between long shots and closeups without changing angle. With a cut, the viewer expects a substantial change in visual information but doesn't get it.

DO

Change angle about 30° when cutting. Adds variety and smooths transition. Builds a fuller perception of subject.

Figure 11-7
Don't and do's of editing

DON'T	DO

Change angle without changing shot size. Twists performer without apparent purpose.	Change shot size as you change angle to add variety and interest.

Figure 11-7 *(Cont.)*

edited. Such JUMP CUTS (or SNAP CUTS or CAMERA MAGIC, as they're called) are obtrusive and disconcerting to the viewer.

When a shot changes, there should be a reason for the change, a look at something new, a look at the interviewer, a look at the subject being talked about, a different camera angle and a closer or farther shot of the talent. By showing something from a different perspective from shot to shot, you provide the viewer with a "reason" for the change in scene, making your edit less obtrusive. You also make your program more enjoyable to watch.

To review, change the kind of shot when you change shots. Don't go from a long shot to another long shot, or from a medium shot to another medium shot. Change from a close-up to a medium shot and then back to a close-up, or use some other varied combination of shots.

A similar method of edit-hiding can be applied when recording something over. Say the talent goofed a line and you have to back up and start the scene over at the last pause in conversation. Using the same shot you left off on (a CLOSE-UP of something in his hand, for instance) would result in the kind of edit which has the object suddenly snapping into another position (because you can never get someone to hold something in exactly the same position as before the edit). To cover this up when making the edit, change the shot to a MEDIUM SHOT of the talent and object, to a CLOSE-UP of the performer's face, or to a TIGHT CLOSE-UP of the person's fingers manipulating part of the object. The viewer, upon seeing the change of shots, will just think you changed camera angles and be unaware that you were editing out an error.

DISPARATE LIGHT LEVELS

Unless you're switching scenes from night to day, where scene brightness is *meant* to change, keep the light levels in your picture the same. Light scenes to an equal brightness so that the switch doesn't call attention to itself.

WHEN TO CHANGE SHOTS

How often should you change shots and what kind of shots should you change to? The main object is to follow the action, keeping it near the center of the screen (unless for dramatic effect you're hiding the action from the audience). If keeping the action on the scene means frequently alternating shots, then do it. But if you have a good view of the action, then keep the shot. Don't change just because the PAUSE button is there on your camera. As long as the audience sees what they want and need to see, they won't care how many shots were used in the process of displaying it to them.

Some amateur directors change shots just to add variety to the show. The wisdom of this procedure depends on the creativity and savvy of the director, the objectives of the program, and the content of the particular shots. Switching to a side shot of the news anchorman, revealing the busy newsroom in the background, adds variety and style to the end of the newscast. Doing this in the midst of his news presentation would detract from it. Cutting to a close-up of an interviewee's nervous, wringing hands adds variety (and insight) to a show. Displaying this shot while the interviewee is making an important point, however, would be distracting to the audience. Showing this shot while the *interviewer* was asking a question may even confuse the audience—they might assume they were watching the *interviewer's* hands.

Perching your camera up in the rafters offers a great opportunity to spice up your presentation of a square dance. This shot adds variety. Since you've gone to the trouble of hoisting the camera up there, why not also use it to view the poetry reading coming up next? This unusual shot should add plenty of variety, right? Ridiculous as it sounds, this is one of the toughest decisions facing a fledgling director: whether to get mileage out of something in one's production arsenal or to disregard the dazzling things one *can* do and instead do what's best for the total production. The production should come first. Try not to get pizazz-happy. Skip the fancy overhead shot if it doesn't add to the drama of the poem.

Sometimes the shots you use imply what's coming next on the screen. The switch from a medium shot to a long shot of a performer indicates that a performer is about to move or be joined by someone. Cutting to a close-up of the performer's face readies the audience to catch his expression. A gesture will be expected if you now switch to a medium shot. And a shot of the door prepares the viewers for an entry. Making any of these shot changes without a specific purpose will not add variety to your show; it will only confuse your viewers. Therefore, change shots for a purpose, not for idle variety.

One outgrowth of the above philosophy supports the general rule: Avoid going from a two-shot to another two-shot, or going from a long shot to another long shot, or from a close-up to another close-up. Switch from one thing to something else—from a long shot to a medium shot. The main idea is to offer a *different* view of something when changing shots. The medium shots of the same thing reveal nothing new and are purposeless.

Still, what do you do with a 45-minute oration made by a long-winded friend who insists on your recording *all* of his after-dinner speech for posterity? Perhaps you slip Ex-Lax into his coffee in the hopes of inspiring brevity. Perhaps you tell everyone to stand and applaud after 5 minutes. He'll think he is done and was a great success. Given no options of editing or shooting related scenes or visuals, you're stuck with Old Talkiepuss for 45 long minutes. Try zooming (or dollying) in from a medium to a close shot, staying there for a minute, then pulling back. This at least puts movement on the screen. Zoom to a long shot, once in a while, again varying the visual. Pedestal up, or down, or truck to the side somewhere along the way for another angle. This indeed isn't (yawn) great television, and runs contrary to some of the aforementioned "rules," but it may be the best you can do with what you've got. Remember, though, that the camera moves which spice up something dull *can* distract from something interesting. Know when not to intrude.

180° RULE

You almost never shoot any subject from opposite sides. The opposing shots can easily confuse the audience because what was on the left in one shot is suddenly on the right in the other. The action that a moment ago was flowing to "stage left" suddenly is moving "stage right." Consider, for example, a shot of a race car as viewed from the grandstands. Switching to a shot of the same car viewed from the island at the center of the track results in a picture on the TV screen of a car first speeding right and then speeding left. Did the driver turn around? Is this another car and is a collision about to occur? Where am I (the viewer)? Similarly, as a performer exits through a doorway walking to the right, when the camera outside the room picks her up, the performer must still be traveling to the right.

Even though the performer may be walking mostly away from you in the first shot and mostly toward you in the second shot, still the flow of travel in both cases is essentially to the right. To have the transition appear otherwise could momentarily confuse the viewers.

One way to avoid perplexing camera viewpoints is to think in terms of angles (see Figure 11-8). Two cameras can shoot a subject from 60° apart, from 90° apart, or even from 120° apart. Beyond that, as the angle between the cameras stretches to 180°, the risk of confounding the audience's sense of direction increases.

Another way to avoid opposing shots is to draw a mental line straight through the center of the stage or performance area (again, see Figure 11-8). The camera can work on one side of that line only, never on the other. Better yet, the camera shouldn't even come near the imaginary line while shooting the performance, if the shots are to be edited together.

What do you do in those rare cases when you *have* to shoot a subject from all different angles? First you could move the camera while it's "on-the-air," dollying it (and your audience) around in an arc to the other side of the subject. Second, you could have the subject move past the camera as the camera pans, keeping the subject in sight. Third, you could use more than two camera angles to shoot the scene. For example, in Figure 11-8, if camera position #1 was first used in shooting the scene, you could switch to camera angle #3 or #4. From either of those you could safely switch to #5. From #5 you could switch to position #6, and later from #6 switch back home to #1. Each camera angle was within 120° of the last, but taken in steps, the angles totaled to a full 360° sweep.

CONTINUITY

Al sits down at the table, picks up his knife and fork and begins to cut his asparagus.

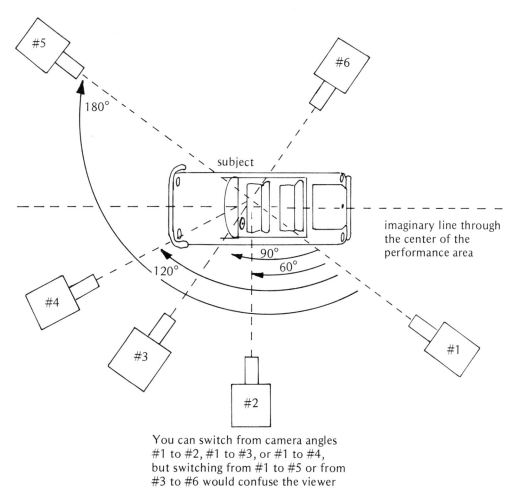

You can switch from camera angles #1 to #2, #1 to #3, or #1 to #4, but switching from #1 to #5 or from #3 to #6 would confuse the viewer

Figure 11-8
Camera angles (as seen from above)

"Cut," you call out, and he feverishly speeds up his sawing until he realizes the scene is over. You continue the scene with the camera in another position for a close-up of his plate. "Okay, roll," you call out, and when Al realizes you're not calling for his buns, he again picks up his fork and dips into his mashed potatoes. Somebody is in for a surprise when editing time comes; somebody who didn't take notes or didn't examine the tail of the preceding scene to assure matched shots. How many viewers will you entertain with your little slip-up?

One way to call the least attention to the exact position of something, is to change shots while it's moving. *Cutting on the action* (as it's called) also makes the movement part of *both* shots, fusing the continuity between them.

For example, someone is about to walk out through a door. You shoot the scene first from the inside, hitting PAUSE *as the door begins to open.* Have the talent memorize which hand was on the knob and which foot was forward. Now run outside, compare your door shot, and have the talent repeat the process of opening the door, hands and feet equivalently placed.

Don'ts and Do's of Editing 317

UNPAUSE as the door opens. If recording RAW FOOTAGE for later editing, the process is similar. Instead of PAUSING, let the talent complete the exit. Have the talent repeat the process while you shoot it again from the outside angle. Start your tape before the talent even begins to approach the door. During editing you will now be able to carefully control exactly where you switch from one scene to the next (before the door opens, as it opens, or as he passed through). As long as the action is matched in both "takes," you should have a number of choices of where to make the switch from the indoor view to the outdoor.

Cutting on the action is a science. If you know you will have to edit scenes, contrive action to cut on. Have your performer turn to a blackboard, sit on the edge of a desk, or simply turn to address a new camera angle. If preparing RAW FOOTAGE, remember to repeat one scene's ending action at the beginning of the next scene, overlapping the action. Pay attention to detail—hand, foot, body, and face positions. Maintain screen direction: Someone exiting the screen to the left should be moving to the left in the next shot (it could be towards the camera or away, but still slightly to the left). If she exits the left side of one screen, she should enter the right side of the other, still moving in the same direction. As mentioned earlier, each new shot should change *both* the closeness of the shot (image size) and the angle of the shot. This change in size and angle will help cover minor flaws in matching the action in the two sequences. And last, feel free to rehearse an action a few times before recording it. It will help everybody relax and perform smoothly.

This rule of "cutting on the action" may seem contrary to the earlier dictum of editing on pauses. Let me clarify: If you are doing STOP edits, INSERT edits, or recording a passage over, your edits are likely to have a substantial glitch. You don't want to lose an important part of your show during a glitch, so you put this messy edit where it will hurt the least, at a pause in the conversation or action. If, however, you're doing PAUSE edits, which are fairly "clean," you don't lose much during the edit, affording you the luxury of applying some of the higher laws of dramatic editing. In short, "cutting on the action" is a more refined technique; however, editing during lapses in conversation and action may be necessary for purely technical reasons.

SOUND COLORATION

Use the same microphone and sound setup (if possible) as you shoot your various scenes. Different mikes impart different fidelity or "color" to the sound. The change may be obtrusive. Be also aware of differing room noise if you shoot scenes on different days. Are the windows open? Is an air conditioner going? Is there traffic, music, or wind in the background? If each scene is clearly in a different place, then it's okay for the background sounds and echoes to be somewhat different. But if the shots are supposed to be of the same person in the same place, you must endeavor to make it sound that way.

Similarly, if you make an AUDIO DUB for a scene, to have the sound match the original sound, record it in the same room with the same mike as the original.

Transitions

There's reading, and there's reading between the lines. There's declaration and there's implication. There are the words, and then there is the mood. The power of television lies in its ability to communicate on various levels simultaneously, and not all of these have to be conscious levels. Beyond the simple sights and sounds there are camera angles and overtones. One more dimension to exploit is the transition, the art of changing from shot to shot.

Setting up smooth transitions from scene to scene requires planning. There's not too much you can do once the RAW FOOTAGE has been shot. If you can visualize beforehand how your scenes will flow together, you can shoot them with the transitions in mind. You can finish one scene with an effect and start the next with the same effect, uniting the shots with a common bond. And yes, you can easily overdo it too. The transition should leave an *unconscious* message. If it intrudes, your transition becomes an eyesore.

THE CUT

CUTS are easiest and quickest to do, requiring only the push of the PAUSE button. CUTTING from one camera angle to another is the least obtrusive way of showing *different views of the same thing.* Changing from a close-up to a medium shot is best done with a CUT. Changing from a performer's face to what the performer is seeing or handling is appropriate for a CUT. Switching from a shot of the football kicker to a long shot of both teams on the field and then to a shot of the receiver, is best accomplished with CUTS. A montage of rapid-fire shots is best handled with CUTS, which in themselves can by their rapidity add tension and excitement to a scene.

THE FADE

A FADE-OUT is performed by closing the camera's lens iris at the end of a scene before hitting PAUSE. *After* UNPAUSING, you open the iris to FADE-IN. When played back, the edited scene goes black and a new scene appears from the blackness. The pros informally call this transition "kiss black." Some cameras have a special button for it.

The FADE-IN is a good way to start a show, the FADE-OUT a good way to end. In the middle of a program, the "kiss black" implies time passing. It may also indicate a change of locale, or a change in subject—like starting a new chapter in a book.

THE DISSOLVE

The DISSOLVE or LAP DISSOLVE is the melting of one picture into another. Actually, one picture FADES OUT while another simultaneously FADES IN. Figure 11-9 shows an example. The process requires two cameras (one of which has special sync circuits in it) and other optional devices called FADERS, GENLOCK, or SPECIAL EFFECT GENERATORS. This equipment, until recently, was very expensive and available only to the industrial, educational, or professional user. It is now finding its way into the more advanced home setups.

Unlike the FADE, which goes to black before it goes to another picture, the DISSOLVE goes directly to the other picture, not passing black, not collecting $200.

The DISSOLVE is used to bridge action smoothly from one scene to another, to indicate a change in time or locale, or to show a strong relationship between two images. Here are some examples:

Bridge action You're shooting a long shot of a ballet. You DISSOLVE to a close-up of the dancer to blend the action without interrupting it.

Workers are erecting a building. You DISSOLVE from a long shot of several to the face of one hammering, or DISSOLVE to a super-long shot to display the flurry of activity surrounding your main character.

Going from your regular scene to a graphic or title (or vice versa), like in Figure 11-9, is handled more smoothly and less obtrusively with DISSOLVES.

Changing scenes to music or displaying slides or still pictures to music is handled smoothly with DISSOLVES. CUTS would be too abrupt unless the music was hard-hitting, such as a march or rock.

Figure 11-9
DISSOLVE

Change in time or locale Time passing is implied as we DISSOLVE from someone sitting to the same person lying down asleep, or a task starting and a task nearing completion.

A DISSOLVE can also say "Meanwhile, back at the ranch..." as you melt from one locale to another.

Strong relationship A DISSOLVE from the chatty din of the secretarial pool to a cackling hen house implies a relationship between the two. A slow DISSOLVE from a close-up of a person rowing to a long long shot of a lonesome boat on the sea reveals great effort pitted against hopeless odds.

Why is there so much emphasis here on an effect most home videophiles can't produce without extra equipment? Because the next paragraph will show you how to get a similar effect without the fancy gear.

THE DEFOCUS—FOCUS

You need a special camera and a special FADER option to DISSOLVE. With your home VCR you can only make CUTS as you edit. This is very restrictive if you desire variety in going from scene to scene.

One method of making a CUT look smoother while adding variety to your edits is to DEFOCUS at the end of a scene, make your edit at another defocused shot, and then have the next shot REFOCUS. As an example of what this could look like, imagine a close-up of a guitar player's hand becoming fuzzy and then magically reforming into a singer's face.

Or how about a zoom-in and DEFOCUS on a burning candle, a calendar, a clock, or a baby's bottle. After the edit, zoom out and back into FOCUS with the image of a burnt out candle, an updated calendar, a reset clock, or an old man's wine bottle. Notice in this case how the DEFOCUS-FOCUS implies the passage of time? It works a lot like the DISSOLVE mentioned earlier, only it doesn't require elaborate equipment. Granted, the DISSOLVE is less obtrusive than this home-made version, but the DEFOCUS-FOCUS can create the desired impression—*if* not overdone.

Two things to keep in mind when you apply this method are: First, it is far easier to get a close-up way out of focus than it is to get a medium or long shot out of focus, so direct your attention to things you can get close-ups of if you want to DEFOCUS easily.

Second, use this method sparingly or else the method won't add variety any more.

LEADING THE ACTION

It's your kid's birthday party with the usual bedlam. As you tape the proceedings, one child suddenly looks "stage right" and in a few moments, all eyes are looking to the right. Now is the time to CUT to Mom carrying the blazing birthday cake into the room. The children's looks made a perfect lead-in to your next shot. The viewer *expected* to see a new shot.

Seated performers shuffle in preparation to stand. There's your excuse to CUT to a long shot of them getting up and strolling off.

Somebody is holding a gem up to his eye. It's time for a close-up on the gem.

The runner lifts her hips in preparation for the gun. It's time for a long shot of the takeoff.

In each case above, the action prepares the viewer for a shot change. The transition from one shot to the other becomes natural and comfortable to view.

REVERSE ANGLES

No one wants to look at the back of somebody's head. But people usually face each other when speaking, and unless you like profiles, you're going to be shooting somebody's behind. Now there are lots of way to get a dialog with two people facing the camera, like people on a park bench, sitting in a car, watching TV or a fireplace, or with one in front of a mirror. In fact, you'd be surprised to see how many natural looking scenes there are in the soaps showing someone facing *away* from the person they're talking with. Nevertheless, at some point you're going to have to figure out how to shoot faces facing faces.

One way is the over-the-shoulder shot, as shown in Figure 11-10. Shooting over one person's shoulder, we get a close-up of the second person's face as he speaks. When the first person speaks, we swap everything around, taking a medium shot of the first person over the shoulder of the second.

If you're editing as you go, you'll find yourself getting a lot of exercise traveling back and forth to get a face shot for each actor's lines. If shooting RAW FOOTAGE for later editing, you could save steps by shooting the entire sequence twice, once facing one actor (or actress) and once facing the other. In fact, each sequence could be recorded on a separate cassette. When editing later, all you do is change tapes to change camera angle and get a face shot of the person speaking.

THE CUTAWAY OR COVER SHOT

When you edit parts of a speech together you run the risk of creating JUMP CUTS if two desired parts of the speech happen to have been shot with close-ups of the speechmaker. Seeing her snap to a new position at the edit point will jolt the viewer and will leave the impression that something was left out. What would be nice is to have, say, a real long shot of the speaker, one where you can't see her lips moving, or maybe a shot of the audience, or an interviewer or an object being discussed. This way, you'd switch from the close shot of the speechmaker to a shot of something else, to another close shot of her. No one may realize you edited-out something; they may think you're trying to show them *even more*. Watch a newscast and notice how the camera jumps from the reporter to the things he or she is talking about. Observe how presidential press conferences are interspersed with shots of the reporters and photographers. They're covering their edits. These press club shots, audience shots, and long shots are called COVER SHOTS or CUTAWAYS.

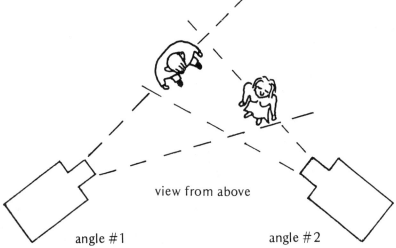

Figure 11-10
Over-the-shoulder shots
Make sure both shots aren't too closely matched (i.e., both medium shots of both close-ups) and still obey the 180° rule

You can do them too, but it's harder for you because what happens to your sound while you are showing your CUTAWAY? You get a gap in the conversation, right? Well, not if you connive a way to cover the gap. Here are some techniques:

Editing as you shoot If you're ASSEMBLE editing scenes as you shoot (sometimes called editing "in the camera"), and conditions permit, you could:

1. Shoot the talent talking to the camera. Then hit PAUSE.

2. Aim the camera at the subject of the discussion or at an interviewer nodding his head. UNPAUSE, and have the talent resume speaking into the camera's mike.

3. PAUSE again at some point, reaim the camera at the talent, and UNPAUSE to continue the speech.

Editing two tapes together You've recorded the graduation speeches (what a waste of tape) and will wish later to edit passages together (so at least you're not wasting *much* tape). While you're there at graduation, shoot some long shots of the

322 Editing a Video Tape

speakers and of the crowd and perhaps a few graduates' faces as they listen intently (with Walkman radios under their robes, earplugs in their ears, and feet a-tappin'). These shots become your COVER SHOTS.

Later, as you edit portions of the oration together, some scenes will be close-ups and others will be medium shots and will match together nicely as-is. When they don't and are likely to make JUMP CUTS, insert a short COVER SHOT between them—but before you do, take the steps to insure that you'll have appropriate sound during the CUTAWAY:

1. Find one or two lines of the person's speech which seem cogent even when taken out of context. Time the sentences and note their location on the MASTER.
2. Now edit-in your CUTAWAY. Make it last as long as those sentences you found.
3. Next go back and AUDIO DUB those sentences over the CUTAWAY.
4. Assemble your next scene onto the end of this CUTAWAY.

When played back, you'll see and hear the old coot droning on, then view the audience as he drivels some more, and then CUT back to him and his concluding statements.

These processes work for more than just speeches, but conversation is the most difficult to handle because viewers notice when people's lips don't match their words. They're less likely to notice when background noises or general chatter aren't synchronized with the scene. Such material is therefore easier to edit.

Professional VCRs make CUTAWAYS super easy by allowing you to INSERT EDIT *just the video*, leaving the sound untouched. Thus you might edit together all the tape segments you wanted without regard for JUMP CUTS, and then go back and cover up all those gauche edits with CUTAWAYS (video only). They do it on the news this way all the time. Outside of professional equipment, a few manufacturers are offering home VCRs which will make VIDEO INSERT EDITS without touching the sound. Hitachi calls this feature "video dub" on their VT-6500A and VT-6800A VCRs. If you modified your VCR with the ERASE BYPASS mentioned earlier, perhaps it too will make VIDEO INSERT EDITS without erasing the sound if you hold the BYPASS button down during the edit. Try it.

THE WALK-PAST

Following people around as they walk from place to place is always hard to condense. One useful trick is to allow the performers to stride toward and past the camera (and out of the picture) and then to edit to another scene of them coming into view from alongside the camera. Also, performers may be allowed to turn corners or pass through doors, leaving the camera viewing an empty set. The next edit begins with another empty set, with the performers entering a moment after the edit.

Condensing a long distance drive into a few seconds is possible by having the vehicle start its journey in one location and drive into or over the viewer. Edit-out (if you survive). The next edit starts with the vehicle going away from (or out from over) the viewer into the new location.

THE BLANK SURFACE

Here, the angry wife exits a room, slamming the door in the viewer's face, leaving the blank surface of the door filling the screen. Edit-out at this point and start your next sequence on another blank surface, perhaps another door which later opens. Better yet, the second surface could be a sheet which disappears as the househusband pulls it from the line. Notice how this transition compresses time as it carries us from the scene of the wife leaving to the scene revealing her husband's plight.

To travel from one place to another, one can tilt up past the treetops into the blue sky and edit-out. Later, starting with the blue sky, one can tilt down past the tops of city buildings into a new scene.

Similar to the walk past is the walk-through, where the talent walks toward the camera, right into it, obliterating the picture. The next scene can begin with the camera close-up on the talent's back, and have the talent proceed away.

There are esoteric transitions which, if not overused, can add intensity or pizazz to a program. Flex your creative muscles and you may come up with more of your own.

THE SWISH PAN

In order to show a move from one location to another (as if to say "meanwhile, across town"), pan the camera rapidly to a blur. Edit-out during this pan. Edit-in the next scene, starting with a fast pan that stops on the next subject to be viewed. The result looks like a hectic pan from one scene directly into another. SWISH PAN edits work beautifully only if they are planned. If the two adjoining scenes are not properly planned and the scene jumps from a pan or a zoom to a static shot, the result will look wretched. So if you can't plan your SWISH PAN edits to come together, avoid even the possibility of getting stuck in one by: (1) if panning or zooming the camera in a scene, come to a static shot before ending the scene, and (2) start scenes with a static shot before you pan or zoom. Thus, all scenes begin and end with static shots and can be edited together in any order.

THE STILL FRAME

One transition you see frequently in movies and TV is used at the end of a scene. The performer looks up and suddenly freezes. It takes special movie and video equipment to make glitch-free STILL FRAMES, but you can get surprisingly good results at home depending on the kind of equipment you're using. If you have a NOISELESS STILL FRAME feature on your VCR, you're in luck.

Say you're editing from one tape onto another. To create the STILL FRAME at the end of a scene, merely press PAUSE on the MASTER VCP. Your SLAVE VCR will record the result. If the STILL FRAME happens to be noisy or unstable, your SLAVE may end up recording useless mush. But if the VCP's STILL FRAME is rock-solid, it may copy okay.

You can often copy other special features like FAST SCAN and SINGLE FRAME ADVANCE from the MASTER player too. Again, the "cleaner" these features perform, the "cleaner" and more stable your edited copy will be.

If your MASTER player *didn't* make a stable enough picture to copy directly but you really wanted a STILL FRAME, there's another alternative: Usually STILL FRAMES show pretty well on a TV set. When you edit the scene which is to contain the STILL FRAME, use the SCAN CONVERSION method of aiming your TV camera at a TV screen which displays the output from the MASTER. The sync instabilities in the STILL FRAME will disappear as the camera rerecords the image off the screen (assuming your TV holds a steady, clean picture—most will).

An extension of this idea can be woven into your transition strategies. Picture this:

Joe Traveler (who won an all-expense paid trip to Secaucus, NJ) is describing his travels and showing his scenic videotapes. The opening shot shows him sitting next to his VCR and his TV. A STILL FRAME of a refinery is on the TV. He introduces the travelogue and turns his VCR to PLAY as he finishes his last line and as you pan away from Joe and zoom in on the TV screen. Once you have a full shot of the TV, there's a tiny blip on the screen as the picture and sound become sharper. The scene could end with this process reversed, or could simply be a CUT to a medium shot of Joe switching his VCR to PAUSE (TV set showing in the background of the shot) and turning to your camera to continue his face-to-face discussion. This process makes for very easy transitions, easy editing, and trouble-free STILL FRAMES. What caused that blip and the sharper picture? Instead of SCAN CONVERTING all his RAW FOOTAGE, you switched to a direct video (and audio, if you wished) feed from his VCR. In other words, his MASTER videocassette player sent its signals to the TV and *also* to your SLAVE VCR. Your VCR started the scene "listening" to your camera input, but when you reached your full shot of the TV set, you switched your VCR's input selector to "listen" to his VCP directly. Thus a better copy resulted. The blip is where you switched. This process works in reverse also, starting with a straight video dub and switching to your camera.

TRICK SHOTS

You can make people disappear by hitting PAUSE, having them walk off the scene, then UNPAUSING. You'll need a steady tripod to keep the scene from changing slightly which ruins the illusion. Things can be moved as well as people, making them pop into and out of a scene.

Although time-consuming, and a bit beyond the range of home editing capability, one could attempt a sort of animation. For instance, someone could drive a chair across the room by sitting still in it, holding their arms and feet in the driving position while you shoot half a second of tape. The friend, while you're PAUSED, would then slide her chair forward 6 inches and reestablish her pose while you clicked off another half second. Clay figures, drawings, and objects can all be animated this way by assembling long series of half-second edits. The process is *slow*. A day's work may take only a minute to play. And if only a quarter of your edits have glitches, then that's a glitch every two seconds when you play your masterpiece.

Editing Strategies

It is easier to plan than it is to reshoot. It is easier to erase pencil notes than it is to erase videotape segments. Plan how your shots will fall together first, then shoot. Here are some things to consider when planning:

ESTABLISHING SHOTS

An ESTABLISHING shot is usually a LONG SHOT introducing the viewer to the setting where the action is about to take place. It may be a shot of the village before we see the inn, or a shot of the inn before we see the lobby, or a long shot of the lobby before we meet the innkeeper. It may be a shot of the patient on the operating table before we see the tonsillectomy. Whatever the ESTABLISHING shot, it sets the stage for the program so that viewers have an idea of what to expect. Without the ESTABLISHING shots of the inn or its lobby, the viewers may be asking themselves: Is this a home? Is this a tavern? Is this a city hotel or a country motel? Is this today or 40 years ago? instead of absorbing the content of the opening scene. In the case of the tonsillectomy, the audience of medical students may be wondering: Is this patient a child or an adult? What are the operating conditions? The answers to all of these questions *may* become apparent as the show progresses, but it's more efficient to sweep them all out of the way with a moment's ESTABLISHING shot.

REVELATION SHOTS

A REVELATION does just the opposite. Something is purposely omitted from the scene and is later revealed to surprise the audience. For example, we're watching the end of a guided tour of the city zoo. The camera slowly zooms out from a close-up of the tourguide's talking face to reveal that he's behind bars, and on the cage door the sign reads "Tourguide—Please Do Not Feed." Another example would be an opening close-up of a news reporter picking up a teddy bear from beneath a wooden plank. As she begins to speak, the camera draws back to a long shot revealing earthquake desolation in the background. Here, for dramatic purposes, visual curiosity is piqued and then satisfied with a revelation.

PARALLEL CUTTING

If you want viewers to feel like they've been presented an overwhelming amount of data supporting one particular point of view, link together interviews or testimonials all showing the same responses. This technique has a powerful effect in commercials where several individuals discuss their delight in a particular product.

INTERCUTTING

Say you wish to present an idea through interview and testimony. Then, as you shoot your RAW FOOTAGE, you discover that everybody is essentially in agreement, giving similar stories. What do you do: Show the second person repeating the first person's story, or just tell the story once with one interview, throwing away the others? One possible answer would be to combine the testimonies to tell the whole story. For example, use one subject's testimony for background, another's testimony to describe the problem, a third's statement on alternatives, and a fourth's to make recommendations. Although each person probably covered all four areas, each may have a more lucid statement on one particular area, so that's the one to use.

JUXTAPOSITION

To create an air of controversy, one may do the opposite of parallel cutting. Record the disparate views of several interviewees, then link together their differing responses to the same question.

TRUE-TO-LIFE INTERVIEWS

In the list of no-nos earlier was the JUMP CUT (or SNAP CUT or CAMERA MAGIC). Strangely, the necessary use of same-angle edits in news and documentary television has trained viewers to associate this style with veracity. Using this phenomenon to your advantage, you could shoot a contrived consumer interview or whatever, loaded with JUMP CUTS. This documentary style of editing can add credibility to your message.

Editing Techniques

EDIT VIDEO FIRST AND AUDIO LATER

Sometimes it is easier to put all the visual scenes together and then go back and DUB in the sound (assuming lip synchronization is not required). This method, called a VOICE-OVER, works best when you find it hard to judge (or don't wish to take the time to calculate) how long each scene will be, and when you do not wish to abridge any of the visual scenes to make them fit any sound passages. So you make a "silent movie" and then go back to DUB in music, narration, or nothing (leaving the original sound track intact). With this method the picture is most important and the sound is a slave to the visual timing. The narrator reads, watches, pauses, reads again; and, if the scene is too short for the narrator's explanation of it, the narrator's script must be abbreviated to fit the scene. During long pauses in narration, light music or background sounds may be employed to fill in the silence.

Sometimes it may be necessary to have lip synchronization, perhaps as a process is explained by a factory worker at his machine. This case calls for a combination of editing techniques:

1. You may wish to shoot the first sequence in "silent movie" style. ASSEMBLE edit these scenes. Afterwards, go back and DUB-IN the sound or VOICE-OVER.

2. For the next sequence, shoot picture and sound together. ASSEMBLE edit the scenes of the factory worker discussing a particular process.

3. The next sequence may again be shot in the "silent movie" style with the narration DUBBED IN afterward.

EDIT ON
CONVERSATIONAL PAUSES

When possible, edit on pauses in conversation. This leaves more margin for error. If you expect to have to edit at certain places, build pauses into the script at those points to make things easier on yourself later.

COVER OVER REMADE EDITS

Sometimes you goof and you have to edit a scene over using the Recording Something Over technique described earlier. The object is to replace the bad edit with a good one, leaving behind no traces of the bad edit, and not cutting into your previous, good scene.

Doing edits over is tricky, and it is one time when you especially appreciate having pauses at the edit points. When redoing an edit, you must start the edit a fraction of a second earlier than the last edit started and if making an INSERT edit, finish it a moment later than the old edit finished. If you begin the edit too late, you may fail to erase the old edit. When played back, you'll see a double edit—the scene will start twice. If you finish an INSERT edit too early, you will not have erased the end of the old edit, leaving you with a double edit when you play back this part of the tape. In short, when redoing an edit, cover up your old one.

Electronic Editors

There are professional BETA and VHS videocassette recorders designed especially for editing. These decks make "perfect" edits, suitable for professional applications. I'll say a few words about professional editors just so you know what exists and what it does.

Professional videocassette editors make perfect edits by electronically changing from PLAY to RECORD (or from RECORD to PLAY) *while the tape is moving.* Since the tape never stops there is no discontinuity in the sync signal recorded on the tape at the edit point. The VCR's electronics synchronize the sync of the already-recorded material with the sync of the about-to-be-recorded material and when the EDIT button is pressed, the machine switches from PLAY to RECORD. Even the switchover is done with electronic precision. Good electronic editing VCRs wait a few thousandths of a second before carrying out the EDIT command so that they make the switch during the VERTICAL INTERVAL (the black bar at the bottom of your TV picture when your TV's vertical hold is misadjusted). Thus the switch never shows in a *visible* part of your TV picture.

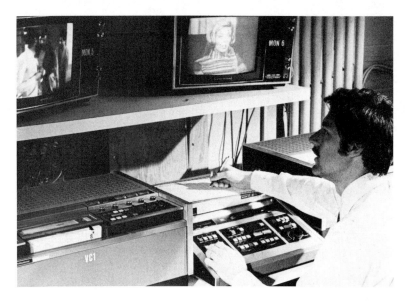

Professional editing VCRs and EDITOR CONTROLLER

And remember the ERASE HEAD problems mentioned earlier in the Recording Something Over and INSERT Editing sections? Unerased tape or erased-against-our-wishes tape caused glitches at the edit-in and edit-out points. All these problems are solved in the professional VCRs because they have separate FLYING ERASE HEADS, hitched to the spinning VIDEO HEADS just upstream of them. They wipe the signal off the tape just before the VIDEO HEADS lay a new signal onto the tape. They stop erasing just a few thousandths of a second before the VIDEO HEAD stops its INSERT edit. In short, the FLYING ERASE HEADS erase only what *needs* to be erased.

An adjunct feature to this is that now you can erase video and replace it while leaving the audio intact. This is called a VIDEO INSERT ONLY. It's especially handy with CUTAWAYS. Say you edited together two medium shots of someone's speech, and in doing so, created an unsightly JUMP CUT. If among your RAW FOOTAGE you have some crowd shots, long shots of the speakers, or shots of an interviewer, you can cover over this unsightly JUMP CUT by making a VIDEO INSERT ONLY lasting about 2 seconds. The result will be a medium shot of a speechmaker, a shot of the crowd with the speaker's voice continuing uninterruptedly. Then the scene cuts back to the speaker, in a medium shot, still talking. No one ever realizes an edit was made.

One difficulty with professional editing is that the tape has to be played *before* the edit. To edit two tapes together, both must be BACKSPACED an equal distance, played simultaneously, and if the timing was right, the SLAVE switched from PLAY to RECORD while the MASTER played on. This precise BACKSPACING is handled nowadays by small computers and VCRs designed to run backwards. You select the edit-in and edit-out points, tell the EDITOR CONTROLLER or BACKSPACER, and it takes care of the rest. It backs up both machines an equal distance, starts them playing, and at the chosen moment, switches the SLAVE to RECORD.

As director of your TV epic, you plan the shots, consider the camera angles, select which shots are to be used, and decide how they are to be assembled as a final product. The director must continually judge the quality of the product and, when necessary, redo a scene (if possible) until it's perfected. For even the most skillful director, there's one job that's probably hardest of all: calmly and open-mindedly accepting criticism about the show from a viewer (or reviewer) who hasn't the faintest idea of how much effort went into producing it!

Editing in a Nutshell

To truly edit video tape, you need a special EDITING VIDEO TAPE RECORDER which is able to switch electronically from PLAY to RECORD with the tape still moving and without causing sync problems. Professional EDITING VCRs also have FLYING ERASE HEADS which remove old video just a millisecond before the new video goes onto the tape. All this results in a glitchless edit.

Home VCRs make edits with glitches. The severity of the glitch depends on how the edit is made.

The cleanest edits are PAUSE edits, where your VCR remains in the RECORD/PLAY mode while you PAUSE and UNPAUSE to record just the material you want. If preparations between scenes will take over three minutes, you will damage your tape if you leave the VCR in PAUSE. You might try shutting off the VCR power while in PAUSE to stop the spinning heads.

Sometimes you goof up during a recording, and have to record a scene over. You would REWIND a ways, PLAY, and watch for a place to break in. When you find it, hit PAUSE, RECORD/PLAY, then UNPAUSE to commence recording. This edit will have a messier glitch than PAUSE edits do.

ASSEMBLE edits tack things onto the *end* of what you've already recorded. INSERT edits replace material in the *middle* of what you've already recorded. You INSERT edit by finding the scene you wish to replace, noting the edit-in and edit-out points, and recording new material over the old (erasing it). You edit-in by playing the tape up to the point where you wish to begin your INSERT, pressing PAUSE, then pressing RECORD/PLAY and then UNPAUSING as the action begins. At the edit-out point, press PAUSE, then STOP. The edit-out may have a glitch that lasts 5–8 seconds. It is possible to modify home VCRs to make neater INSERT edits.

By using SCAN CONVERSION (copying a tape by playing it on a TV screen, aiming a TV camera at the screen, and recording the image) you can fade the TV (or camera) to black to hide the glitches when copying a glitchy taped program.

AUDIO DUBBING allows you to erase an old sound track on a tape and add a new one, perhaps containing narration or music. By recording and editing together scenes, "silent movie" style, you can neatly bridge the visual transitions from scene to scene using the sound track as a common element.

Editing from another tape has the same disadvantages of copying from another tape—your picture and sound quality go down one generation. What you gain is the freedom to shoot your production out-of-sequence and to shoot scenes over and later pick the best "take" for editing.

Keeping track of your RAW FOOTAGE is best done with a SHOT SHEET, which lists and indexes all your "takes." Using an EDITING SHEET, you can lay out all your editing plans before you begin. The EDITING SHEET speeds up the actual editing process by listing the location of all the scenes you plan to use.

Some of the "rules" of editing aesthetics are:

1. Change camera angle and shot closeness when you change shots. Avoid JUMP

CUTS, where your subject snaps into a new position but the rest of the shot remains unchanged.

2. Keep light levels the same from shot to shot.

3. A variety of shots is good but don't change from a good shot to a worse shot just to provide variety. Try to imagine what the audience will want to see and use that shot.

4. Imagine a straight line drawn through your performance area. Always keep your camera on one side of that line. See that performers traveling to the right in one scene are still traveling to the right when you cut to your next scene.

5. Pay attention to hand, foot, and face position and other details as you edit so that the scenes match when put together. If you edit during action sequences it's easier to mask minor inconsistencies between scenes.

6. Different mikes and different rooms add their own coloration to the sound. When editing and AUDIO DUBBING, use the same mikes and locations so the sound remains the same from scene to scene.

Transitions from scene to scene can leave unconscious messages. When shooting your RAW FOOTAGE plan for these transitions so your scenes will fit together smoothly. CUTS are easy and unobtrusive. FADES, DISSOLVES, and DEFOCUS-FOCUS shots imply time passing or change of locale. REVERSE ANGLE shooting allows you to shoot over one person's shoulder into the face of the speaker, then swap positions when the other person speaks. The CUTAWAY is a shot of something other than your main subject. By inserting CUTAWAYS between shots of the subject, you can avoid JUMP CUTS and also disguise your edits, making them less obtrusive. Leading the action is a technique for preparing the audience for the next shot. For example, the performers look to the side before you CUT to a shot in that direction to reveal what they saw. Following your performers around as they travel from place to place can be handled by having them exit one scene and appear in another, or by having them walk toward and by the camera in one scene and away from the camera into the next. The camera could also pan, tilt, or zoom to fill the screen with a blank surface from one scene, then you could edit to a similar blank surface in the next scene, then pan, tilt, or zoom the camera to take in the action of *that* scene. A SWISH PAN moves the viewer from place to place during the blur of a fast pan. You edit during the pan.

You can copy many special effects as they play from the MASTER VCR. STILL FRAME is a stylish effect when editing from another tape.

Things can be made to disappear and reappear if you move them while the VCR is PAUSED and your camera remains stationary on a tripod.

ESTABLISHING SHOTS tell your viewers where they are, "setting the stage" for action to come. REVELATION SHOTS do the opposite by involving viewers in the action, then piquing their curiosity about the circumstances or surroundings of a scene. Here the element of surprise is used for dramatic effect.

By editing a tape a certain way you can create the impression of overwhelming evidence, or discord, or credibility.

Index

NOTE: Page numbers 1–330 appear in paperback edition, Volume I. Pages 331–554 appear in paperback edition, Volume II.

A

Accessories, buying, 495, 496
Accessories, video:
 ad eliminator, 396, 397
 audio noise reduction devices, 384, 385
 bulk tape eraser, 400
 cables, 383, 384
 captioning equipment, 401, 402
 carrying cases, 401
 cigarette lighter power adapter, 400, 401
 commercial squelcher, 396, 397
 demodulator, 398, 399
 detailer, 377–379
 distribution amplifier, 382, 383
 enhancer, 377
 faders, 394–396
 filters, noise, 401
 framus, 403
 microwave antenna, 403
 modulator, 398
 monitors, 374, 375
 multistandard adapters, 402, 403
 patch bay, 380, 381
 processing amplifier, 375
 pulse code modulators, 386, 387
 RF generator, 398
 routing switcher, 381, 382
 special effects generators, 394–396
 stabilizer, 375–377
 stereo sound adapters, 385, 386
 switchers, 394–396
 tape demagnetiser, 400
 telecine adapter, 399, 400
 tuner, 398, 399
 tuner/timer, 399
 up converter, 397, 398
 VCR modifications, 387, 388
Acoustic coupler, definition, 434
Ad eliminator (see Commercial squelcher)
Adapter cable, camera-to-VCR, 516

Adapters, video, 15, 16
Addresses of manufacturers and suppliers, 542–549
Adjacent channel rejection, TV specs, 500
Ads, time taken up by, 301
Aesthetics (see camera angles)
AFT (see also Automatic fine tuning), definition, 5
AGC (see Automatic gain control)
Ailments (see specific device's ailments, i.e., VCR ailments)
All channel antenna, 30–31, 39
Amplifiers, TV antenna, 41
Analog, definition, 433, 434
Antenna, microwave, 403
Antenna, TV:
 accessories, 36–41, 43, 47
 ailments, 42–48
 aiming, 32, 33, 34
 bow tie, 3, 26, 27
 connecting accessories, 58–64
 connections to VCR, 79
 dipole, 25
 how they work, 29, 30
 impedance, 36, 37
 improvised, 28, 29
 installing, 32, 37
 kinds of, 30–32
 loop, 26, 27
 monopole, 25
 portable, 27, 28
 power line, 28
 rabbit ear, 3, 25
 specifications, 502–504
 wire, 34–36
Antenna rotor (or rotator), 30
Antenna sensitivity:
 TV projector specs, 524
 TV specs, 500
Antiope, 442, 443
Anti-piracy signals, 291, 375–377
Artwork for TV (see Graphics)
Aspect ratio, graphics, 264, 265
Assemble edit, 300–302, 308
 definition, 299, 302

Attenuator, audio, 83, 229
Attenuator, TV antenna, 41
Audio:
 adapters, 215–217
 ailments, 237–241
 basics, 4–6, 210, 211
 cable handling, 242
 coloration and room tone, 318
 connectors, 213, 214
 cueing a record or tape, 233–235
 definition of, 4
 dolby, 235, 236
 dubbing, 223
 editing, 223, 306–308
 graphic equalizer, 236, 237
 hi fi from VCRs, 135, 241, 243
 how microphones work, 214
 impedance, 212
 making a sound check, 223
 microphone selection and use, 217–223
 microphone types, 211, 212
 mixers, 225–231
 mixing techniques, 231, 232, 233
 modifications of VCR, 388
 music recording, 221, 241
 noise reduction devices, 384, 385
 pulse code modulators, 386, 387
 related terms, 211–214, 228, 229
 simulcast recording, 224, 225
 sound coloration, 223
 stereo adapters, 385, 386
Audio connections on a VCR, 81, 106
Audio dub, 306–308
 control on a VCR, 87
 definition, 299
Audio feedback (see Feedback, audio)
Audio frequencies compared, 236, 237
 (see also Frequency response)
Audio from VCR into hi fi, 101, 102

Audio in connection to a VCR, 82, 83
Audio levels, 82, 83, 226, 227
Audio out connection to a VCR, 83, 84
Auto-focus lens feature, 143
  definition, 140
Automatic fine tuning (AFT), TV set, 2, 7
Automatic gain control (AGC):
  audio, 223
  TV camera, 144
Automatic iris, TV camera, 144
Automatic white balance, 146

B

Backgrounds for titles, 277, 278
Back light, 251–254
  definition, 249
Backspacer:
  definition, 299
  editing, 298, 327–329
Balanced line, 214, 215
  unbalancing, 216
Balun (see Matching transformer)
Band pass, VCR specs, 507
Band separator (or band splitter), TV antenna, 37
Barn doors, 254
  definition, 249
Battery care, 369–371
Battery power for VCR, 108, 109, 110
Beach taping, 111–113
Beam control on TV camera, 153–156
Beamwidth, TV antenna specs, 503
Beta format, 92
Beta format transfer, 532, 533
Block converter (see Up converter)
BNC connector, 11–12, 15, 16
Books, video, 539–541
Boom microphone, 218, 219
Booster, TV antenna, 41
Bow-tie antenna, 3, 26, 27
Bridged inputs (see Looped-through inputs)
Brightness:
  lighting for TV camera, 248
  TV screen, 2, 8
Bulbs, lighting, 259–261
Bulk tape eraser, 400
Burn in, 174, 460
Buttonhook feed, 470
Buying video equipment:
  general hints, 491–497
  sources, 549–554
Byte, definition, 434

C

Cable care, 371
Cable compatibility for TV cameras, 170, 171
Cable handling tips, 104
Cable ready:
  TV sets, 56, 57
  TV specs, 501
Cables, making video, 383, 384
Cable TV:
  computer games (Playcable), 455, 456
  connections, 53, 55, 56–64
  frequencies, 52, 53
  how it works, 49–53
  programming, 54
  two-way, 65–67
Camcorder, 134, 135
Camera, TV
  adjustments to picture and color, 144–146, 152–156
  ailments, 171–175
  angles, 180, 181
  automatic iris, 144
  backgrounds, 186–189
  black & white, 150–157
  burn in, 173, 174
  buying considerations, 510–512
  cable compatibility, 170–172
  close-up shooting, 161–163, 185
  color temperature, 144, 145
  connections, 80, 82, 103, 141, 151–153
  controls, 144–156
  depth of field, 158, 159
  features, 144–150, 511
  focal length, 159, 160, 168
  focusing, 141, 160, 162, 184, 185
  focusing on graphics, 274, 275
  how they work, 138, 139
  how to operate, 141–143
  industrial, 150–157
  iris adjustment, 141, 157–159, 247, 248
  kinds of, 143, 150–153
  lens care, 166–168
  lenses, 140–143, 157–168, 511
  light specifications, 514
  microphones, 149, 150
  placement (see Camera, TV, backgrounds)
  portable color, 143
  specifications, 512–516
  steadying, 182
  techniques for shooting, 207, 208 (see also Camera angles)
  titler built-in, 273, 274
  tripods and dollies, 168–170
  vidicon tube, 138
  viewfinders, 146–149
Camera angles, 189–202, 316, 321–325
Camera connections to a VCR, 82
Camera in, VCR connection, 81
Camera light, portable, 246
Camera tricks, 200–207, 306 (see also Camera angles and Mood creation)
Cannon plug, 213, 214
Capacitance, definition, 420
Capacitance electronic disc, 423–425
  definition, 421
  specs, 524–526
Captioning equipment, 401, 402
Cardioid microphone, 212, 221, 222
Care of tape (see Video tape, care)
Cartivision format transfer, 532, 533
Case, carrying, 401
Cassegrain feed, 470
Cassette (see Videocassette)
Cassette storage boxes, 126, 127
CAV (see Constant angular velocity)
CB interference on TV, 42, 43
CED (see Capacitance electronic disc)
CED videodisc player, specs, 524–526
Ceefax, 442
Center focus lens attachment, 205, 206
Channel selector, TV set, 2
Channels, TV specs, 500, 501
Chrominance, 139
Cigarette lighter power adapter, 400, 401
Cinching, tape, 363
Circular polarization, TV antenna, 34
Cleaning (see Maintenance)
Clogged video head (see Videocassette recorder, head cleaning)
Closed captions, 401, 402
Close up, 179
Close up lenses, 161–163
Close up shooting, 161–163, 185, 186
Clubs, tape swapping and rental, 535–537
CLV (see Constant linear velocity)
C-mount, camera lens, 167, 168
Coax (or coaxial) cable, 13, 36, 43
Color:
  adjustments on TV sets, 8, 9
  how color TVs work, 6, 7
  lighting for, 258
Color camera adjustments (see Camera, TV, adjustments or specific adjustments)
Color intensity, TV set, 2, 8, 9
Color temperature, 144, 145, 246, 258, 259
Comb filter, 23
  TV projector specs, 524
  TV specs, 499

Commercial squelcher, 396, 397
Community access cable TV, 50
Compatibility:
    between countries, 533, 534
    between VCRs, 532, 533
    between VCR and camera, 170–172
    foreign standards, 402, 403
    teletext, 447
    TV and computer, 458–460
    videotext, 448
Compuserve's Micronet, 446
Computer:
    compatibility with TV, 458–460
    connections, 435, 436, 439, 440, 456–460
    games (Playcable), 455, 456
    how it puts words on TV screen, 437–439
    selecting for titling and graphics, 462, 463
    teletext, 440–443, 447
    text on a TV screen, 436, 437
    TV adjustments when using computer, 459, 460
    used for lettering TV tiles, 273
    using data banks and host computers, 458
    videotex, 443–455
Connectors, 11–17, 213, 214
Constant angular velocity disc, 422
    definition, 421
Constant linear velocity disc, 422
    definition, 421
Contrast, TV screen, 2, 8, 20
Convenience outlet on a VCR, 85
Convergence, TV projector specs, 524
Converter, cable TV, 50–53, 60, 61
Copyguard, 291
Copyguard stabilizer, 375–377
Copying film and slides, 399, 400
Copying movies and slides, 278–283
Copying video tapes:
    basics, 285, 286
    connecting equipment, 287–289
    copyright law, 290, 291
    from a TV screen, 291–293, 305, 306, 324, 325
    photographically, 293–297
    quality of result, 286, 287
    technique, 288, 290
    using a distribution amplifier, 382, 383
Copyright law, 290, 291, 536
Corner insert, definition, 392
Cost, video vs disc vs film, 426
Counter (see Index counter)
Counterfeit blank tape, 122
Cover shot, 321
CPU (Central Processing Unit), definition, 434, 435
Crawl text on screen, 275–277

Credits, 300
    definition, 299
Crispener, picture, 377–379
Crystal microphone, 212
Cueing a record or tape, 233–235
Customs hints, traveling with video equipment, 345
Cut:
    definition, 387
    video transition, 319
Cutaway, 321, 338
CVC format, 91
CVC format transfer, 532, 533

D

Dampness (see Dew indicator)
DBS (Direct Broadcast Satellite), 488, 489
    definition, 465
Deaf, captioning equipment for, 401, 402
Decibel, definition, 228
Decoder, cable TV (see Converter, cable TV, or descrambler, cable TV)
Definitions relating to:
    audio, 211–214, 228, 229
    cameras, 140, 182
    computers, 433–435
    editing, 299
    lighting, 248, 249
    satellite receivers, 465, 466
    switchers and special effects, 389–394
    TVs, 2–5, 10
    VTRs, 70, 73
    videodiscs, 420, 421
Defocus, video transitions, 320
Degaussing heads (see Videocassette recorder, demagnetising heads)
Demagnetiser, videotape, 400
Demagnetising VCR heads (see Videocassette recorder, demagnetising heads)
Demodulator, 50, 398, 399
Depth of field, 158, 159
Descrambler, cable TV, 55–57, 63, 64
Detail, 2
Detailer video, 377–379
Detail loss from photo to video medium, 269
Dew indicator, 86, 100, 101, 110
Digital, definition, 433
Digital audio recording, 386, 387
Diopter, lens attachment, 162, 163
Dipole antenna, 26
Direct Broadcast Satellite (see DBS)
Directional microphone, 212
Directionality, TV antenna specs, 502, 503

Director's duties, 331, 332
Dish, antenna (see Satellite dish antenna)
Dish, satellite (see Satellite dish antenna)
Dissolve, definition, 390
    video transition, 320
Distant/Local TV set control, 2
Distribution amplifier, 382, 383
Documentary script, 340
Dolby, 90, 235, 236, 284, 285
Dollies (see Tripods)
Dolly movement of camera, 179, 182, 184
Double speed play on a VCR (see Search)
Downlink, satellite (see TVRO)
Dramatic script, 339, 341
Dropout compensator on a VCR, 87
Dropouts:
    specs, 520
    video tape, 119, 121, 301, 364
Dub, definition, 299
Dubbing a video tape (see Copying a video tape)
Duplicating video tape (see Copying a video tape)
Dynamic microphone, 212
Dynamic noise reducer, 384, 385

E

Editing a video tape:
    assemble, 300–302, 308
    definitions, 299
    electronic editors, 298, 301
    from another tape, 308–313
    insert, 302–305, 308
    organizing yourself, 311, 312
    pause, 89, 102, 103, 300
    planning strategies, 325–327
    raw footage, 309–312
    techniques and rules, 313–327
    transitions (cut, fade, etc.) 318–325
Editing controller, 298, 327–329
Editing sheet, 312
EIAJ format, 91
EIAJ format transfer, 532, 533
8-pin plug, 16–19
Electret condensor microphone, definition, 211
Electrical focus of an older black & white TV camera, 156
Electronic editing, 298, 301, 327–329
    definition, 299
Electronic viewfinder, 148, 149
    definition, 140
Enhancer, video, 377
Equipment sources, 549–554
Erasing tape:
    avoiding accidental, 122, 123
    while recording, 74

Establishing shot, 325

F

Fade, video transition, 319
Fade feature on cameras, 146
Fader, 394–396
   definition, 391
F connector, 11–13, 16, 39
Feed, antenna, 469–472
Feedback:
   audio, 131, 222
   video, 206, 207
Field, video, 294
Fill light, 250, 251–254
   definition, 248
Film-to-tape transfer, 278–283
Filter:
   audio, 229
   lens, 163–166
   noise, 401
Fine tuning, TV set, 2, 7, 8
Fixed focus lens:
   definition, 140
   focusing, 140, 141
Flagwaving, 124, 133
Flying erase head, 301
FM radio signals:
   from antenna, 38, 47
   from cable TV, 52, 53
Focal length, lens, 157
Focus, editing transition, 320, 321
Focusing on graphics, 274, 275
Focusing techniques, camera lens, 160, 161, 184, 185
Folded dipole antenna:
   how to make, 29
   yagi, 33
Foot-candles, 514
Foot-lamberts, TV projector specs, 524
Foreign TV standards, 93
Foreign TV standards adapters, 403
Format, type of program, 333, 336
Format, VCR:
   compared, 118, 119
   copying from one to another, 532, 533, 534
   definition, 71
Format, videodisc:
   CED, 423–425
   compared, 428–431
   definition, 420
   LV (laservision), 421–423
   VHD, 425
Frame, video, 294
Frame advance feature on a VCR, 88
Framus, 403
Frequency response:
   microphone specs, 517
   tape specs, 519
   TV specs, 500
   VCR audio specs, 509
   videodisc specs, 526

Front screen projection of graphics, 279, 280
Front-to-back ratio, TV antenna specs, 504
F-stops (see Iris, camera lens)

G

Gain:
   satellite receiver specs, 528
   TV antenna specs, 502, 503
Games, computer, 455–457
Gel/cell battery, 369, 370
Genlock, definition, 394
Geosynchrous satellite, 466
Ghost eliminator, TV antenna, 41
Gigahertz, definition, 465
Graphic equalizer, audio, 236
Graphics:
   adding movement to, 278
   aspect ratio, 264, 265
   backgrounds for titles, 277, 278
   basics, 263
   changing, 275–277
   computer, 460–462
   crawl, 275, 276
   film to tape transfer, 278–283
   focusing on, 274, 275
   lettering and titling techniques, 267–274
   lighting, 274
   roll, 275, 276
   safe title area, 265–267
   slides, 267

H

Hardware, computer definition, 434
Harmonic distortion, VCR specs, 509
HBO, 54
Head (see Tripod)
Head cleaning (see Videocassette recorder, head cleaning)
High band TV:
   antenna, 37
   signals, 53
High pass filter (see Trap)
Hi level audio input (see Audio levels)
Hi-Z input, 14
Horizontal hold, TV set, 2, 10, 20
Hue, TV set, 2

I

Impedance:
   audio, 212, 213, 215, 227, 229
   TV antenna, 36, 37
Impedance matching transformer, audio, 213, 215

Index counter on a VCR, 86, 93–98
Input selector, 76, 78, 103, 104
Insert edit, 302–305
   definition, 299, 301
   modifying VCR for, 304, 305
Insertion loss, switcher, specs, 529
Interactive TV, 65–67
Interactive videodisc and VCR systems, 441
Interference:
   audio, 240
   TV, 42–48
Interlace, video, 294
Insertion loss, switcher, specs, 529
Interactive TV, 65–67
Interactive videodisc and VCR systems, 441
Interference:
   audio, 240
   TV, 42–48
Interlace, video, 294
International standards (see Compatibility)
Interviews, editing, 326
I/O (Input/Output), computer definition, 434
IRE, TV camera specs, 514, 515
Iris, camera lens:
   adjusting, 141, 157–159, 247, 248
   definition, 140
   depth of field, 158, 159, 160
Isolation, switcher specs, 529

J

Joiners, TV antenna, 37
Jump cut, editing, 313, 314, 321–323

K

Key effect, definition, 393
Keyfax, 446
Key light, 250, 251–254
   definition, 248
Keystone effect on graphics, 276, 279
Kilohertz, 52

L

Lag, TV camera image, 153–156, 173, 203
Lapel mike, 217, 218
Laservision (LV) videodisc, 421–423
   definition, 420
Laservision (LV) videodisc player, specs, 524–526
Lavalier mike, 217, 218

Legal aspects of tape:
  copying, 290, 291
  trading, 536
  (see also Copyright law)
Lens, TV camera:
  attachments, 162–168, 205, 206
  automatic iris, 144
  care, 166–169
  cleaning, 369
  close up, 161–163
  definitions, 140
  depth of field, 158–159
  filters, 163–166
  focal length, 159, 160, 168
  focusing, 141, 160, 161, 184, 185
  focusing on graphics, 274, 275
  iris adjustments, 157–161
  macro, 162
  mounting, 167, 168
  specs, 515
Lens iris, (see Iris)
Lettering for TV, 267, 269–274
Lighting:
  basics, 244–247
  brightness ratio, 248
  bulb types, 259
  care of fixtures, 260, 261
  color, 258, 259
  graphics, 274
  indoor, 245–247
  outdoor, 247
  placement, 245–247
  techniques, 249–259
  terms, 248, 249
Lighting and camera placement, 186–188
Line equalizer, TV antenna, 41
Line filter, 401
Line matching transformer, audio (see Impedance matching transformer)
LNA (Low Noise Amplifier), 472–473
  definition, 466
Locating segments on a tape (see Index Counter)
Lo level audio input (see Audio levels)
Long shot, 179
Loop antenna, 26, 27
Loop through inputs, 14
Low band TV, 53
Low band TV antenna, 37
Low Noise Amplifier (see LNA), specs, 528
Low pass filter (see Trap)
Luminance, 139
Lux, 514
LV videodisc (see Laservision)

M

Macro lens, 162

Magazines, video, 538, 539
Mailing videocassettes, tips on, 534, 535
Maintenance:
  ailments, 351–353
  batteries, 369–371
  demagnetising heads, 361, 362
  head cleaning cassettes, 354, 355
  lens, 166–169
  repair service costs, 347–351
  saltwater soaked VCR, 113
  tape splicing, 364–369
  TV camera burn-ins, 173, 174
  VCR head cleaning, 353–361
Manufacturer's addresses, 542–549
Manufacturers of video equipment, general groups, 494
Master tape, best quality, 286, 287
Matching transformer, 36–39
  transformer/splitter, 39
Mechanical focus of TV camera, 152, 153, 175
Medium shot, 179
Megahertz, 52
Memory, computer definition, 434
Memory, index counter on VCR, 86, 89, 93–98
Microphone:
  input to VCR, 82
  selection and use, 217–223
  sound check, 223
  specs, 517
  types, 212
Microwave, 466–468
  antenna, 403
  definition, 465
  TV signals, 51
Microwave receiver, definition, 465
Mid band TV, 53
Mini plug, 213, 214
Mixer, audio:
  adjusting volumes, 225, 226
  ailments, 237–240
  impedance, 227, 229
  inputs to, 226, 227, 237
  mixing without a, 230, 231
  monitoring sound, 227, 228
  outputs, 227
Modem, 440, 443
  connecting, 457
  definition, 434
Modifications (see Videocassette recorder modifications)
Modulator (see RF generator)
Monitor, TV, 10, 11–14, 374, 375
  connecting up, 11
  definition, 10
Monitor/receiver, TV, 16–19
Monopod, 170
Monopole antenna, 25
Mood creation, 198–200 (see also Camera angles)

Movies copied onto video tape, 278–283

N

NABTS (North American Broadcast Teletext Specification) standards, 447
Narration techniques, 335–338
NiCad battery, 370, 371
Noiseless still frame feature on a VCR, 87, 88
Noise temperature, LNA specs, 528
NTSC standards, 533

O

Ohm (Ω), 34, 36
Omnidirectional microphone, 212, 218
On-location shooting (see Portable VCR setup)
Optical sight viewfinder, 147
  definition, 140
Oracle, 442
Output selector control on a VCR, 75
Overscan TV, 459

P

Pad, audio (see Attenuator)
PAL standard, 533
Pan (movement of camera), 179, 182–184
Parabolic antenna, 468
Parallax, 147
Parallel cutting, 326
Patch bay, 380, 381
  specs, 529
Patch cords, 13
Pause control on VCR, 73
Pause editing, 300, 301
Pay TV, 51, 54, 55
  programming, 54
  theft of, 57
Peak level indicators, 228
Pedestal control on tripod, 168, 169
Pedestal control on TV camera, 156, 157
Pedestal movement of camera, 179, 182
Periodicals, video, 538, 539
Phone plug, 213, 214
Phono plug, 213, 214
Photoflood lamp, 246, 259
Photographing a TV screen, 293–297
Picture composition (see Camera angles)
Picture control on TV set, 8
Pictures for use on TV, (see Graphics)
Picture tube, TV specs, 498

Pirate, pay TV, 57
Pirate video tape, 522, 523
Planning a production, 331–346
Planning and scripting, 332
   organizing pictures, 283
Play, control on VCR, 72, 73
PL259 connector, 11–12, 15
Plugs (see Connectors)
Polacoat rear screen material, 280
Polarization, satellite signals, 470–472
Polarizing filter for lens, 165
Polar mount (see Satellite TV, antenna mounts)
Portable VCRs (see Videocassette recorder, portable)
Portable VCR setup, planning a shoot, 343–345
Power adapter, cigarette lighter, 400, 401
Power focus lens feature:
   definition, 140
   focusing, 142, 143
Power line antenna, 28
Power needed for lighting, 261
Power saver feature on a VCR, 109
Power supply for VCR, 108, 109, 110
Preamplifier, audio, 82, 83
Preamplifiers, TV antenna, 41
Prestel, 444
Prime focus feed, 269, 270
Problems (see specific device ailments, i.e., VCR ailments)
Processing amplifier, 375
Programmable timer (see Timer)
Program select control on VCR, 73
Pull focus, 200, 201
Pulse code modulator, 386, 387

Q

Quartz iodine lamp, 259, 260
Quasar 1000 format transfer, 532, 533
Qube cable TV, 65–76, 446

R

Rabbit ear, antenna, 3, 25
Radio Frequency (see RF)
Raw footage, 308–312, 318, 319
   definition, 299
RBG (red, blue, green) output, 458, 459
RCA connector, 11–13
RCA plug for audio, 213, 214
Rear screen projection of graphics, 280
Receiver, satellite, (see Satellite TV)

Receiver, TV:
   component, 23
   controls, 1–10
   definitions, 2, 3, 10
   how it works, 4–7, 294, 295
   how to operate, 1–19
   large screen (see Projectors, TV)
   miniature, 21, 22
   placement, 4, 27
   specifications, 497–502
Recording TV programs off the air, 74, 77, 78, 106, 107
Remote control of a VCR:
   pause connection, 84
   wireless, 84
Rental clubs, tape, 535–537
Repair (see Maintenance)
Resolution:
   tape specs, 519
   TV cameras, 513
   TV projectors, 524
   TV specs, 498, 499
   VCR specs, 507
   videodiscs, 526
Revelation shot, 326
Rewind control on VCR, 72, 73
RF (Radio Frequency), 4, 6, 18, 19
   definition, 5
RF converter, 79
RF generator (or modulator), 107, 398, 399
RF out, 79, 80, 106, 107
Roll titles on screen, 275–277
Routing switcher, 381, 382
Rush, editing, 308–312
   definition, 299

S

Safe title area, graphics, 265–267
Satcom 3R satellite, 479–482
Satellite dish antenna:
   aiming, 477, 478
   definition, 465
   mounts, 474–477
   types, 468, 469
Satellite TV:
   advantages and disadvantages, 487, 488
   antenna aiming, 477, 478
   antenna mounts, 474–477
   buying and building info, 489
   DBS (Direct Broadcast Satellite), 488, 489
   definitions, 465, 466
   dish antenna types, 468, 469
   feed antenna, 469–472
   getting best signal, 478, 479
   how it works, 466, 467
   legal considerations, 485, 486
   licensing, 485
   positions of satellites, 475–477
   receivers, 473, 474

Satellite TV (Cont.)
   scrambling of signals, 486, 487
   SPACE, 485, 486
   stations, 479–485
   transponder, 467
Satellite TV receiver:
   selection, 526–528
   specs, 528
Scan conversion, 291–293, 305, 306, 324, 325
Scan picture feature on a VCR (see Search)
Scrambling satellite TV signals, 486, 487
Script:
   documentary, 340
   dramatic, 339, 341
   narration, 335–338
   planning, 332, 333
   rundown sheet, 341
   storyboard, 341–343
   writing mechanics, 338–341
   writing strategy, 333–335
Search feature on a VCR, 88, 94
SECAM standard, 533
Secret recording (see Surreptitious recording)
Segue:
   audio, 231
   definition, 229
Set light, 251–255
   definition, 249
Seventy-five ohm (75Ω), 14
   coax, 36
Sharpness (see Detail)
Shielded twin lead (see Twin lead)
Shotgun microphone, 149, 212, 218–220
Shots, types of transitions, 319–324
Shot sheet, editing, 312
Shutter bar, 294, 295
Shutter bar when video recording movies, 282
Shuttle search on a VCR (see Search)
Signal splitter, TV antenna, 38, 39
Signal-to-noise ratio:
   satellite receiver, 528
   tape, 121
   tape specs, 519–521
   TV camera specs, 514
   VCR specs, 507, 508
   videodisc specs, 526
Sign-on definition, 434
Simulcasts, recording stereo, 224, 225
Single channel antenna, 30, 31
Skew error (see Flagwaving)
Slave, 285–289
Slides for TV, 267, 278–281
Slow motion on a VCR, 88
Snap cut (see Jump cut)
S/N ratio (see Signal-to-Noise ratio)
Software, computer definition, 434

Solenoid controls on a VCR, 88, 89
Sound coloration, 223
Source (data base), 446
SPACE (Society for Private and Commercial Earth Stations), 485, 486
Special effect generator, 394–396
    definition, 394
Special effects, video, 389–396
    (see also Camera tricks)
Specifications:
    microphone, 517
    patch bays and switchers, 529
    satellite receiver, 528
    TV antenna, 502–504
    TV camera, 510–516
    TV projectors, 523, 524
    TV receiver/monitor, 497–502
    VCR, 504–510
    videodisc players, 524–526
    video tape, 518–521
Specs (see Specifications)
Speed, VCR tape, 76, 86, 91–93, 118, 120, 286
Splicing videotape, 364–369
Split screen, 391
    definition, 390
Sports, shooting, 208
Stabilizer, 291, 375–377
Standard lens:
    definition, 140
    focusing, 141, 142
Standards:
    converting from PAL or SECAM to NTSC, 291–293
    foreign, 402, 403
    (see also Compatibility)
Standby lamp, 86
Standoffs for antenna wire, 35, 36
Star pattern lens attachment, 205, 206
Stereo sound adapters, 385, 386
Still control (see Pause control)
Still frame, editing effect, 324, 325
Stop, editing method, 300
Stop control on VCR, 73
Storage of cassettes (see Video tape storage)
Stores, kinds of video, 494–497
Storyboard, 341–343
Summary (see end of each chapter for review)
Super band TV, 53
Superimposition (or Super), 396
    definition, 390
Surreptitious recording, 207
Swish pan, 324
Switcher, 394–396
    definition, 389
    RF, 529
Sync, 4, 5, 6
    definition, 4
    external, 389, 394, 395

Sync (Cont.)
    TV camera spec, 513
Sync generator, 152

T

Tai chi camera-holding position, 183
Tape, video (see Videotape)
Tape care (see Video tape care)
Tape path, VCR, 358–361
Tape speed on a VCR (see Speed, VCR)
Tapes on the subject of video, 541
Target control on TV camera, 153–156
T connector, 15
Telecine adapter, 280, 281, 399, 400
Telematique, 446
Telephoto lens converter, 163, 164
Teletext, 440–443, 447
    definition, 435
    disadvantages of, 455
Television receive only (see TVRO)
Television standards (see Foreign TV standards)
Television text systems, 448–455
Telidon, 443, 444
Terminator, 14, 15
Terminator, 75Ω, 40
Test tape, 351
THD (see Total harmonic distortion)
Threading, video tape, 118
Three-dimensional TV, 404
Three hundred ohm (300Ω), 34
Through-the-lens viewfinder, 147, 148
    definition, 140
Tilt movement of camera, 181, 182–184
Time constant, TV spec, 500
Timer, VCR:
    how to operate, 77, 99, 100
    programmable, 90, 134
Tint (see Hue)
Titles by computer, 460–462
Titling, lettering for TV (see Graphics)
Tone, TV set, 2, 4
Total harmonic distortion:
    tape specs, 520
    VCR specs, 509
    videodisc specs, 526
Touch tone teletext, 443
Tracking control on a VCR, 86, 87
Transponder, satellite, 467
    definition, 465
    stations, 479–485
Trap, 54, 55
    TV interference, 43, 44
Travel hints with VCR, 345
Trick shots, 325

Triple speed play on a VCR (see Search)
Tripods, 168–170
Troubleshooting (see specific device ailments, i.e., VCR ailments)
Truck movement of camera, 179, 182, 184
Tri-standard TVs and VCRs, 533, 534
TTL (see Through-the-lens)
Tubes, TV camera compared, 512, 513
Tuner, TV, 398, 399
Tuner/timer, 399
TV ailments, 19–21
TV antenna (see Antennas, TV)
TV antenna ailments (see Antenna, TV, ailments)
TV cameras (see Cameras, TV)
TV coupler, 38, 40
    passive and amplified, 40
TV monitor (see Monitor, TV)
TV monitor/receiver (see Monitor/receiver, TV)
TV projectors:
    advantages and features of, 416
    buying, 523
    disadvantages of, 413–416
    future developments, 416, 417
    how they work, 407–410
    light valve type, 410
    multiple tube, 410
    one piece vs two piece, 410–413
    single tube, 407–409
    specs, 524
    viewing area, 414
TV receiver (see Receiver, TV)
TVRO (TV, receiver only)
    definition, 465
TV screen cleaning, 369
Twin lead, 29, 34–36
    shielded, 35, 43
Typing TV titles, 271–273
Typography (see Lettering)

U

UHF:
    antennas, 26, 27
    band splitter, 37, 38
    cable TV use, 52, 53
    definition, 5
Unbalanced lines (see Balanced lines)
Underscan TV, 459
Uni-directional microphone, 212
Up converter, 56, 62, 63, 64, 397, 398
Up converter/VCR connections, 113–115
Used video equipment, buying, 496, 497

V

Variable focus light, definition, 249
Variable speed slow motion on a VCR, 88
Variactor tuning, 88
V-Cord II:
   cassettes, 533
   format transfer, 532
VCR (see Videocassette recorder)
Vertical hold, TV set, 2, 9
VHD videodisc player, specs, 524–526
   (see also Video high density)
VHF:
   antennas, 90
   band splitter, 37, 38
   cable TV use, 52, 53
   definition, 5
VHS-C format, 92
VHS:
   format, 91
   format transfer, 532, 533
Video, 4, 5, 6
   definition, 4
Video adapters (see Adapter, video)
Videocassette:
   definition, 70, 71
   running times, 120
   splicing and repair, 366–369
Videocassette recorder:
   ailments, 127–134, 351–353
   battery care, 369–371
   battery power, 108, 109, 110
   beach taping, 111–113
   buying considerations, 505, 506
   cable care, 371
   camera compatibility, 170–172
   care, 109–113
   compatibility, 90–93
   connecting up, 79–85, 103, 105–108, 113–115 (see also Cable TV connections)
   definition, 71
   features, 86–90, 109, 505, 506
   formats, 91, 92, 504, 505
   function controls, 72, 73
   future developments, 134, 135
   head cleaning, 353–361
   head problems, durability, care, 352, 353
   how it works, 294, 295
   how to operate, 71–79, 100–102, 105

Videocassette recorder (Cont.)
   improving picture and sound, 101, 102
   input and output controls, 76
   modifications, 387, 388
   portable, 104–113
   specifications, 504–510
   stereo, 231
   taken apart, 356–359
   tape, 115–127
   threading, 117
   winter taping, 109, 110
Videocassettes on the subject of video, 541
Videocassette tape (see Video tape)
Videodisc:
   advantages and disadvantages, 425–429
   basic description, 419
   CED, 423–425
   connecting up player, 429, 430
   definition, 420
   durability, 429
   kinds of, 419–425
   laservision (LV), 421–423
   VHD, 425
Videodisc player:
   buying, 524–526
   specs, 526
Video feedback (see Feedback, video)
Video head (see Videocassette recorder, head cleaning)
Video high density (VHD) disc, 425
   definition, 421
Video in, VCR connection, 79, 80
Video out, VCR connection, 80, 81
Video projectors (see Television projectors)
Video repair (see Maintenance)
Video reverse feature on TV camera, 157
Video tape, 115–127
   care, 122–125, 363–369
   comparisons, 521
   definition, 70
   how it works, 115, 116, 117
   kinds, 116
   renting, 521–523
   selecting, 120–122
   specs, 518–521
   storage, 124–127
Video tape recorder, definition, 71
Video tape splices (see Splicing video tape)

Videotex, 443–455
   applications, 445
   definition, 434
   disadvantages, 455
Video User's Handbook, 541
Vidicon tube, 138
   cleaning, 369
   definition, 140
   mechanical focus of, 152
   versus other types, 512, 513
Viewdata (see Videotex)
Viewfinder, TV camera, 146, 147
   definitions, 140
   electronic, 148, 149
   optical sight, 147
   through-the-lens (TTL), 147, 148
Viewtron, 444
Vignetting, camera lens, 168
VIR (Vertical Interval Reference), TV specs, 501
V lock adjust, control on a VCR, 88
VLP, disc format, definition, 421
Voice over, audio, 231, 232
VTR (see Video tape recorder)

W

Wall Street Journal videotex, 447
White balance, 246
White balance, on color TV camera, 145, 146, 204
Wide angle lens converter, 163, 164
Winter taping, 109, 110
Words to know (see Definitions)
Writing for TV (see Script, writing)

X

XLR plug, 213, 214

Y

Y adapter, 101, 102, 230
Yagi antenna, 30, 31

Z

Zoom lens:
   definition, 140
   focusing, 141, 142
   using, 182, 184, 185